AF350088

BACK TO
REALITY

A REVISION OF
THE SCIENTIFIC WORLDVIEW

ARTO ANNILA

PRIVUS PRESS

PRIVUS PRESS
privuspress.com
Copyright © 2020 Arto Annila
artoannila.com

Publisher's Cataloging-in-Publication Data

Names: Annila, Arto, 1962- author.
Title: Back to reality, a revision of the scientific worldview / Arto Annila.
Description: New York: Privus Press, 2020 | Including bibliographical references and index.
Identifiers: LCCN 2020909194 | ISBN 978-952-94-3364-3 (hardcover) | ISBN 978-952-94-3365-0 (ebook).
Subjects: Physics. | Biology. | Economics. | Philosophy.
Classification: LCC QC174.8 2020 (hardcover) | LCC QC174.8 (ebook) | DDC 530.13

Printed in the United States of America
ingramspark.com

Original title in Finnish: *Kaiken maailman kvantit, luonnontieteen todellisuuskäsityksen tarkistus*
Copyright © 2019 Vastapaino, Arto Annila
ISBN 978-951-768-747-8
Publisher Vastapaino
Yliopistonkatu 60 A, 33100 Tampere, Finland
vastapaino.fi

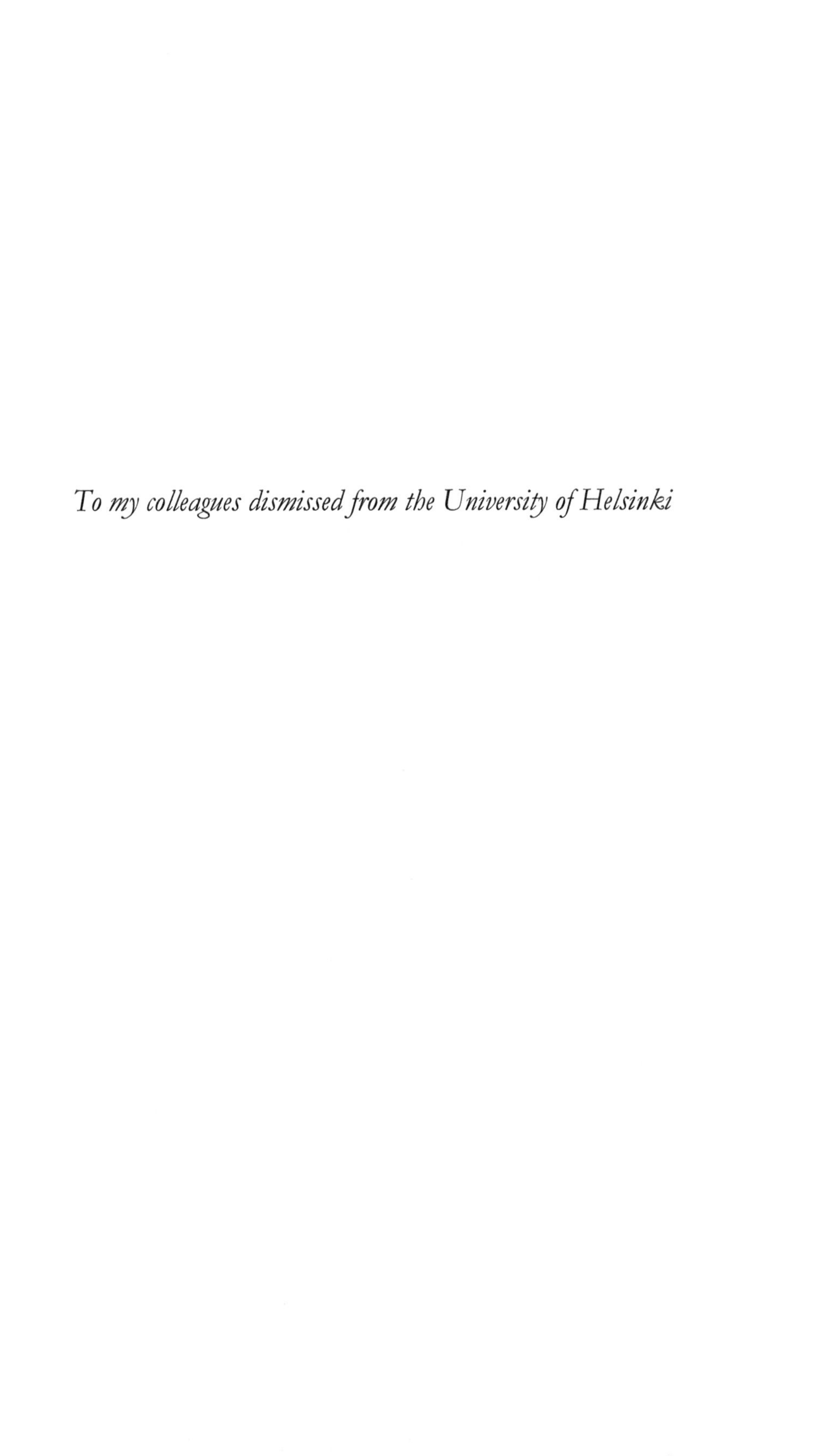

To my colleagues dismissed from the University of Helsinki

Contents

PREFACE

A worldview is an all-encompassing set of beliefs. It is not a static doctrine but has evolved through the ages. Once, the Earth-centered stance held sway but had to give way to the Sun-centered view. Most people also pictured the world fundamentally as timeless, eternal, and unchanging before Darwin's theory of evolution opened their eyes to the endless transformations of Nature. There is thus no guarantee that our current comprehension is accurate, either.

While most of the time science refines our conception of the world step by step, at certain times, stunning new panoramas have opened up. Such a rare moment was described by Ludwig Boltzmann in 1886: "Thus natural science appears completely to lose from sight the large and general questions; but all the more splendid is the success when, groping in the thicket of special questions, we suddenly find a small opening that allows a hitherto undreamt of outlook on the whole."[1]

Reality appears to us as a coherent whole. Nonetheless, scientists tend to tackle fundamental problems about time, space, matter, life, and consciousness as if independent of one another. While a path to a unified worldview may not be apparent, we have learned from the history of science that asking questions – challenging ground-laying assumptions – has led to revisions of the mindset. This age-old method can also work in our time.

In 2001 I embarked upon a search for a comprehensive view of Nature, as a newly appointed professor of biophysics at the University of Helsinki. This discipline aims at understanding those principles of physics that explain how biological systems work. I began by asking myself: Why Charles Darwin's theory of evolution, the basic tenet of biology, has remained a mere narrative? Why is the evolutionary theory not formulated as a law of physics? After all, physics endeavors to account for everything with mathematical rigor. So I thought that if evolution were put in the definitive mathematical form of a natural law, it could reveal what *natural selection* truly is and perhaps also explain the origin of life. What could be a more meaningful goal for a professor of biophysics?

The idea that evolution could be written as a natural law may seem far-fetched. However, ever since Galileo, physics has proved to be a successful method for showing that seemingly complex facets of reality conform to simple laws.

After a few years of exploration, I found the evolutionary equation with surprising ease when adopting the old idea that everything comprises quanta of light, the basic building blocks of Nature. All of a sudden, a straight path opened up to a broader understanding and insight to make sense of all kinds of processes extending well beyond biological evolution. This wide-ranging result is to be expected, for everything changes through time; logically, all processes contribute to the evolution of the whole universe. Nonetheless, I had not anticipated that the implications of this general principle would force me to question some of the most established doctrines as well.

When exploring this theory of *nonequilibrium thermodynamics*, I realized that a few eminent scientists from the past had already known about it. Although their insights were lost as the scope of this dynamic tenet was narrowed down, even distorted into mere mathematical models of equilibrium, the original principle explains many recent findings and puzzling observations. This is not surprising, for thermodynamics is considered a universal theory.

One's worldview is part and parcel of one's identity. Therefore, when one's own closely held beliefs are challenged, it is common to become emotional and defensive. It is unpleasant to acknowledge that one's convictions are unwarranted, even outright discordant. However, in the long run, all realistic views must be welcomed. Let's face it: for all its achievements, the theories of modern physics are mathematical models of static systems. They do not explain the world in evolution, the process of transformation. That is why one's worldview will inevitably change upon understanding evolution in its essentials.

In the past, scientific revolutions have been preceded by observations deviating from predictions, inexplicable coincidences, and disconnected disciplines. All these hallmarks of impending change are evident today. However, history also tells us about disturbing first re-actions to what were initially viewed as seditious ideas, followed by a rational re-evaluation and, ultimately, the adoption of a more realistic revised worldview. Today, we ought to see the world in a way that is consistent with reality if disastrous outcomes on a global scale are to be alleviated or even avoided.

STRUCTURE OF THE BOOK

This book examines the fundamental questions of science. Such questions point out the limitations of our knowledge, the inconsistencies in our thinking, and even our misunderstandings; otherwise, we wouldn't keep on asking, would we?

Since nothing is beyond doubt, we must consider all learning falli-ble. I, therefore, go beyond merely laying out the facts to challenging contemporary truths and putting together a unified worldview from the inferences of many thinkers known from the history of science and from more recent scientific publications, including my own. Chapter by chapter and question by question, I argue for a coherent worldview by comparing its conclusions to precise measurements and unambiguous observations, as well as to prevailing assumptions and potential objections.

The first part of the book examines the ultimate nature of existence as Philip W. Anderson described Richard Feynman doing: "... the key to understanding nature's reality is not anything 'magical', but the right attitude, the focus on asking the right questions, the willingness to try (and to discard) unconventional answers, the sensitive ear for phoni-ness, self-deception, bombast, and conventional but unproven as-sumptions."[2]

The first chapter guides the reader through an examination of data from wide-ranging phenomena, leading us to consider the possibility that all phenomena might display the same basic pattern because the data without labels and headers look the same irrespective of scale and scope. Since a pattern implies a rule, the question arises: what natural law could explain this universality, coined as *Grand Regularity*, across all kinds of processes?

The second chapter argues that all processes are necessarily alike because the flow of time is physical; it is a flow of fundamental elemental constituents, known as quanta. In the third chapter, the structures of all the substances that exist are understood in terms of quanta. From this all-inclusive perspective, problems of elementary particle physics are also tackled. The fourth chapter addresses the evolution of the universe, as all processes are part of it. The deep questions of cosmology, including imperceptible dark matter and mysterious dark energy, are also unraveled. The fifth chapter discusses mathematics as the language of the expression of natural laws as well as interpreting mathematical models as reality.

The second part of the book deals with life, economy, and especially we human beings. Might all these expressions of reality ultimately be only about quanta re-distributing energetically ever more favorably in the form of matter and space? Undeniably, many mechanisms of occurrence are complicated. Might their underlying organizing principle nevertheless be simple and readily comprehended?

In the sixth chapter, life is understood as the chain of events from molecules to the biosphere. From that viewpoint, evolution is causal, teleological, purposeful, but not in the sense of a previously known or predetermined goal. In the seventh chapter, this naturalistic theory exposes the concept that subjectivity, nondeterminism, and intentionality are characteristics not only of consciousness but of all processes. In the eighth chapter, humanity's future is examined as a nondeterminate process where we face waning natural resources and a warming climate.

The book's last chapter deals with the significance of the worldview and attitudes toward reforming the prevailing one. How we see reality and how we opt to act is not predestined – it is all in our hands.

It should be noted that the holistic worldview, the atomistic tenet, sees things through its own lens, just as any other tenet will have its own particular perspective. That which is left without explanation – if anything – the thermodynamic theory does not encompass, and this book thus does not discuss them.

THE LOST LOGIC

Given that fundamental scientific questions about time, space, matter, life, and consciousness remain unanswered today, more precise measurements will not help. Instead, we need to unearth and re-examine the beliefs from which the questions stem.

I find Lee Smolin's and Robert B. Laughlin's views on the essence of time and the substance of the void insightful as we strive to understand reality more profoundly. I also concur with cutting comments on theoretical physics made by Jim Baggott, Philip Ball, Sabine Hossenfelder, Tim Maudlin, Thomas Neil Neubert, Alexander Unzicker, and Peter Woit. We need to explain phenomena rather than model data. Similarly, Stacy McGaugh, David Merritt, Marcel Pawlowski, and Paul Steinhardt have made uncompromising conclusions about contemporary cosmology. Everything is evolving: not just living organisms but the entire universe. Ergo, we need a valid theory to bring both the details and the whole into complementary correspondence.

The materialistic worldview has been both debunked and defended in debates about the origin of life and the quintessence of consciousness. While many commentators with opposing views talk past each other, Thomas Nagel does not choose sides but concludes in his book, *Mind and Cosmos* (2012), that evolution is not random but a teleological process, yet without a preset goal. I now see that the conclusion could not be more accurate. However, this kind of logic would have been lost on me earlier in my career. I had received a contemporary education in physics, and thus the essence of time was beyond my knowledge, as was the true nature of causality.

Initially, I did not have the faintest idea about how to express the evolution of systems within systems. Stanley Salthe's book *Evolving Hierarchical Systems* (1985) put me on the right track. It is difficult to be aware of the dogmas of one's own discipline unless one is open to learning about other perspectives. Change is the prominent characteristic of biology, whereas constancy, or invariance, is the assumed and imposed attribute of physics. So recognizing the change in an invariant was crucial to grasping the essence of evolution.

As the first physics, Galileo's method expresses experience as a law of nature. One is easily fooled into regarding such a genuinely empirical but primitive approach as ambiguous and amateurish. Nevertheless, this still-living source of science remains open to draw understanding. And this nonfiction book is a natural way of telling how we may come to have a deeper understanding of Nature by returning to that spring. The supporting mathematics is exemplified in the appendices and quantitative analyses are available in the references.

The essence of matter and space, as well as the relationship between cause and effect, have intrigued physicists and philosophers throughout history. Today, the mysteries of modern physics, albeit seemingly remote to common sense, have influenced how we weigh our ability to understand the world through the popularization of science. As early as 1923, George Bernard Shaw was lashing out at what he saw as the preposterous scientific ethos in that "… modern science has convinced us that nothing that is obvious is true and that everything that is magical, improbable, extraordinary, gigantic, microscopic, heartless, or outrageous is scientific."[3]

This book seeks to restore confidence in our innate reasoning and reconnect theory to experience. The same pursuit once distinguished modern science from Renaissance magic. Today, we should demand the same transparency of open public debate and reject experts' obscure credo and cliquish consensus. Science is not free of social influences and value judgments, as it is a profoundly human activity. From the history of science, we are all too familiar with the tension between a progressive individual who ventures to think outside the box and a conservative community, nonetheless, as it seems, overly obeisant to the scientific authority to oppose. To think is to differ.

Edmund Husserl's book from 1936 is a relevant analysis and penetrating critique unmasking the foibles of modern science too.[4] Specifically, when we express our reasoning in the language of mathematics, we often set conditions that weaken, even sever, verifiable connections to reality. Consequently, fundamental questions arise but remain open. Husserl recognized the deep historical causes and far-reaching consequences of this profound problem. The philosopher pointed out the nature and necessity of explanations, and above all, the major but often unrecognized obstacles to obtaining them.

Without further philosophizing, I lay out what can be understood solely by requiring concreteness and insisting upon consistency. In this way, we obtain some distance from the nebulousness of modern science and attain a clear connection to the history of science. We will not just marvel at the technical excellence of modern physics and revel in its achievements but will have the chance to discuss problems and share ideas, to know science in its most authentic form.

Many have seen that the current scientific problems stem from the disposition of contemporary science itself. In turn, I have certainly missed and therefore have not cited many meritorious works arguing for a more realistic worldview. I did not recognize the worth of such criticism before I awoke to the fact that the problems of science are not so much about Nature itself but rather about our own thoughts about Nature.

Preface to the English Edition

In the spirit of modern science, the book, now also in lingua franca, is intended to be available for everyone keen on comprehending reality.

PART I
THE NATURE OF EXISTENCE

We must begin with the mistake
and find out the truth in it.
That is, we must uncover the
source of the error;
otherwise hearing what is true
won't help us.
It cannot penetrate
when something is taking its place.

Ludwig Wittgenstein

1. Why?

The world is complicated
but regular.

"Why?" is the question we ask when looking for a cause. For example, an investigation into an accident aims to uncover the particular events that led to the incident. When causal connections between these antecedents and the coincidence are established, we are said to understand the course of events. The world is an arena of causes and consequences.

The relationship between cause and effect is generally recognized as a central law of nature, perhaps its most important one. However, I cannot recall an explanation of causality from any lesson or lecture. Ignorance is, of course, not a problem; insofar as when there is no understanding, there can be no misunderstanding either, which can be misleading, not to mention hard to eliminate. Back in my student days, I did not even think about the essence of cause and effect. But we all should be familiar with such a basic relationship, for it provides the necessary foundation for comprehending reality.

What, then, do we know about the law of cause and effect? Surprisingly little. Events follow one another in time, and yet we do not understand why time goes by and why things happen. When such a central issue is unclear, what kind of certainty do we have about the truth of contemporary knowledge in the first place? How do we not know what time *is*?

Starting point: We have in front of us a grand mystery waiting to be solved like a murder in a detective story. There is a lot of evidence about the march of time but a shortage of inference. What is the agent of time? What is its motive?

We are not the first to be hunting down the natural law that relates causes to consequences. Throughout history, it has seemed clear that the course of events cannot be random. There must be a governing rule, since the same patterns emerge from a wide range of processes:[1] a spiral galaxy looks like a cyclone; a neuronal network is much like a telecommunication network; a shrub with branches resembles lungs with bronchi; bacterial colonies and urban areas spread in matching ways. This *Grand Regularity* of Nature is newly on display in vast archives of data, but the idea of the unity of everything is ancient.

WHAT IS THE CAUSE?

The dream of comprehending the world through a single principle was reawakened during the Enlightenment. Notably, the work of Sir Isaac Newton pointed toward a unified worldview. In the preface to *Principia* (1687), the natural philosopher introduced forces and motions. A force whatsoever is *a cause* of a change in motion, and a change whatsoever in motion is *a consequence* of a force. Causes relate to consequences through Newton's second law of motion.

In the mid-1700s, the French polymath Pierre-Louis Moreau de Maupertuis used the same Newtonian principle, formulated in energetic terms, to explain both the passage of light and the motion of celestial bodies, as well as the proliferation of life, the essence of consciousness, and the imperative of economic growth.[2,3] Likewise, at the beginning of the 19th century, Sadi Carnot, the founder of thermodynamics, showed that machines also operate following the same simple principle.[4]

It was revolutionary to realize that the whole of Nature complies with the law given in a mathematical form. Today, we know more about atomic structure, cellular metabolism, connections in neuronal networks, and transactions in the global economy, but our knowledge is fragmenting into discipline-specific descriptions. However, do the different phenomena differ in principle? Isn't it a force that causes a stone to fall, a plant to grow, a signal to transmit, and a company to prosper? So why did we abandon the old but general law of causality?

Might it be that this universal principle of the Enlightenment, while beautiful, perhaps offers too perspicuous an explanation? On the one hand, complexity in itself should not pose a problem. Contemporary physicists handle massive datasets and even model the expansion of the whole universe. On the other hand, there is a problem if a theory does not match the data. And there is, for sure, a welter of issues. For example, we have not been able to directly detect dark matter or dark energy, even though they are thought to encompass more than 95% of the universe.[5] Nor can we precisely explain why there is so much excess material in our DNA, with over 95% of the genome of most organisms being seemingly useless.[6] Moreover, why does the world economy not obey our economic theories, but instead, frequent crises take us by surprise? Could these disparities only stem from our failure to measure numbers to enough decimal places, or do they originate from our misunderstanding of the leading digit?

Perhaps there is no universal law at all, contrary to the beliefs of the Enlightenment. Isn't the whole idea that events are guided by a natural law implausible? Wouldn't that imply some ultimate objective, a final cause, as understood by Aristotle? Science does not recognize or acknowledge such a teleological explanation, an intention, a purpose in Nature. But instead, it relies only on detailed observations and precise measurements to draw its conclusions. Indeed, do we have a shred of evidence that all processes result in regularity by complying with a general principle?

Had someone asked me this twenty years ago, I would not have even understood what regularity we might be seeking with this line of reasoning. At that time, I studied the structures of protein molecules, the building blocks of life. Yet I should have had a clue, knowing that these molecules of life have a common origin. Biochemistry is not a hit-and-miss affair: proteins are mutually related, much like organisms are relatives of one another. As such, I was well aware that the structures of complex biomolecules were also generated through molecular evolution.

Evolution is not random; it is a law-like process. In Darwin's words, viable molecules, cells, and organisms are *naturally selected* from variation. Of course, I knew this all along. Even so, I did not grasp that evolution is just sequences of events in which causes give rise to effects. That is all there is to it.

It is high time to examine this worldview-shaking tenet that evolution does not make a distinction between the living and the lifeless, the microscopic and the cosmic, or the simple and the complex, but that all courses of events follow natural law instead of being the result of a random walk.

ARE THERE SIGNS OF REGULARITY?

Today, the spectrum of our knowledge extends from elementary particles to enormous galaxies and from the richness of genes to the abundance of species. We know a whole lot about cellular regulation, as well as about social relationships. We know a good deal about the nexuses of neurons in the brain as well as about the connections of companies in the global economy.

As startling as it is, these data are highly similar, regardless of what we look at. Universal characteristics[7,8] are evident in immense masses of information called *big data*. The world is clearly not random but regular. Could it be consistent with just one single rule?

Unless headings and units are labeled in each descriptor of different datasets, we cannot say when just looking at the data from where the data originates. As an illustration, the length distribution of genes in a genome looks much like the length distribution of words in a book. The lengths of words vary from language to language just as those of genes differ between organisms, but these scale-free distributions are skewed alike. Medium-length words are the most frequently used. A short word may be deft, but a few sounds cannot be combined into many unique words. Conversely, as long words are laborious to use, exceedingly long words are rare. Does that mean that *survival of the fittest* is a decisive factor, perhaps a universal criterion, not only of length but also any attribute?

The lengths of genes vary like the lengths of words. A short piece of DNA is long enough to instruct the synthesis of many a small hormone. However, making the actual building blocks of life, the proteins, requires lengthier blueprints, but not at any cost, as there are very few extremely large proteins. The situation at your local library is analogous: there are a lot of ordinary-size books but very few lengthy tomes. The reason is apparent: such an assortment meets the readers' needs. Does this equivalence of the distribution shape imply some ultimate purpose or profound principle?

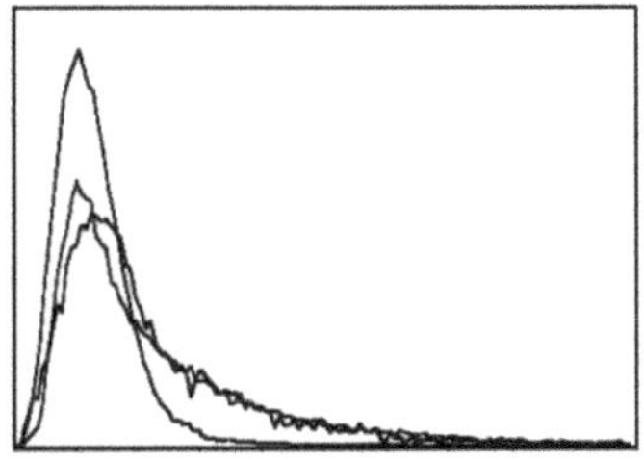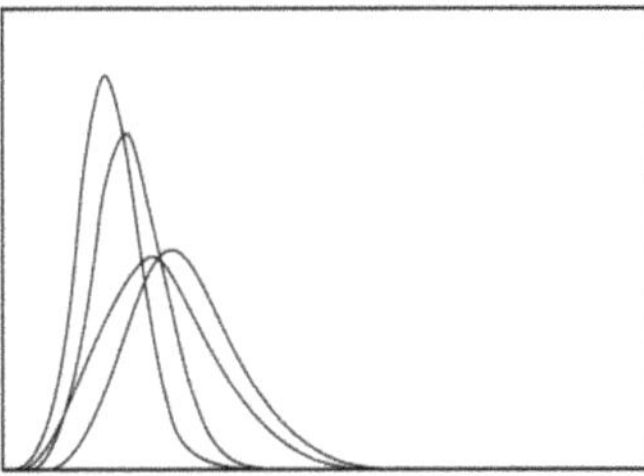

Length distributions of genes[9] (left) and words[10] (right) are skewed. Relatively few long genes or long words exist. When there are no headers and axis labels, the data's provenance is shrouded. Thus, it becomes apparent how the names and measures we have given to various things can kindle in us an illusion of fundamental differences between them.

The distributions of animal and plant populations in an ecosystem are skewed like genes in a genome and words in a book. There are many small fish and tree saplings, whereas Chinook salmon are rare, as are giant redwoods. Distributions of wages and wealth are also skewed: many are quite poor, very few are super-rich. The size distribution of earthquakes looks like that of the activated cortical areas in the brain,[8] with a huge quake being as rare as an immense sensation. Conversely, a slight shivering of the ground is as ordinary as a minor stimulation of the senses.

Similarities are found everywhere. There are more and more animal and plant species in larger and larger areas. For example, small islets serve as habitats for but a few bird species, whereas larger islands are home to many more species of birds. The number of vocations, too, increases as the economy develops over time; technological progress has created digital careers.

When zooming into the depths of the night sky with a powerful telescope, galaxies pass by[11] at a similar relative frequency to junctions when driving on a highway.[12] In the center of a cluster of galaxies, neighbors are close to each other; in the suburbs, road-crossings are near one another. At the edges of the cluster, as in a trackless wilderness, there is a lot more space.

What is it that underlies this *Grand Regularity* that is evident in our heredity and language and apparent in the food webs of ecosystems and the structures of human societies? When similarity ranges from the fine details of matter to the vast structures of the cosmos, could it be that all processes follow one and the same law of nature?

Yet another example of *Grand Regularity* is the branching of a nerve cell, which is similar to the branching of a tree.[13] The trunk forks here and there, while the branching quickens and ends in many leaves at the top. The distribution of branch lengths from the base to the ends is skewed in a universal manner. The units and scales vary from system to system, yet the form is ubiquitous regardless of the source.

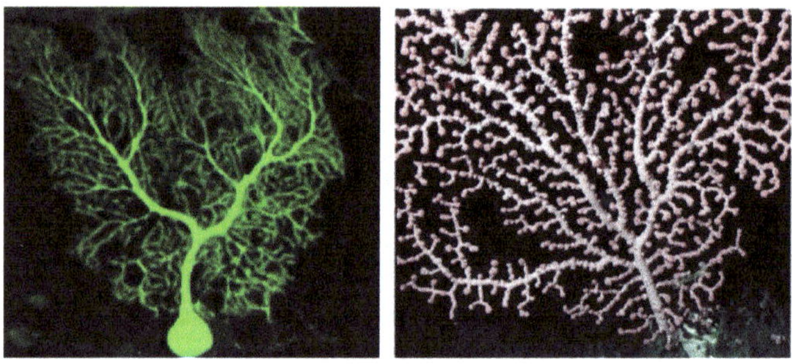

The similar branching of a nerve cell[14] and a coral[15] suggests that their principle of organization is the same.

Natural spirals, such as clamshells, the heads of flowers, hurricanes, and galaxies, all whirl in a similar manner.[16] The dense center curls tightly, whereas the sparse outskirts swirl widely. This skewed distribution of matter is evident to us directly, without any analysis or theorizing.

The similarity of the data across scales is inconceivably broad. It must be regarded as incredible unless we can see a common cause. The greater the number of different phenomena that share the same shape, the more general the explanation we should seek. Newton was likewise after the same explanation for similar natural phenomena in his rules of scientific reasoning.[17] If any system behaves in the same manner as any other, then everything should be of the same content, fundamentally commensurable at the basic level. Thus, we are led to track down the fundamental universal law of nature.

The similarity between galaxies, hurricanes (left), heads of flowers (middle), leaf positioning, and molluscan (lower right) spirals suggest the same governing principle.[18]

At first, it may seem rather absurd to compare arbitrary data with no common unit of measurement. Nonetheless, this is how we break free from the barriers of fragmented knowledge to an awareness that the world is amazingly similar everywhere. Conversely, our view of reality would be incoherent if we were to describe some particular system as profoundly different from everything else. Yet scientists do just that today. They seem unable, for instance, to relate dark matter or dark energy to anything that we already know.

Moreover, since Einstein, physicists have come to the conclusion that space is devoid of any substance, despite our sensing something that causes gravitational and inertial effects. Biologists, in turn, tend to think that there is some difference between the living and the non-living but are unable to define it. Likewise, neuroscientists wonder about the essence of consciousness because they fail to recognize its characteristics elsewhere. In contrast to these divided views, *Grand Regularity* suggests a deep unity among the void and matter, living and non-living, conscious and unconscious.

Sequences of events range from orderly oscillations to chaotic courses. Atoms vibrate in a molecule as signals oscillate in the central nervous system.[19] The economy fluctuates in the same way as predator and prey populations vary from year to year. There is chaos in market turmoil as in atmospheric turbulence. Chaotic processes are not altogether random either; they, too, exhibit *Grand Regularity* since significant events are rare and insignificant ones frequent.

This recurrence of patterns is not new or numbing. On the contrary, we use metaphors to talk about sameness, but we haven't determined the cause of the similarity. We have modeled the regular forms in mathematical terms, but we haven't explained the cause of the regularity. The narrative in words and data in graphs give us descriptions, not explanations. We need a universal theory in a mathematical form for quantifiable accounts of data. Such a valid theory is not based on data but on a fundamental assumption, a postulate, an axiom from which the interpretation of data follows.

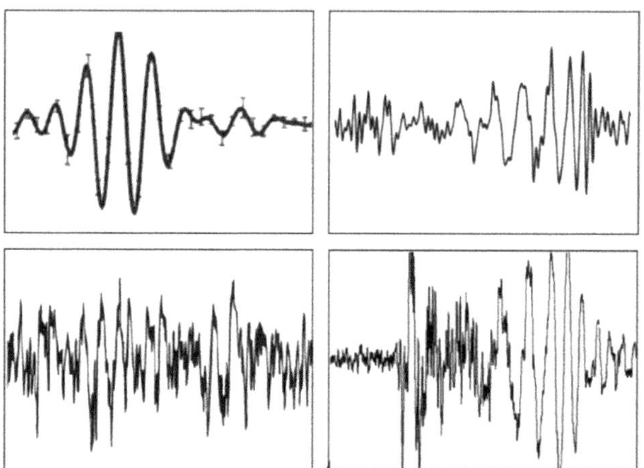

Abrupt changes cause ripples. A pulse of laser light agitates electrons[20] (top left). Gravitational waves arise when two black holes merge[21] (top right). Stock prices fluctuate unpredictably[22] but not all arbitrarily (bottom left). A 5.8 magnitude earthquake was recorded on August 23, 2011, in Virginia (bottom right) (WVGES).

Does this *Grand Regularity* emerge across processes because the same universal law of nature governs them all? The idea is astounding. Even so, could it be true? Water finds its way to the sea; a plant turns toward the light; an animal seeks food; a company pursues profit. Do we also display in our behavior nothing but one inexorable natural law?

So it seems. Regularity is also apparent in our cultural habits. We shake hands with our right hand, except for members of particular groups, e.g., the Scouts and Guides. A right-handed convention for vehicular traffic is the rule in many countries, with notable exceptions. Furthermore, screws and nuts are usually right-handed. Presumably, the right-handed majority set the standard. Nonetheless, counter-clockwise threads, too, remain useful for particular purposes. Not only screws and nuts but also numerous industrial components are standardized, compatible, as they say.

At the core of existence, rules are more stringent than standards in industry and norms in a society. Atomic nuclei are positive, and electrons are negative. Antimatter elements, where positrons circulate negative nuclei, are almost nonexistent in the universe. Similarly, the chemical structure of natural amino acids is left-handed. Their right-handed mirror-image compounds are almost absent in the biosphere.

Standards are helpful, for they help to make things happen. For example, a conventional measurement system is a pragmatic agreement, and a common currency is a convenient means of payment, if nothing else. We understand this compatibility: an incorrect component jams the assembly line as a poison blocks the metabolism. It seems that the higher the degree of standardization across a system, the more profuse its interactions. Could it be that the cause of standardization is the same for matter as for habits?

WHAT EXPLAINS *GRAND REGULARITY*?

Many growth curves follow the form of the letter "*S*", i.e., they are sigmoidal. For example, a bacterial population grows in this way. The growth spurts of children and young people are also sigmoidal. Chemical reactions proceed and economies progress likewise. The French sociologist Gabriel Tarde discovered that innovations spread similarly to epidemics.[23] The universal patterns have been noticed and modeled but not yet explained.[24]

In the middle of the 19th century, the Belgian mathematician Pierre François Verhulst found a mathematical function that matches many datasets of growth.[25] Verhulst's logistic curve, however, does not say why growth is sigmoidal; it only follows the data. Moreover, variation in fossil diversity in geological strata shows that speciation bursts as the growth curve shoots up.[26] Subsequently, evolution comes close to stalling for eons. However, why the course of events first soars and then almost stops is still without a clear explanation.

The extreme values of many datasets extend far beyond the arithmetic average.[27] To give an example, there are only a few large islands, as there are only a few super-rich people. As the English chemist Francis Galton and the Scottish doctor Donald MacAlister realized, the long tail of the skewed distribution can be squeezed when the plot axes are marked at even intervals with orders of magnitude (i.e., 1, 2, 3) instead of their numerical values (i.e., 10, 100, 1000). After this mathematical transformation, known as the logarithm, the distribution looks almost like a normal distribution. In other words, natural distributions are nearly *lognormal* but not *normal*. The Gaussian curve, already familiar from our school days, is not found in Nature, only in books. This normally distributed curve, symmetrical about the arithmetic average, is certainly *abnormal* in Nature, where the outcomes of natural processes extend beyond the spread of sheer coincidence.[24]

Nevertheless, as Gabriel Lippmann said, "Everyone believes in the normal law, the experimenters because they imagine that it is a mathematical theorem, and the mathematicians because they think it is an experimental fact."[28] This Luxembourgian physicist reminded us that repetition makes a thing familiar. Soon the familiar notion is taken as the truth, while it may not be true – merely a convention.

The world is statistical. Yet Nature's statistics are not the random variation of the normal distribution but the regularity of skewed distributions.[29] So, what is the causal law from which *Grand Regularity* follows? Mathematically speaking, what is the law of nature that underlies the statistical law of large numbers and the central limit theorem? Jacobus Kapteyn was looking for the answer. In 1903, the Dutch astronomer, who had an interest in biology, asked, "What is the reason for the widespread occurrence of just this [lognormal] curve?"[30]

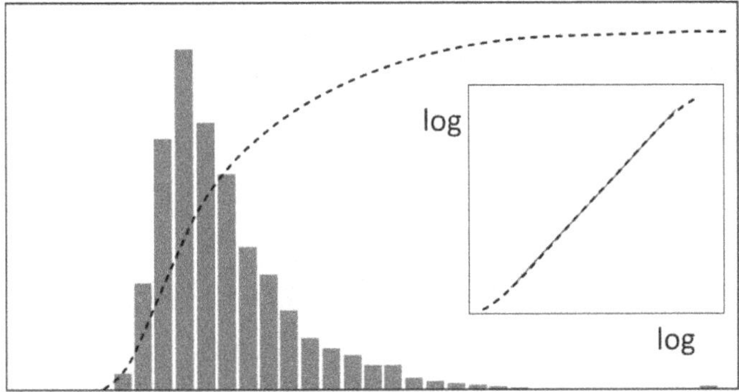

Natural distribution is skewed with long tails. The histogram columns extend far above the average, not the arithmetic but the geometric mean. This implies that the unexpected is to be expected. The nearly lognormal distribution accumulates in a sigmoidal manner (dotted S-shape). This cumulative curve, in turn, closely follows a straight line when the logarithm is taken from the horizontal and vertical axis values (inset). Thus, the different formats of data display the same regularity.

The skewed distributions are alike; so are their S-shape sum curves. These curves cumulate such that all the preceding values of the distribution are added together at each point. At its final score, e.g., all fish caught in a fishing net are tallied up from the smallest to the largest. Thus, regardless of the subject matter, the cumulative curve of a nearly lognormal distribution climbs up in an S-shape and therefore follows mostly a mathematical form known as the power law. It is a straight line on a log-log plot. The representations of the *Grand Regularity* are thus convergent.

At the beginning of the 20th century, the Italian social scientist Vilfredo Pareto[31] and the American linguist George Kingsley Zipf[32] realized that the power law is ubiquitous. The rule of thumb is, for example, that 20% of game company customers bring in 80% of the income, and 20% of accidents cause 80% of injuries. This ballpark figure is a handy approximation of the sigmoid curve. The 80/20 rule, the law of the vital few, agrees well with the outcomes of many natural phenomena and human activity.

Besides the power law, say, Pareto distribution, there are also other mathematical models of the data, yet they are only models. Instead of merely modeling natural phenomena, we are looking for a natural law that explains these ubiquitous patterns. Is the leitmotif, the *Grand*

Regularity, a manifestation of a physical principle? Is it the solid ground upon which we could build a scientific worldview?

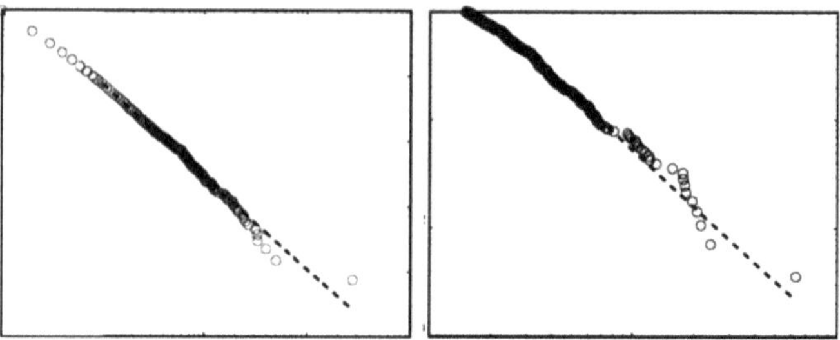

Severe accidents happen rarely, small ones all the time. When the full dataset is presented on a logarithmic-logarithmic graph, it mostly follows a straight line (left). There are many small caverns in the Earth's crust but only a few extensive cave systems. On a log-log graph, the data is chiefly on a straight line (right). When there are no headers or axis labels, the data do not show whether accidents or cavities or something else entirely is being displayed.[33]

The American astronomer Simon Newcomb noticed in 1881 and the American physicist Frank Benford again in 1934[34] that the first or most significant digit is the number one in about 30% of cases and the number nine in less than 5% of cases. This rule applies to a number sequence, such as the Fibonacci series, and the value of a physical constant, such as Boltzmann's constant. The incidence is not random but tends to follow the power law. Why?

Regularity is also reflected in the size, form, anatomy, physiology, and behavior of animals. This pattern was noted by, among others, Galileo Galilei[35] in 1638, and subsequently scientists Otto Snell[36] in 1892, D'Arcy Thompson[37] in 1917, and Julian Huxley[38] in 1932. The bones of an elephant are, of course, much thicker than those of a mouse. But isn't it remarkable that the relationship between body weight and bone thickness abides by the same mathematical law for all mammals? This isometric scaling, also known as allometry, results from a chain of events where each stage of development follows from all the previous steps, from history. Why are these path-dependent passages, such as proportionate growth, similar across species? That is what we seek to explain.

In 1926, the American biophysicist Alfred Lotka noticed that most scientists publish relatively few papers in any given period of time while only a few publish many, such that the number of scientific publications per author closely follows a power law.[39] Derek John de Solla Price, a British physicist, advocated 1965 a similar model for the growth of citation networks.[40] Today, we know that likes per post and tweets per person distribute in the same skewed manner. The English mathematician, physicist, and peace activist Lewis Richardson noted in his 1948 book that the destructiveness of wars also follows the power law.[41] These data are worth pondering. What are the forces that are pulling us? What are we being attracted to?

In the late 1980s, the Danish physicist Per Bak stirred up a vibrant discussion by pointing out that the most complex systems show the same simple regularity, regardless of the details.[42] Stephen Wolfram, the creator of the Mathematica software, demonstrated that primitive computer programs, so-called cellular automata, suffice to generate complex but nonetheless regular patterns.[43] Cellular automata are also familiar from the English mathematician John Conway's computer game *The Game of Life* in the 1970s.[44]

By the turn of the millennium, the physicist Albert-László Barabási, mathematician Steven Strogatz, and sociologist Duncan Watts had shown that the World Wide Web, cell regulatory and metabolic networks, and social networks are also nearly scale-free.[45] In other words, the system looks the same even when we zoom into smaller details. Within each node of the net, there are, on the one hand, many nodes with very few links and, on the other hand, few nodes with many links. A case in point, we are highly socially integrated, as everybody on this planet is, on average, only six connections away from anyone else. This independence of scale is in line with the power law, the characteristic of *Grand Regularity*.

Events may take their unique courses. Nevertheless, they are very much alike. Technology moves from one innovation to another. Prototypes are followed by first-generation products, then second, third, etc. Ultimately, the mature product becomes so ordinary that no one is interested in what generation it represents. An ecosystem evolves likewise from one species to the next during ecological succession. In a fire-cleared area, mosses grow first, then grasses, soon shrubs, and finally trees. The growth curves are alike. These similar curves imply a universally applicable principle, a *constructral law*, as Adrian Bejan, a professor at Duke University, refers to it.[46]

Complex systems science is the new discipline that models this type of scale-free similarity across subjects.[47] For instance, cities swell into the surroundings in the same manner as fluids percolate into rocks.[24] The mathematical models of lognormal distributions, S-curves, and power laws approximate various growth forms irrespective of the field. For example, Robert Gibrat, a French engineer, proposed as early as 1931 that most firms, independent of their sizes, grow at a proportional rate, yielding the approximate lognormal size distribution of firms.[48] In turn, the physicist Eugene Stanley with his collaborators found in 1996 that the growth rate of firms follows a power law.[49] Company lifespans also exhibit the same universal pattern, as Geoffrey West explains in his book *Scale: The Universal Laws of Growth, Innovation, Sustainability, and the Pace of Life in Organisms, Cities, Economies, and Companies* (2017).[50a] This theoretical physicist admits that although the consequences of these occurrences are everywhere, the cause of *Grand Regularity* is not known.[50b] He hungers for a grand unified theory.[50c] Maybe such an understanding could help us redirect our way of life into a more sustainable mode of existence.

When the provenance of the data is not indicated and the scale is not specified, the plots of different datasets become interchangeable by stretching or shrinking the axes. Obviously, the way in which the data is shown does not change the data itself. Nature does not distinguish between the animate and the inanimate, the minuscule and the gigantic, the basic and the complicated. Scales and other labels are our inventions and conventions. The United States uses inches, pounds, and gallons for the same things that people in most countries talk about in metric dimensions.

While *Grand Regularity* is on display all the time, its ubiquity may cause us to pay it little attention. Its universality was not fully realized until a wide array of self-organizing, spontaneously assembling, organically evolving systems were studied. The electrical activity of neurons is synchronized like the pulsing glow of a swarm of fireflies. Small robots flock like birds and fish.[51] Emperor penguins move about in breeding colonies like particles in fluids.[52] We walk smoothly, even though nerve impulses flow in our muscles spasmodically, or – more likely – because of that. Society works efficiently, even though the tasks and chores of people differ somewhat from day to day, or – more likely – because of that.

In the 1980s, when I was an undergraduate, all sorts of digital data were beginning to be amassed. Already then, I knew from scientific journals that simple mathematics accounts for diverse data astonishingly well. However, I did not yet crave an explanation for this *Grand Regularity*. It did not occur to me that the various phenomena could, after all, have something in common or perhaps a deep connection.

If you ask for the reason behind something so obvious yet overlooked, the question itself may already point to the answer. The query itself makes you aware of what to look for. In general, science focuses on those unknown phenomena thought to be knowable. But the cause of *Grand Regularity* does not seem to be contained within our theories, which suggests we need to think differently. That is why we should start from scratch and progress from personal experience to scientific thinking.

The *Grand Regularity* stems from the same root as Galileo's idea that every phenomenon in Nature can be represented in mathematical form. Husserl reminded us that the Pythagoreans already knew that the length of an instrument's string determines the pitch of the sound it can produce and other mathematical dependences. Still, the generalization of these connections into mathematical laws had to wait for Galileo.[53] To him, once represented mathematically, the fall of an object was an example of a universal law. To us, a set of observations is now an example of regularity, a general rule that we wish to find.

We have now gone through the facts. The ubiquitous patterns of skewed distributions, sigmoid curves, spirals, power laws, and even chaos accumulate from processes over periods of time. This weight of the evidence points to *time* as the culprit for *Grand Regularity*. Next, we must choose the line of inquiry that will allow us to catch the carrier of time and understand the driving force that makes things happen.

HOW ARE THE LAWS OF NATURE FOUND?

The regularity found in the datasets is salient. But does it hold good in reality? What if *Grand Regularity* is only a figment of our imagination? Surely the pervasive patterns and similar shapes in themselves mean nothing and prove even less.

Every observation would indeed remain meaningless without some form of interpretation. Invariably "something is understood as something". This adage of the German philosopher Martin Heidegger motivates the theory of interpretation (hermeneutics). Professing the regularity does not mean anything significant is also an interpretation. If it were a pure coincidence, that would be incredible, as events would have no connection whatsoever to each other.

Then again, is it merely our ordering of things, from the largest to the smallest, from the fastest to the slowest, et cetera, that produces the regularity? Is the similarity we observe due solely to the mere fact of our putting things into a serial order? We may have our doubts, and yet such an interpretation implies that our subjective sorting of observations deviates from other natural processes. Instead of accepting such an inconsistency, we should reason logically that all events involve subjects. Ultimately, is the idea of a purely objective view, in fact, a delusion?

We do not have to agree about the significance of *Grand Regularity*, for unanimity is not the goal of science. Truth is. As the philosopher of science, Karl Popper, pointed out, "The growth of knowledge depends entirely upon disagreement."[54]

We can debate, for example, how similar datasets really are, as we tend to think that a law of nature means a course of events without alternatives. However, the world does not seem to be deterministic, or for that matter, indeterministic. Random processes lead nowhere and therefore deliver symmetrical distributions, whereas natural processes produce skewed distributions.

It is rather remarkable that even today, the character of the natural law remains ambiguous.[55] While physicists reason that gravity, electromagnetism, and the weak and strong nuclear force were once united and branched out from a common stem at an early stage of the evolving universe, wouldn't it be more reasonable to suppose that the forces are contingent upon causality itself? Philosophers, in turn, argue that there must be a sufficient reason for causality, too.

We tend to think that small deviations in data are due to random fluctuations or measurement errors. However, the variation is not random but has its causes, however small and momentary. Since neither determinism nor indeterminism explains the universality of patterns, we must look for a nondeterministic, unpredictable, yet causal law that accounts for historical contingency. According to such a law, there would be no random deviations from an average, so to say, from an ideal course of events, but all courses of events would be relevant because even the slightest consequence has its cause.

We know this by experience. A measurement is inaccurate when the object under inspection moves about. Often, many factors affect the result; for example, an individual's height is influenced notably by nutrition and genes shaped by past generations' diets. Therefore, the statistical variation is not normally distributed random variation but skewed, stemming from causes and consequences.

Admittedly, it may seem strange that the law of nature could describe regular yet unique processes. It is difficult for us to doubt the truths of our time, just as it was for past generations to doubt the truths of their time. From antiquity to the turn of the 17th century, it was believed that the more massive an object was, the faster it would fall. Even before Galileo, the Dutch scientists Simon Stevin and Jan Cornets de Groot, by letting two lead balls, big and small, fall from the bell tower of Delft's new church (*The Beghinselen of Weeghconst*, 1586), showed that Aristotle was wrong in maintaining that motion requires force. Similarly, we must consider whether the present doctrines agree with the observations. Steven Strogatz pointed out that theorizing can cut both ways: "The art of abstraction lies in knowing what is essential and what is minutia, what is signal and what is noise, what is trend and what is wiggle. It's an art because such choices always involve an element of danger; they come close to wishful thinking and intellectual dishonesty."[56]

We have categorized phenomena, created concepts, and set up disciplinary boundaries, but instead of splitting the unity of Nature, we should discuss phenomena across disciplines in an integrated manner. Newton did this by overturning the old dogma that the mechanics of the heavens differed in kind from the mechanics of the Earth.

As we seek the explanation of *Grand Regularity*, rather than new scientific results, general education helps us unify things. But, in this quest, we must also prepare ourselves for an unexpected explanation. We know from the history of science that a revolutionary perspective has often been diametrically opposed to the prevailing opinion.

Ptolemy assumed that the Sun revolves around the Earth, while Nicolaus Copernicus inferred the exact opposite. Aristotle thought that force maintains motion, whereas Isaac Newton understood that motion continues in the absence of forces. The English natural philosopher Joseph Priestley assumed oxygen was air from which an imaginary substance, phlogiston, released in combustion, had been removed. The French chemist Antoine Lavoisier realized that there is no phlogiston. Instead, oxygen in the atmosphere reacts in combustion. Revolutionary views have often been more straightforward and comprehensive than their predecessors. We should expect the same.

Many geniuses have been polymaths rather than specialists. People with broad interests are disposed to be the first to discern similarities, which are then used to organize observations into a theory. Against the backdrop of history, we should proceed likewise.

Observations

> Many changes to the worldview have begun by noticing a regularity. For example, the German astronomer Johannes Kepler realized that planetary orbits are proper ellipses. Now we see that regular patterns are the outcomes of all kinds of processes.

Concepts

> At one time, John Dalton, an English chemist, physicist, meteorologist, understood the atomistic idea underlying the regular weight ratios of chemical compounds. Now we should find the axiom underlying *Grand Regularity*.

Law

A law of nature states a relationship. For example, Newton's law of universal gravitation relates the gravitational force to masses and the distance between them. Now we should seek out the law of nature that relates causes to consequences and gives rise to *Grand Regularity*.

Explanations

A natural law explains a broad spectrum of observations. For example, the Scottish scientist James Clerk Maxwell's theory applies from X-rays to radio waves. Likewise, we should expect to find a universal law of nature that explains *Grand Regularity* beyond our own experiences.

The English philosopher Francis Bacon introduced a scientific method to distill a common rule from systematic observations. Natural laws have been found in other ways, too. Newton used Galileo's approach, in which one's own experience is mathematized into a general law. According to the myth, Newton saw an apple fall from a tree and generalized the falling apple to the universal law of gravitation. Einstein, on the other hand, mathematized the idea rather than his experience of free fall into general relativity. Maxwell, in turn, constructed the theory of electromagnetism by insisting on consistency among the equations that had already been found. In the same manner, the German physicist Max Planck discovered the law of radiation.

When equations are derived from equations without a connection to the underlying experience, it is not apparent what the theory means, and the mathematical becomes mystical. Equations are then interpreted as reality instead of reality being structured in equations. The mathematics of quantum mechanics has been taken to connote that there are universes parallel to our own. Since antiquity, natural philosophers have pondered whether mathematics can be interpreted as real or only as describing reality.

The mathematical equation expresses the axiom or the postulate as equivalence. According to Einstein, the forces experienced as gravity and acceleration are equivalent. Unless the equivalence is understood and hence respected, general relativity will turn in the hands of successors into a malleable mathematical model with pliable parameters, such as dark matter and dark energy, in order for the calculations to match the observations. Physicists take this constructed correspondence between astronomical data and the cosmological model as evidence for the existence of dark matter and dark energy, whereas for philosophers, dark matter and dark energy, as long as they remain without any substance, remain only parameters of the model.

Galileo's method seems to yield a theory of physics firmly related to reality. Hence, let us use it to find the law of causality and its axiomatic basis by structuring our experience of time into a mathematical form. Let us look for meaningful explanations rather than intellectual challenges. Let us rely on common sense rather than established theories when interpreting observations. When we construct a theory in this way, there is a certain familiarity and unquestioned confidence.[53]

Honesty suffices. Let us not embrace an explanation unless we understand it, however celebrated it may be. Let us not accept an intellectual achievement, although it may be marvelous in its intricacy, unless we understand how it was made. For an explanation to qualify as such, it must tell us what the thing is, what it does, what its structure is, and what follows from it. By not legitimating anything incomprehensible, our worldview shall remain a sound guideline.

Naturally, we hesitate to begin the undertaking. Why should we figure out the relationship between cause and effect? Are such grand missions not rather the domain of leading scientific teams? Paradoxically, the big questions at the heart of reality are open for anyone to explore. Otherwise, they would not be big. Often, the very thing directly under our eyes is the most difficult to perceive. In the past, the foundational questions pointed to the essential truths; they might do it again, so let us be brave enough to ask them anew.

KEY POINTS

- The similarity of numerous datasets suggests that all processes follow the same law of nature.
- Our own experience is a credible way of finding this law.

2. WHAT IS TIME?

Time is seen as an enigma
but comprises periods of quanta.

Time is not only a riddle in its own right but of everything that emerges with it. For instance, the history of humankind accumulates from the unique lives of each person, the evolutionary tree grows over eons from the lives of every single species, and the course of our home galaxy contributes to the overall evolution of the universe. The associated ubiquitous patterns, skewed distributions, spirals, and S-shaped curves come into sight with time in structures on Earth, from minuscule molecules to magnificent mountains, as well as in structures of the cosmos, from dwarf galaxies to vast voids.

When we examine the course of events in the finest detail, we realize that the flow of time is synonymous with the flow of the fundamental elemental constituents of Nature. These are quanta of light. They carry time because time *is* the period of a quantum.[1] The concrete conclusion is straightforward, but the trail past the abstractions to its revelation has been devious.

THOUGHTS ABOUT TIME

We experience time passing, but the experience itself lacks a theoretical formulation. Every process involves a passage of time, yet the essence of time is unclear in the equations of contemporary physics. There is thus an enormous blind spot in our scientific worldview. For us to see clearly, the flow of time should be brought into the form of a natural law.

We readily use notions of time: just in time; time flies; only time will tell. Yet, we have a devil of a time defining time itself. We use time to relate events to one another, but we are not quite able to relate the concept itself to anything. So, why is time instinctively felt on the one hand but beyond our ken on the other?

Some thirty years ago, it did not even cross my mind to think about the essence of time. I did not know how to get hold of something that I had no grip on at all. My arrogant attitude was also an obstacle. It paralleled the saying, customarily credited to Richard Feynman, "The philosophy of science is as useful to scientists as ornithology is to birds." The American physicist, known for his lively stories, was a pragmatic thinker and understood the nature of our thought process: "The first principle is that you must not fool yourself – and you are the easiest person to fool."[2]

Wise words. But how do we make sense of seemingly insubstantial concepts that underlie physics, such as time? As Aarne Oja, my supervisor, once said: "You can't formulate a tenable theory without first getting your hands dirty." You can think big as long as you get a good handle on your thoughts. So, how do we get a hold on time?

Heraclitus' well-known verse conveys the ancient ideas about time: "Everything changes and nothing stands still." The philosopher expressed the irreversible passage of time: "No man ever steps in the same river twice, for it's not the same river and he's not the same man." This remark that nothing can change unless everything else changes is both plain and profound. The evolution of life depends on the evolving universe. The elements, sunshine, the cold night sky, and other prerequisites for life are fruits of the evolving universe.

Aristotle, too, sensed his existence in relation to the past and the future. Today, our comprehension of time has become obscure, as our understanding of the wholeness of Nature has been shattered across disciplines, each expressing its own points of interest with specific concepts. Be that as it may, experience is the mother of wisdom. Each of us is somehow intuitively aware of time. This experience is what we should express in scientific terms.

Newton considered time as ideal and absolute, not, strictly speaking, part of the physical but the mathematical world. This stance is reflected in this key passage from the *Principia*: "Absolute, true, and mathematical time, of itself, and from its own nature, flows equably without relation to anything external, and by another name is called duration: relative, apparent, and common time, is some sensible and external (whether accurate or unequable) measure of duration by means of motion, which is commonly used instead of true time; such as an hour, a day, a month, a year."[3] Absolute time and space are thought to be the basis of Newtonian mechanics. However, for Newton himself, following Galileo, time and space were inexplicable axioms of the *Principia Mathematica* for geometrizing Nature, establishing theorems, and calculating the future positions of celestial bodies. Words can be confusing, and the original meaning between natural phenomena and their mathematical idealizations can easily be lost.

Newton's understanding that each object has an absolutely *unique* position in conjunction with the rest of the universe aligns with the current conception: the universe is expanding away from every locus. Each of us is thus at the center of the universe. While no one's position is too modest, no place in the cosmos is special, either. Apparently, the same laws reign everywhere. This inference, the so-called Copernican principle, implies the unity of the universe.[4]

WHAT IS THE PROBLEM OF TIME?

Although one of the most mundane matters, time is a big problem for physics. While we experience time to have a direction, the laws of physics, as we know them today, do not make a difference whether time flows from the past to the future or from the future to the past. So, "Where does the arrow of time come from?"[5] asked Arthur Eddington. The English astrophysicist became famous in 1919 for measuring how much a beam of light bends in the Sun's gravity.[6] The result agreed with Albert Einstein's general relativity, becoming the first proof thereof.

Theoretically speaking, everybody in the universe is immersed in space-time. Yet, general relativity explains neither the substance of space nor the flow of time. So there is a serious lacuna in our learning, given that history is on display everywhere. "Of all obstacles to a thoroughly penetrating account of existence, none looms up more dismayingly than 'time',"[7] wrote John Wheeler, a well-known physicist, a great figure of the bygone golden age of general relativity.

Contemporary physics is thus stuck with equations that yield no clue about what the flow of time *is*.

Time is expressly a problem of physics, as Ray Monk, a professor of philosophy, pointed out in his review of Lee Smolin's book *Time Reborn* (2013):

> The problem here is that the philosophical view, for which Smolin is arguing, is not one that many non-physicists would find particularly controversial. It is that time is real, a position that Smolin describes as a 'revolutionary view', but which, for most people, is just common sense. Of course, time is real! For most of us, casting anxious glances at the mirror as the effects of time reveal themselves in the aging process, it is all too real. To understand why this unexceptional, common sense assertion is regarded as revolutionary, one must, to some extent at least, understand how the world looks to modern physicists.[8]

Let us first clarify the nature of this problem for physicists and then focus on expressing our unproblematic everyday experience of time using the concepts of physics. This is how Galileo structured observations as mathematical laws.[10]

THE TROUBLE WITH PHYSICS

More than a hundred years ago, physicists encountered difficulties in explaining with their usual methods, which are now called classical physics, the results of some simple experiments. These conceptual problems led to modern physics. Nowadays, we have another issue: although calculations do match measurements, we do not understand how equations relate to reality. This paradox is at the heart of the whole trouble with modern physics.

Smolin, well-known for his nonfiction books, argues that physicists, dazzled by the beauty and success of mathematics, have rejected the true nature of time. In modern physics, time is a variable without substance. Smolin, a professor of physics at the Perimeter Institute in Waterloo, Canada, questions this stance and hence the whole of modern physics. Such a posture is revolutionary, indeed Copernican.

Smolin expects that when we find out what time *is*, we will also get answers to many other questions. The revolution will not be limited to physics but will revise our whole worldview because all processes embody time. Revelations of comparable magnitude have happened before. This is why we remember Galileo, Newton, and Einstein. Physics is of significance. What, then, is the true significance of time?

Einstein once noted: "The distinction between the past, present, and future is only a stubbornly persistent illusion."[i] In general relativity, as well as in quantum mechanics, the flow of time is without cause, so there are no consequences, either. Bodies move along their optimal paths; the planets orbit the Sun one cycle after the other; comets come and go. Within modern physics, it does not even make sense to ask why things happen.

Such a view of the world is strange to the man in the street. We don't understand any effect without some cause. In fact, the central tenet of science itself is that every single phenomenon in the universe can be shown to have a natural cause. Since the flow of time is a natural phenomenon, it seems reasonable that it, too, should be shown to have a natural cause. If a field of science stops looking for causes, it stops advancing.

The calculations by quantum mechanics reproduce the properties of systems, but the connections between these mathematical equations and physical reality are obscure, shaky at best. Erwin Schrödinger, one of the foremost architects of quantum mechanics, lectured once that the equation named after him does not outline alternative events but rather all possible events superposed.[11] We do not have any personal experience of such superposition. I am experiencing only the present, not a combination of all imaginable versions of the present. I exist only here and now, not as a superposition of all conceivable places. This is why quantum mechanics goes beyond our comprehension. It does not seem real.

As physicists and philosophers know, no experimental proof of superposition or entanglement, i.e., correlated superposed states, exists as a phenomenon in Nature itself. Quantum mechanics, like any

---

[i] Quoted from Einstein's condolences to the family of Michele Besso.[10]

other theory, only provides a framework for interpreting data. Are the prevailing interpretations true? Assessing the truth of the relationship between theory and observations is difficult. The argument is circular: scientists interpret the experimental data based on the theory and design new experiments based on the interpretation.

Since we cannot free ourselves from this hermeneutic circle, we must be skeptical about all observations, experiments, and theories. Given such a challenge, Robert B. Laughlin finds it lamentable that physics students are thrilled by the quirks of quantum mechanics. The Nobel Laureate of physics reminds us: "In science, one becomes enlightened not by discovering ways to believe in things that make no sense but by identifying things that one does not understand and doing experiments to clarify them."[12] Being fascinated with conceptual conundrums arising from theory and taking existing interpretations at face value is ultimately being unfaithful to science. As a young student, the weirdness of quantum mechanics enticed me, too. But later, I realized that initiation into mystics molds an apprentice into a magician rather than a physicist.[13]

According to quantum mechanics, superposed states collapse into one state instantaneously at the event of observation. At that moment, the outcome of an experiment is not a consequence of causes but is claimed to be a probabilistic event without any cause. Einstein's famous criticism of quantum theory, "God does not play dice,"[14] captures the foolishness of a belief that any consequence could result from mere chance without any proximate cause. Nature is causal in its ways, not mysterious. A phenomenon may appear random, but there is no guarantee that this is truly the case. Science does not have criteria for proving a phenomenon to be arbitrary.

While criticizing it, Einstein was unable to demonstrate that quantum mechanics is wrong. He drifted away from mainstream physics — the stream he had unleashed. A few years before his death, Einstein wrote to Maurice Solovin, an old friend: "You imagine that I look back on my life's work with calm satisfaction. But from nearby, it looks quite different. There is not a single concept of which I am convinced that it will stand firm, and I feel uncertain whether I am in general on the right track."[15] Thus, it is unclear how Einstein himself ultimately rated his own work. The implications are far-reaching because modern physics emerged in the wake of Einstein, and our view of the world transformed. Today, a century later, we may still wonder whether or not we are on the right track.

UP IN THE AIR

Imagine that the Accident Investigation Board announces that there is no cause behind an airplane crash.[16] The members of the authority just go on restating that their calculations agree closely with the data, and let us say, on average, one flight out of a million gets into an accident. The professionals ground their argument in technical terms of space and time, fields, and functions. They maintain that it is not even sensible to look for reasons behind the airplane crash because any particular crash is a totally random event. (Members who have argued against this doctrine have been dismissed.) Would you accept this?

This state of affairs, unbelievable as it may seem, is modern physics. The winged words of superposition and entanglement correspond to but do not explain data. Even worse, physicists don't seem to want to understand the theories they use.[17] By all appearances, the scientific community accepts this situation as canon. Hence scientists cannot pilot into investigating causes, that is, do what science is supposed to do. We need a realistic view of the world to cruise safely and not perish in catastrophe. "Roger, over."

TAKE IT OR LEAVE IT

The mystery of time has only deepened since the early 20th century. Quantum entanglement is not bounded by the microcosm of particles but leads logically to the existence of parallel universes. In 1958 Hugh Everett asked: "What if the Schrödinger equation always applies and applies to everything – objects and observers alike? What would such a world appear like to us?"[18] The American mathematician proved that quantum mechanics implies many worlds, parallel universes, the multiverse, if the wave functions are real and reality is independent of the observer. Take it or leave it.

If you take quantum theory seriously, your worldview is inevitably remote from the reality we know from our own experience. However, we cannot present any evidence of a parallel existence. If the theory cannot be put to the test, does it qualify as science? On the other hand, if you abandon quantum mechanics as absurd, how would you then calculate the results of some simple experiments? It was out of the question for the physicists to discard this productive even if a strange way of doing science, known as instrumentalism.[19] This stance has no ambition to explain phenomena but regards a theory merely as a means to model data after the fact. Our modern scientific worldview rests on this soft footing.

We cannot high-handedly label one outcome of modern physics as suitable, such as an accurately calculated atomic spectrum, and another as unsuitable, such as the multiverse. The oddities must not be ignored but ought to be illuminated. Even where we deem our reasoning to be right, our thinking might be wrong. Feynman emphasized that "the unknown must be recognized as being unknown in order to be explored."[20] Time without substance is an impenetrable abstraction. You need to name it to know it.

We experience the passage of time. Even so, textbook physics states that in the microscopic world of particles, the laws of physics are independent of time's direction. However, it is a thin line between the microscopic and the macroscopic. So, could it be that so long as nothing is happening, time just doesn't point anywhere? In other words, have we simply chosen to define the laws of physics to be free from time's arrow to attain the maximum precision of a steady state? Have we thus excluded the flow of time from our theories?

We have a hard time comprehending theories that provide numbers in agreement with measurements but no explanation of causality, time's arrow, for we experience vividly that the past is irreversibly distinct from the future. So, what's amiss with modern physics?

When modern physics is taken as real, reality appears fundamentally random, unfathomable. And yet, many are just enthralled by these enigmatic creeds. As a result, the popularization of physics muddles up fact and fiction. Is this the purpose of science?

Jim Baggott asks this question in his book *Farewell to Reality: How Fairytale Physics Betrays the Search for Scientific Truth* (2013). The physicist and science journalist thinks physicists underestimate people's reasoning ability.[21a] Trust in science is jeopardized as we can never test theories with such bizarre concepts. The truth is no longer sought. On the contrary, stories are concocted by building on the assumptions of modern physics. Intellectual vacuity expands in the face of imaginative parallel universes, supersymmetry between particles, curled-up dimensions, cosmic shortcuts known as wormholes, dark energy, dark matter, etc. Words mean nothing when they don't relate to anything real; only mansplaining elevates fiction to fact.

RIGHT FROM THE START

How can we tell the difference between science fact and science fiction? Whom can we rely on if not erudite experts? Carl Sagan's words are food for thought. "Arguments from authority carry little weight – authorities have made mistakes in the past. They will do so again in the future."[22] Should we then rely on our own reasoning and experience?

We expect theories to tally with reality – and the calculations of modern physics indeed match the data. But despite their consistency, the theories seem inexplicable. The problem of time might very well lie beneath this disconnect.

It is shocking that we do not understand what time *is*.[23] After all, causality is at the core of reality. "The world is not a collection of things, it is a collection of events," writes Carlo Rovelli, a physicist known for quantum gravity, in his book *The Order of Time* (2018).[24] What else do we not understand when we fail to grasp the quiddity of time, its essential quality?

Our nescience also threatens the certainty of what we think we know. This is nothing new. When the Scottish physicist Maxwell amalgamated electricity and magnetism, our vision of light widened.[25] It was thus made clear that the spectrum ranges from X-rays to radio waves. When Darwin made us see the common origin of species, we understood that we are part of Nature and not her lords. Presumably, when we realize what time *is*, we will also think differently about what we presume to know – about light and evolution too.

Such a suggestion may seem exaggerated. Despite the trouble with time, our concepts of reality cannot all be awry, can they? Not really, as the problem with time challenges our theories rather than our observations, unlike, say, in 1820 when Hans Ørsted noticed, to his surprise, that a compass needle swayed when he plugged electric current into a nearby wire. The phenomenon appeared almost supernatural to the students attending the renowned physicist's lecture. By contrast, we find the passing of time entirely natural. It is just that we are not fully conscious of time, for our understanding of time has not yet been structured in the form of a scientific theory.

When a problem, such as that of time, has been bothering us for so long, we tend to deem it difficult to solve. However, perhaps we ought to be thinking differently. It may be that the relevant observations have already been made, but we have not interpreted them correctly. Maybe the data have already proven our theories false, but we do not realize this because the calculations match the data.

The history of science reminds us that some of the foremost problems were at last cracked simply by interpreting the findings differently. At one time, the Earth-centered model of the universe was refined over and again to better match observations, but at last, the Sun-centered model of simple elliptical orbits was able to replace the complicated epicycles of planetary motion. Neither today, adding more and more decimals, so to speak, suffice if we have not even gotten the first digit right. In contemporary physics, time is instrumental, insubstantial, incomprehensible, whereas it should be concrete, causal, comprehensible. That is why we need to go back to reality and once again bring Galileo's method to bear.

IS TIME AN ATTRIBUTE?

A clear, frosty night under a starry sky is a great experience, except that it feels cold with time. Heat does not escape by itself but together *with time*. The observation is obvious, but that is precisely why it is precious. Can we thus infer that the passing of time is associated with a flow of energy? What *is* it that moves when energy and time flow?

This trivial reasoning about the quintessence of time may seem quite amateurish. How could it possibly lead to a breakthrough? After all, the nature of time is a world-class mystery.

It is good to become aware of such prejudices, although it is almost impossible to avoid their influence even then. It is also good to know about the first physics, Galileo's method, which is about mathematizing experiences in the form of natural law. By contrast, modeling only alienates us from reality, reducing mathematics to a habit of thought. Many textbooks show how one equation is derived from another, but very few show the experience from which the law was originally sourced. When the equation for time cannot be found in textbooks, we must draw understanding from experience. This does not mean discarding the achievements of physics but rather making sense of them.[9a]

What *is* time? It is an open question. Can we face it with an open mind? We have measured and we have theorized, but we have not figured out what time *is*. Therefore, how we sense reality is a prudent starting point for grasping the substance of time.[26a] "Pure logical thinking cannot yield us any knowledge of the empirical world; all knowledge of reality starts from experience and ends in it."[27] Einstein's opinion parallels philosophers' outlook on experience as an invaluable source of knowledge.[9c,28]

WITH TIME

Under the starry firmament, we feel cold because heat escapes from the warm skin to cold space. This experience exhibits causality. The difference in temperature is the cause, and the loss of heat is the effect. Conversely, intense radiation from the Sun feels unbearable. Whether moving away from or toward us, heat is carried by photons.

The photon is a quantum of light, or, as Newton said, a corpuscle of light. The human eye is sensitive enough to register even a single quantum of light. However, it takes two simultaneous photons to produce sensory perception because our brains perceive such a confluence of events as noteworthy.[29]

Since we see and sense the quanta of light, we know them through experience. According to Bertrand Russell, the underlying elements of reality must be objects known through experience.[30] The philosopher advocating empiricism advised eschewing concepts without connection to our experience as meaningless.

From the whole spectrum of light, we sense infrared as heat; ultra-violet tans and burns our skin; X-rays penetrate our bones. Clearly, the photon carries energy. *But does the photon carry time, too?* This is an essential question.

The union between time and energy is really a matter of everyday life. It is apparent, for example, at sea when a heavy swell first lifts and then drops the boat. This happens over the wave's intrinsic period of time. Likewise, the photon carries energy on its period. Thus, sunlight does not bring about a change as energy alone but also as time. Aristotle had already concluded that time measures change.

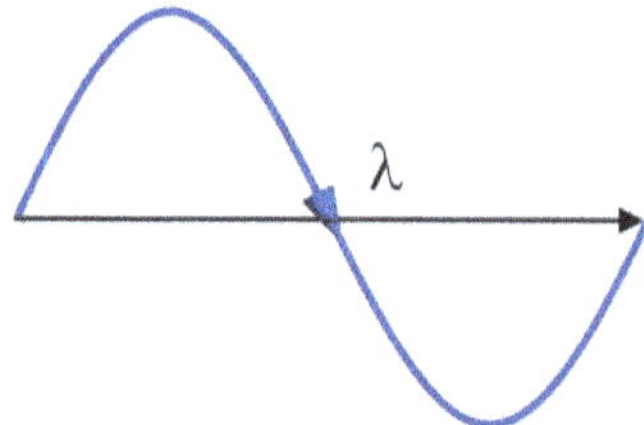

The photon is a wavelet that carries energy on its period. So, as the photon propagates, time and energy move. The speed of light is $c = \lambda / t$ for all wavelengths, λ, and periods, t.

Energy and period are inseparable properties of a photon, as Max Planck exposed in 1900. The fame following this great discovery did not make Planck haughty: though the equation works, he admitted to not understanding the meaning of a constant that the spectrum of light exposed. The photon energy, $E = hf$, can be calculated by multiplying the frequency of oscillation, f, corresponding to the color of light, by Planck's constant, h. This is among the most important equations of physics. But what does it mean?

The frequency indicates how many periods fit in one second. Many. The period of visible light is so short that there are a hundred trillion periods in one second. The period, t, is very short, but not nought. Time is real. It is a property of the photon, like energy.[31]

Time *is* the period of the light quantum. This is a new viewpoint, not a new finding. The second is *defined* as 9 192 631 770 multiples of the photon period, whose energy makes the cesium-133 atom oscillate. As such, time is not illusory. It is tangible, even visible; the period of a red photon is longer than that of a blue one.

This seemingly trivial conclusion about the essence of time will lead to a significant change in worldview, as envisioned by Ludwig Wittgenstein: "The problems are solved, not by giving new information, but by arranging what we have always known."[32] According to the new view, time and energy do not exist as such. For example, there is no energy, as such, in the car's fuel tank, but rather the fuel contains energy. Time is not an illusion, either, but a concrete quality of the quantum. As the quanta move, both energy and time flow.[33]

The period, $t = 1/f$, is the inverse of the oscillation frequency, f. Thus, Planck's constant, $h = Et$, is equal to energy multiplied by time. This rearranged form is mathematically equivalent to the one in the textbook, but the meaning is different. Here, Planck's constant is not just a coefficient of proportionality but the measure of the photon, the quantum of action, as the Indian physicists Chandrasekhara Venkata Raman and Suri Bhagavantam determined from their measurements in the early 1930s.[34] As quanta make up everything, Planck's constant occurs in many contexts and appears in many equations.

Milestone 1: We have now identified the agent of causality. It is the quantum. The fundamental element of Nature carries both energy and time. This revelation is a breakthrough in the investigation. The motive force of occurrence is still missing.

The photon energy decreases as its period increases. However, energy cannot drop to zero, and the period cannot extend to infinity. If this were possible, the photon would vanish, but Planck's constant means that the quantum is an indivisible and eternal constituent, a solid fact that bears the weight of a worldview.

Galileo founded physics as a method for mathematizing experiential knowledge into a universal law. This procedure is what we have followed. The experience of heat escaping from the skin with time renders the concepts of energy and time complementary properties of the quantum. Rather than through such an experience, Planck found the constant by interlacing two equations together. While covering the spectrum's two branches, the low- and high-energy bands, Planck's law of radiation does not actually explain the light. Planck was, therefore, blind to the essence of light.

THE MATTER OF TIME

Planck did not look for but found the natural constant bearing his name. It took a few more years before Einstein suggested that the constant means the corpuscle of light, the light quantum, the photon.[35] Soon, however, the focus of physics shifted from the light quantum to the energy being quantized.[36] Energy in discrete chunks is spurious yet understandable interpretation. While the photon's energy can shift continuously, energy flows quantum by quantum, in bits.

When a system is static, quanta move neither in nor out. Nothing happens. Quantum mechanics works fine at this balance. But imbalance it expresses with impenetrable parlance. The wave function spanning all conceivable states, so to say, collapses at the event of observation. One of the superposed states will become instantly observable, but which one will be determined by pure chance. That interpretation is not only inexplicable but incorrect. Measurements made in 2019 revealed that the quantum processes are both measurable and reversible; they are neither entirely arbitrary nor exactly instantaneous.[37]

Quantum mechanics, in its arbitrariness and non-locality, surpasses our understanding. We have no experience of entangled existence, superposed states, and wave function collapse. It does not make sense that the world would be profoundly probabilistic. "The past is not probabilistic. What's done is done,"[38] as Jeremy Bernstein, a professor emeritus of physics at the Stevens Institute of Technology, pointed out. But mere criticism of modern physics is of no avail. It only emphasizes our need to fathom the essence of time.

Although we do not have and cannot have our own experience of everything, we can equate phenomena with our own experience, provided that the same universal law can parse all processes, as *Grand Regularity* implies. For example, if we did not lose a single quantum, we would not lose any heat. So, a system without a flux of quanta remains as it is, stationary. We understand as well that no system can have zero energy because no system exists without any quanta. So, the energy of the vacuum, space without any matter, is not precisely zero either. One cubic meter holds the energy of about one billionth of a joule.[39] The void is thus not just a geometric space; it is some sort of substance embodying energy and time, i.e., quanta.

A concrete concept of the void as substance prevailed from ancient times until the early 20th century. As late as 1916, Walther Nernst, a German chemist, suggested that space is replete with photons, which constitute electromagnetic radiation with the lowest possible energy.[40] The old idea that the vacuum must be some everlasting substance is not out of date because, ultimately, the quanta that escape from our skin end up in the void.

"Vacuum holds the key to a full understanding of the forces of nature,"[41] philosophized the British physicist Paul Davies from Arizona State University. The ancient philosopher Parmenides had already concluded that everything, including the void, consists of uncuttable and eternal primary particles named *atomos*.[42] So, we may already imagine how central an insight it is to regard time as the quantum's period and the quantum as the element of everything. The atomistic tenet is holistic. The postulate that *everything comprises quanta* transfigures our outlook on reality into a thoroughly tangible worldview. "It is in vain to expect any great progress in the sciences by the superinducing or ingrafting of new matters upon old. An instauration must be made from the very foundations, if we do not wish to revolve forever in a circle, making only some slight and contemptible progress," advised Francis Bacon.[43]

Since the onset of modern physics at the beginning of the 20th century, we have faced many conceptual problems. Admittedly, we may contest this correlation as causal. Is it not our observations that have forced us to ask, for example: What is dark energy? What is dark matter? Likewise, we could not possibly have asked why the universe is made almost exclusively of matter instead of an equal amount of antimatter *before* we had any clue about antimatter. Or how could we have been puzzled about the relationships between the four fundamental forces of nature, that is, gravity, electromagnetic, weak and strong nuclear force, *before* we had measured their strength?

True, but we ask about what we do not understand. What would we have understood by now had we taken the quantum as the fundamental element right from the start rather than quantizing energy?

Speculating about history seems pointless. Would we not have inescapably ended up with modern physics by studying Nature? Possibly, given enough time. However, had we recognized the quantum as the primary element at the outset, we would now regard quantum mechanics and relativity as mathematical models of reality rather than theories of reality. Because of this confusion, we may not even distinguish between a model and a theory. But we should. A model mimics data, whereas a theory explains a phenomenon in unison with data — or if it doesn't, the theory proves insufficient as the phenomenon remains unexplained.

THE ORDER OF TIME

The idea that the fundamental element of time is the period of a quantum is perhaps surprising in its simplicity. The notion's modesty can even be disappointing. There is no more of a mystery hidden in time than in energy.

Planck's constant, the fixed *product* of energy and time ($h = Et$), may not measure up to our expectations of an indivisible unit with the least energy or the shortest period. Still, it makes sense that time and energy are complementary properties, as the founder of quantum mechanics, Niels Bohr, formulated.[44] One cannot be without the other. When the photon energy decreases, the photon period increases. This change is as familiar as the fading flame of charcoal that shifts from a blue-yellow blaze to a red-hot glow. As the color shifts from blue to red, the photon energy per period decreases; that is, the power drops. It becomes chilly. Nonetheless, the light quantum maintains its size as per Planck's constant.

Energy and time also change hand-in-hand when the quantum takes off from a flashlight into the sky. As the wavelet of light dashes away from Earth's gravity into space, its energy lessens, and its period lengthens. The color shifts toward red, albeit only slightly, for Earth's gravitational field is hardly denser than free space. However, the light that departed from the blazing early universe and arrived at the cold present has extended so much that our eyes cannot see it. But our bodies can still feel it. The night sky is cold and dark because the average temperature of the expanding universe has fallen to less than three degrees above absolute zero. This massive change in energy took eons, about 14 billion years. It is pivotal that the photon, unlike a stable particle, is open to change. The universe could not be expanding unless the photon period was increasing and energy decreasing.

Since time and energy, as well as momentum and wavelength, are complementary properties, the steps in a sequence of events are not interchangeable. Mathematically speaking, they are non-commutative. For example, velocity measurement changes the particle's velocity and thereby affects the subsequent position measurement of the particle and vice versa. Since no observation will leave its object intact, the results depend on the order in which the measurements are made. In other words, the order of time is the order in which quanta flow.

By now, I am quite used to thinking that the periods of quanta embody time. Nonetheless, I can still relive how perplexed I was about this logical conclusion, no matter how lucidly it corresponds with observations. That is why I was not surprised by the bursts of disbelief about the period of quantum being time, that students of mine expressed in their Facebook posts. The immediate natural reaction is to defend ingrained beliefs. So, their comments targeted me personally rather than the concept they found to be out of the ordinary. "Thinking is difficult, that's why most people judge," the founder of analytical psychology, Carl Jung, said.

As much as assumptions control our everyday activities, they also rule our scientific studies. And for good reasons, we must hold on to them until proven dead wrong. But we must also be ready to reject a hypothesis that goes against observations, just as we must be open to examining a tenet that is not an oxymoron but merely incongruent with the prevailing thoughts. So, we need to ask what might be better understood by considering time as the quantum's period. And we ought to ask whether any observations are inconsistent with this concrete concept of time.

WHY IS TIME RELATIVE?

Although Einstein realized that time is relative, he did not clue us in to what time *is*. General relativity, as an effective theory,[45] does not explain why clocks run at different rates at different places; it only reproduces the measurements.[46] For instance, the clock up in the attic runs faster than the one in the basement. The difference is minimal, but it has been measured so accurately that there is no doubt of its validity. The rate of the most precise clocks follows the slightest change in the surroundings, perhaps even the rise and fall of a gravitational wave.[47]

Surroundings affect not only the rate of a clock but also other processes so that thermodynamic balance is always attained in the shortest time. For example, in winter, our house will cool down quickly if someone has left the front door wide open, whereas, in summer, there is not much drag because indoors and outdoors are equally warm. In general, the larger the energy difference, the faster the energy flows. In a perfect balance, nothing happens. Time does not pass when no force drives the quanta out to the surroundings or into the system. For the same reason, the clock stops when its battery has run down.

Since quanta carry both time and energy, we understand the rate of a running clock in the same way we do the rate of heat flow. The passage of time in the attic is faster than in the basement because the *energy difference*, the *imbalance* between the clock and the surroundings, is bigger in the attic than in the basement. Since the gravitational field decreases upward, a clock running in the attic gives its quanta away slightly faster than one in the basement.

This experience-based comprehension of the passage of time is mathematically consistent with general relativity.[48] The calculation is just as it should be; only our interpretations are revised.[49] Since energy is relative to the surroundings, so is time.

Milestone 2: We have now solved the mystery of time; we have found the motive. An imbalance of any kind forces the quanta into motion. The flow of quanta is the flow of time because the quanta carry time and energy.

We have made explicit that the smallest event entails the change of one quantum. The change consumes an energy difference, free energy, a force, not in any which way, but in the least time. This is the *principle of least time*.

CALORIC

Following Lavoisier, it was reasoned until the mid-19th century that the flow of heat from one body to another is the motion of a gaseous, massless, and invisible substance, referred to as *caloric*. Are the quanta, therefore, the elements of caloric? No. The caloric theory assumes that heat is conserved as it is assumed that energy is conserved. But energy is not conserved. For example, the energy of a photon traveling from Earth into space decreases.[50]

The caloric theory was discarded by reasoning that heat could not be an indestructible substance because more of it can be created by rubbing pieces of material against each other. So it was concluded that heat is actually the motion of particles, such as atoms. Ever since then, energy has been theoretically treated as insubstantial, immaterial, ephemeral rather than realistically as property, even if its substance is just space. So, what is the substance of the gaseous, massless, and invisible void that embraces everything? What is it that moves in the form of the vacuum?[51] To tackle these basic questions in physics, we need to lay bare the nature of substance.

ON THE FLY

Time as the quantum's attribute explains many oddities of modern physics. We can comprehend, for example, why the clock runs more slowly in motion than when at a standstill. This prediction was verified elegantly in the early 1970s.[52] The physics professor Joseph Hafele and the astronomer Richard Keating spent their research grants on round-the-world fares. To be on the safe side, they took flights in both directions, east and west. Tickets were also bought for the atomic clock. The chronometer was seated affordably to tick in economy class.

At the end of the flight, the clock on the eastward-bound jet was found to be ahead and the westward-bound to be behind the reference clock that stayed on the eastward-rotating ground of the Earth. The amount of time gained or lost on the fly, along or against the Earth's rotation, was as much as had been calculated beforehand.

Such results convince physicists that special relativity is valid, although it does not explain why the clock runs slower in motion than at rest. It is not even meaningful to pin down the causes that would affect the rate since relativity does not aim to explain time. It just provides mathematical transformations from a stationary to a moving coordinate system.

Since the clock runs slower in motion than at rest, it must be that in motion, the imbalance relative to the surroundings is smaller than when at a standstill. If the clock were to move at the speed of light, it would stop. The clock cannot give away any photons in a perfect balance with the surrounding vacuum. By the same token, an oar that falls into a river will drift away only if the boat moves more slowly than the stream. Is the vacuum, therefore, a stream of light?

That is how Maxwell, a leading light of the era, approached the problem. He discovered that the electromagnetic properties of the vacuum determine the speed of light.[25] In 1865, Maxwell wrote that light is a wave of the void, a ripple, an undulation. Since a wave of water is water, could not the substance of the vacuum be the quanta that move at the speed of light? This was a revolutionary suggestion.

In 1940 Bruno Rossi and David Hall showed that when a spontaneously decaying particle moves very fast, almost at the speed of light, its lifespan increases greatly.[53] However, no particle can attain the speed of light and become uncuttable, *atomos*. Only the photon is indivisible and eternal. Light does not age in a constant vacuum, but in expanding space, its period lengthens.[54] Due to enormous expansion, the bright blaze of the nascent universe has shifted to deep red; the sky has become black.

The flow of quanta explains the twin flow of time and energy, but it does not tell us what embodies the energy differences, driving forces. We cannot really draw a parallel between the passage of time and our experience of heat flow unless we know what the void is, what the gravitational field of a body is, and what particles are made of.

We strive for a consistent and comprehensive worldview. With only a single theory, there would be no need to decide which phenomenon should be interpreted by which theory. The *Grand Regularity* suggests that one theory should suffice. To actualize that vision is the age-old goal of humankind.

WHY DOES TIME MOVE FORWARD?

The preconceived idea that ever-increasing disorder is what directs the arrow of time is deeply rooted in contemporary physics.[55] Our own experience is that it also takes considerable time to put things in order. I would say it takes quite a bit of time to tidy up my garage.

We favor a proposition in line with our experience and disfavor those against it. For example, we see that order increases when water freezes, and we see that disorder increases when the ice melts. So, order, just as disorder, emerges as the energy difference between the environment and the system evens out. It is, therefore, not an increasing disorder but an imbalance that directs the arrow of time.

Physics textbooks nevertheless present increasing disorder almost as a law of nature.[56] While disorder does increase when the system evolves toward a balance in an environment where disorder rules, it is also true that order increases when the system becomes synchronized with its environment where order reigns. For example, ecosystems keep to daily and annual rhythms. Societies are locked into the pace of the clock and the course of the calendar year. Atoms synchronize with their environment.[57] An increase in disorder or order is no cause of anything.[58] Both are consequences of moving toward a balance.[59]

When a film is played backward, the spiral of events looks unreal. Shards of glass on the floor just cannot merge into a solid vase and rise back onto the table. Work needs to be done for that to happen, but we see no one doing it. Indeed, time does not step all by itself but by free energy, i.e., force. In many cases, one does not have enough resources to repair the damages. No one can oppose the universal flow of time, all processes, the total consumption of free energy, which amounts to the expansion of the universe.

A textbook considers increasing disorder as a cause, but from experience, it is a consequence. The conflict between authority and reality is blatant. Indeed, Boltzmann's theory was found problematic shortly after publication in the late 19th century. His statistical mechanics applies only to a steady-state system. Nonetheless, we have gotten mired in it for lack of a statistical theory that provides a rationale for the imbalance. Philip Ball, a well-known science writer, phrases this weakness: "[classical thermodynamics] seeks to account for change, but it can't actually say anything about the process of change itself."[60]

To echo teaching does not make what is taught right. To doubt it, however, adds to certainty. It would be irresponsible not to question the solidity of the theoretical grounding. If it will not hold, the whole discipline is in danger of collapsing. That has happened before; it is not ruled out today, either.

THE PRINCIPLE OF LEAST TIME

Time does not move forward all by itself. It takes an imbalance for the quanta to move. The motive force could be a temperature difference, a height difference, or a density difference. Energy differences are also called fields, such as the gravitational field and the electric field. And sometimes, force is termed free energy.

Time flies, time crawls, time stalls. Figures of speech illustrate the pace of change, but that is not enough. The exact analysis calls for an equation. Only when an event is noted down with the mathematical precision of a single quantum are its causes and consequences wholly captured.

Boltzmann sought this equation of motion. He was impressed by Darwin's tenet of evolution by natural selection. Still, he did not see a fundamental distinction between the living and the non-living, envisioning evolution of any kind to follow the same principle.[61b] The evolution of a species on Earth is no different in principle from the cooling of tea in a cup.[62] Be it in temperature, chemical energy, or anything else, differences diminish with time.

Long ago, the biosphere, as a mechanism in its entirety, emerged to consume the energy imbalance between matter on the globe and the hot sunlight.[63] Nowadays, solar panels gain ground for the same reason; they collect photons even more effectively than plants.[64] As such, from the viewpoint of the overarching principle, photovoltaics is a species subject to the same evolutionary process as trees.

Flows of energy *naturally select* their paths through the most powerful mechanisms. The widest jaws or the sharpest elbows are naturally selected from a crowd when these are the means to make a living. Likewise, heat naturally selects an open door rather than a wall for the most voluminous outflow. Temperature differences thereby diminish as quickly as possible. A stream naturally selects its course along the steepest descent. In that way, the differences in water heights even out as quickly as possible. Time does not just run; it runs in the least time.

The principle of least times was known in Persia and Mesopotamia a thousand years ago.[65] Pierre de Fermat made the Europeans aware of it. As a young man, the French lawyer honed his math skills at the gambling table. Besides probability calculus, Fermat advanced the field of geometry. Most notably, he figured out that "Light chooses its way so that time is the shortest."[66] The optimal path is called the *geodetic line*. Photons mediate forces through these lines. The photon itself is a piece in the line of force.

The ubiquitous patterns of big data, discussed in the previous chapter, suggest that optimality is universal from the quantum onwards. For example, the slightly flattened form of the rotating Earth is energetically optimal, having the least-time shape: a clock runs as fast at the Equator as at the North Pole. On the one hand, the clock would run faster at the Equator, where gravity is weaker as the distance to Earth's center is longer than at the pole. On the other hand, the clock would be running slower at the Equator due to the Earth's rotation. These two opposing effects precisely cancel out.[46] The optimum expressed in terms of time and energy is one and the same because time and energy are inseparable attributes of the quantum.

The substance of probability

Boltzmann's vision of evolution is eye-opening. Everything is evolving toward energetically more favorable states, in other words, toward more and more probable states, as quickly as possible. Gottfried Leibniz also said: "This world is the best of all possible worlds."[67] Thus, a comprehensive worldview has been around for quite some time. However, its mathematization in thermodynamic terms had to wait until the present time (Appendix A).

Regarding the role of probability, the French mathematician Pierre-Simon Laplace put it aptly: "It is remarkable that the discipline that emerged from gambling would become the keystone of human knowledge."[68] Probabilities are calculated, but what *is* probability?

The probability of rolling any one face on a die is one-sixth – unless someone has brought in a loaded die. After losing all your money, you might just figure out that the die is loaded. Then you will realize that the real probability is not a mere number but a measure of energy. The heaviest side faces down because that state is energetically the optimum, the most probable. Likewise, among a hatch of eaglets, the one who gets the most food will grow stoutest. It is energetically in the best position.

Boltzmann aimed to formulate the probability of any system in terms of its factors, even without explicitly knowing them. For example, to evaluate the energetic state of an eaglet, we need not know whether its mother has caught a rabbit. It is enough to have that option in the equation. And it is there because it is easy to quantify everything when everything ultimately comprises quanta.

We can readily comprehend an equation that describes our experience. For instance, my existence is a mathematical product of numerous factors because I need at least oxygen and, in the long run, also water, food, and whatnot. If any one factor is absent, my likelihood of existing exactly the way I do now would be zero. The factors define the subject. My dog's existence is a similar product of substances to mine, but the lack of vitamin C, for instance, is no threat to the pet. Its body can produce the vitamin, whereas mine cannot.

The concept of probability articulates in a mathematical form what I am. The phrase "you are what you eat" covers a good part of the probability that expresses my whole identity with the precision of a quantum. When I lose just a single quantum as heat, the loss will change me a bit. With numerous such losses, I feel cold. Similarly, if I go hungry, I lose weight. With time I will change in many ways through various processes that all are basically flows of quanta.

THE MIND OF MATH

Mathematizing one's own experience of existence in the form of the law of nature follows Galileo's method (Appendix A). In this way, the theory is grounded in reality. Of course, we can also formulate equations in other ways, for instance, by modeling data, but there would be less proof of the theory's truthfulness. For example, defining the probability as a normal distribution of random variables does not agree with reality. Data do not conform to the bell curve but spread in a skewed manner with long tails.

To understand physics, we must go to its source. We need to know the experiences or images that were mathematized in the past. The subsequent derivations do not change the perspective on reality. The mathematical analysis cannot change a thing. It can only highlight the implications of theories, such as the multiverse of quantum mechanics and the wormholes of general relativity, which we might have otherwise missed. So, if we come across an unrealistic implication, we should conclude that the theory is not fully in touch with reality elsewhere either. Wittgenstein pointed out that mathematical truth is no guarantee of equivalence with reality.[69]

We have empirical grounds for believing that causality manifests itself as *Grand Regularity* observed throughout Nature. Now we have also a theoretical basis as the equation of time accounts for the S-shaped curves that sum to skewed distributions as well as for spirals, oscillations, cycles, and chaotic courses, and the multiplicative and branching courses of history (Appendix A).[70]

THE GIST OF THE THEORY

Boltzmann was convinced that the universe is composed of atoms. The elements of any system evolve one way or another toward thermodynamic balance, just as gas atoms attain a balance through collisions. At about the same time, the American physicist Willard Gibbs realized that a chemical system evolves toward balance through reactions. Everything is on its way toward balance. Thermodynamics[ii] is thus held as the most comprehensive among theories. It explains why

---

[ii] Thermodynamics stems from the Greek words *thermós* (hot) and *dúnamis* (force).

a stone falls, an eagle catches its prey, and a company serves its customers. Gravity, vital force, purchasing power: forces of all forms are what make things happen.

It is a groundbreaking thought that all events, from a nuclear reaction to the expansion of the universe, from a chemical reaction to the evolution of the biota, and from a small purchase to world trade, are all aspects of the quantized substance moving toward a thermodynamically more favorable state in the least time.

HOW DOES TIME CHOOSE ITS COURSE?

It is only natural that the universe expands everywhere in every direction, a stone falls straight down, a plant grows toward light, and you go for the best price. In this way, balance is pursued in the shortest time.[58] The maxim is, in a sense, a truism. When this quest for balance in the least time is understood as *natural selection*, that is, Nature selects, evolution incorporates not just the living but everything. Temperature difference forces hot tea to cool down, just as food powers the growth of eaglets. These phenomena involve different mechanisms but the same underlying principle. That is why the same patterns are present everywhere.

When forces are small and fluctuating, the ensuing events are insignificant and neutral.[71] They will hardly lead anywhere. The eagle, hovering over its territory, is trying out various options to get prey. Out of variation, flows of energy *naturally select* their courses toward energetically more and more favorable contingencies. The eagle targets a kill. The catch will be energetically rewarding. Likewise, the river varies its course to find the swiftest way. The Colorado River has dug its deep canyon through time, taking the fastest course downward to sea level. Plants have evolved to high effectiveness in harvesting light, sheep in grazing, and wolves in catching prey. In every case, the future will be energetically more favorable, i.e., more probable than the present, which in turn is more probable than the past.[72]

Boltzmann's holistic comprehension of natural selection is startling. Do we humans, too, value everything ultimately in energetic terms? Do we also adapt our behavior to consume free energy in the least time? If the principle is universal, where do differences in execution come from?

The quest for balance is universal, but we go about attaining it in different ways because different forces influence us. As an illustration, my dog enjoys pig's ears, but I don't. There is no unanimous wisdom. Rather, you are moved only by the forces that pertain to you. So it is natural that my daughter went out with the bloke who captivated her, not with the boy who impressed me. Although the criterion for natural selection is simple, it is not easy to navigate in the choppy sea of diverse forces.

Any system keeps evolving until it attains a balance of forces, a steady state. Boltzmann understood this superbly, perhaps too well. He did not bother to look for the equation that captures the tendency toward balance but instead hastened to formalize the equilibrium condition alone. Einstein understood that no one could dictate the balance in advance, but the system itself works its way to it.[73] This insight came too late, for Boltzmann had already taken his own life. Thus, the complete equation of time, where the forces drive the system toward balance, remained unknown until now.

Sometimes when explaining the least-time principle, I am asked why forests do not spontaneously burn down in order to release the energy bound up in them as quickly as possible. The answer is that events do not take place to dissipate energy but to attain balance in the least time. Forests are an expression of the quest for balance because photosynthesis raises the Earth's energy content closer to the energy of sunlight. A reproducing, recurring steady state, such as the biosphere in its entirety, is known as a dissipative structure.

I have likewise been asked, would not the least-time law mean that explosives ought to detonate on the spot, without ignition? The answer is that flows of quanta always need some means of transmission. The spark from ignition is what opens the flow, which multiplies rapidly into an explosion. Similarly, the course of events got out of hand when the water level of Lake Höytiäinen in eastern Finland was lowered in 1859. The outlet expanded beyond its bounds as masses of water eroded the sandy bank away and deluged the downstream city of Joensuu. Priming will also precipitate learning.

ENTROPY

The equation engraved on Boltzmann's gravestone in Vienna's central cemetery, $S = k_B \log W$, shows that the logarithm (log) has been taken of the probability (W, from the German *Wahrscheinlichkeit*) and multiplied by Boltzmann's constant, k_B. The thus obtained entropy, S, is an additive measure of its factors. We are used to adding things up when we measure things. For example, when a market seller measures two pecks of potatoes, they first pour one into the bag and then the other, that is, they add the two together.

Entropy was originally defined without understanding the flow of quanta. Thus, many a physicist still today assumes without proof that increasing entropy equates to increasing disorder. By contrast, many a chemist says right away that reactions tend toward balance where free energy is at its minimum and entropy at its maximum.[74] Thus, the concept of entropy does not reveal anything more about causality than does the concept of energy. Entropy increases as energy differences decrease, irrespective of whether the course of events results in order or disorder, wealth or famine, joy or pain.

While money is a pretty good measure for many things, it cannot buy everything. The best measure for everything is the total energy, the product of the system's temperature and entropy. Any one thing can be compared to any other in energetic terms. The total energy of a system includes the energy already bound into it and the energy that is still free.[72] For example, energy is bound into eaglets, and there is free energy insofar as there is still prey available for feeding the brood. Free energy is in deficit if the brood has already exceeded the environment's carrying capacity. The energy bound into the eaglets has to decrease, so they starve. In one way or another, all systems gain a balance with their surroundings.

A system at balance is getting nowhere. There are no causes for changes. Modern humankind has little experience with stasis, but there have been times where generations followed one another with little change. When a new emperor rose to power, the calendar was restarted, but stagnation persisted because no one gained access to free energy sources, such as fossil fuels. Balance is the natural destination for all processes, but it would be hard for us to adjust to it, were it to occur now, accustomed as we are to living in a changing world.

THE UNIVERSAL PRINCIPLE

The theory of time can be written in the form of a simple equation, as Maupertuis discovered in the mid-18th century. The luminary used it to explain a wide range of phenomena.[75,76] The cause for the orbital motion of planets and the survival of organisms is universal: the quest for balance. Maupertuis also understood that the equation, which he named the *principle of least action*, was consistent with the original form of Newton's second law of motion. And the principle of increasing entropy, the same as the second law of thermodynamics, is equivalent to Maupertuis' equation divided by temperature. Thus, the principle of least time, the second law of thermodynamics, and Newton's second law of motion are one and the same natural law (Appendix A).[77,78] Henceforth, I use these names interchangeably for the theory that I propose explains time. This natural law must also be the law of causality we are looking for, as it accounts for *Grand Regularity*. It relates causes to effects and answers "why" questions.

Such sensemaking, in Boltzmann's view, "opens up a hitherto undreamt of outlook on the whole."[61a] It allows us to comprehend phenomena from the smallest to the largest and from the simplest to the most complex in the same way as we deal with everyday matters of life. Our understanding of what we cannot see or what we cannot reach is not based on a mere metaphor, but the explanation of one phenomenon and another is the same because the governing principle is the same. So we are no longer at the mercy of the acumen of experts. Instead, we can deduce for ourselves and take responsibility for our conclusions. While blind belief in authority often begins where our own thinking ends, we will better appreciate the knowledge when we find it through our own examination.

Paradoxically, the old unified view of reality retreated as science advanced by diverging from the natural philosophy of the Enlightenment and fragmenting into distinct disciplines.

A WAY OF SEEING

The principle of least time is familiar to us, even if we have not recognized and rationalized it in the form of a natural law. The principle is so obvious that we do not pay much heed to it. We are not surprised by a stone falling straight down but would be puzzled if it zigzagged. And we are not amazed that a creek runs down along the steepest descent of a hillside but would be stunned if water were to creep uphill. Neither are we baffled by the spread of aphids, snails, and other pests from one plant to another. However, if a shoot were to escape from the vermin, we would be looking for a reason behind it.

Events are not haphazard but happen in the least time. It is quite ordinary that if a way of doing things turns out to be a blind alley, another way out will soon be found. If one species cannot use a new kind of food, soon another one will be consuming it. If one does not seize the opportunity, someone else shall promptly take it. Unless a company actively seeks to seize market opportunities, competitors will pounce on them, leaving the sluggard to bite the dust. Nature behaves as Darwin's cardinal concept stipulates. It is not the strongest of the species that survives, nor the most intelligent. It is the one that is most adaptable.[79]

In the mid-18th century, Maupertuis had already concluded that there is no principal difference between the animate and the inanimate.[75] Galaxies and elementary particles, as well as societies and individuals, are born, develop, mature, and die in attaining balance with their surroundings. Yet, large or small, a system may experience the steady state only fleetingly, for its circumstances continue to change as everything affects everything else. New opportunities will open, and old ones will close.

In the traditional view, thermodynamics describes flows of energy; here, *thermodynamic theory, the theory of time*, accounts for flows of quanta because quanta carry energy. "A theory is the more impressive," asserted Einstein, "the greater the simplicity of its premises, the more different kinds of things it relates, and the more extended its area of applicability. [Thermodynamics] is the only physical theory of universal content which I am convinced will never be overthrown, within the framework of applicability of its basic concepts."[80] Eddington, too, endorsed thermodynamics: "If someone points out to you that your pet theory of the universe is in disagreement with Maxwell's equations – then so much the worse for Maxwell's equations. If it is found to be contradicted by observation – well, these experimentalists do bungle things sometimes. But if your theory is found to be against the second law of thermodynamics, I can give you no hope; there is nothing for it but to collapse in deepest humiliation."[5a]

Let us compare the straightforward principle of least time with the abstract thinking of modern science. Comparisons open eyes because one way of seeing obstructs other ways of seeing.[81]

FINDING THE LOST

Some 20 years ago, when I was looking for the equation of evolution, I could have found it directly from Maupertuis' work. His name was remotely familiar to me, but his works hardly at all. I knew the principle of least action only through Lagrange's textbook equation. It does not describe events but outlines a fixed trajectory, so I had to work out the evolutionary equation myself. It was not hard to do; I only needed to find the lead.

After taking my professorship, I dabbled in thermodynamics from time to time in the hopes of relating it to evolution, but without a breakthrough. Finally, a discussion with Kari Keinänen, a biochemistry professor at the University of Helsinki, gave the proper impetus. Over a flight from squalls of sleet in Helsinki to the polar night in Ivalo, Lapland, to attend an annual meeting of our Graduate School in late 2005, we posited that evolution must be a *probable* process.

Probability is the central concept of statistical physics. So, I took the textbooks out as soon as I got back home. Boltzmann's theory, however, only applies to the equilibrium system. So the equations needed a revision.

Over the Christmas break, I derived a general expression for probability. I noticed it was higher for a cocktail than for a homogeneous set of molecules. I was excited. For a moment, I thought I had succeeded in writing for the first time a formula that accounts for species diversity, which is a characteristic of life. In truth, I had found nothing new, for Willard Gibbs had derived the entropy of a chemical mixture over a hundred years ago.[74] However, from that result, it was easy to see how to account for photons streaming into a system. They cause things to happen; new molecules emerge from the power of light.

That is how I brought forth the evolutionary equation,[72] on January 6, 2006. It soon dawned on me that the system and its surroundings evolve toward a mutual balance. The eureka moment was somewhat shocking to realize that both animate and inanimate evolution is nothing more than the sequence of events. There is nothing fuzzy or emergent[24] in the flow of time being the flow of quanta. My worldview was changed once and for all.

That day also stuck in my mind because, on my way home, another car drove through a yield sign, crashing into my old jalopy. The distracted driver was relieved at my calm attitude; crumpled metal did not mean much to me compared with the day's discovery.

It is not that many things would be that different from this new perspective; it is about seeing them in a broader context. We already think that evolution has been going on for eons. Now, we only extend the thought from organisms to everything that exists. We already consider matter to consist of atoms. Now, we only refine the scale down to the quantum. While it might be hard to swallow this holistic naturalism without hiccups, we are not digesting the bits and pieces of information about Nature to reach a consensus — nor to end up with a disagreement — but rather to come to a coherent view of reality.

WHY IS THE FUTURE UNPREDICTABLE?

In ancient Greece, *theōría* meant viewing, considering, examining, whereas today, we expect theories to provide precise predictions. Is our expectation legitimate or not?

Once upon a time, astronomers amazed citizens and kings by foretelling when a known comet would show up in the sky. The triumph of such timing went to the heads of the scientists. This rationalistic arrogance radiates from the famous quote: "Sire, I have no need of that hypothesis." So Laplace answered when Napoleon inquired why he had never mentioned the creator of the universe in his books.[82] The supreme ruler may well have been pleased with the reply. But at the end of the 19th century, the tone changed as doubts snuck in. After all, it turned out to be impossible to predict the orbits of heavenly bodies throughout eternity. The concern was grave. Will the Earth even stay in its age-old orbit?

Gösta Mittag-Leffler cashed in on the opportunity.[83] The professor of mathematics, who also served the University of Helsinki for a few years at the end of the 19th century, suggested to the King of Sweden that the problem of three bodies was worth a prize. At that time, mathematicians wondered why it seemed impossible to calculate the trajectories of the Earth, Moon, and Sun forever into the future. Of course, complexity makes it hard to predict, but the intermingling of causes with consequences precludes it.

The prize went to Henri Poincaré. The French polymath showed that there is no guarantee of stability when the orbits of bodies are open. When energy changes, time passes as well. The sought-after calculation cannot be precise until and unless energy is constant. However, the calculation of a closed orbit is not a prediction about the future. It is a disclosure of the unknown trajectory because, in such a system, time does not advance but circulates. The outcome is a paradox: the equation of motion has the elements of the explanation, but at the point of balance, where nothing happens, there are no causes or consequences to be explained.[84]

The inevitable conclusion is that the future is genuinely unpredictable yet bounded by free energy. The faster and greater the change, the harder it is to pin anything down. The values of stocks are hard to predict when prices change a great deal. Earth's orbit is not stable, either. However, our dear orb derailing is not much of a danger because the whole solar system's massive energy shifts only gradually. Earth drifts about 15 cm (6 inches) per year away from the Sun, and the Moon about 4 cm (an inch and a half) from Earth.

Through the ages, tea leaves, molten tin, cards, and the like have been used for divination, but the idea of predicting came to science quite late. Newton and his contemporaries did not yearn for prophecies but for comprehension. *Principia* has relatively few equations. Not many are needed to note down forces and changes in motion, viz., causes and consequences. In the 17th and 18th centuries, mathematics' triumphal march shifted the focus from comprehension to calculation. Time became a mystery when those equations that could not be solved were set aside, as it was presumed that there was something wrong with them. However, "not everything that counts can be counted and not everything that can be counted counts,"[85] as Einstein said. The philosopher Paul Feyerabend also saw that "mathematical reasoning is not only exact; it has its own criteria for reality."[86]

In 1766 Joseph-Louis Lagrange became the president of the Prussian Academy of Sciences. Soon, Maupertuis' evolutionary principle was superseded by the Lagrange equation, in which energy is constant, and so is time. "When the equilibrium of motions is understood abstractly and independently of forces, it offers the tantalizing possibility of disconnecting cause and effect of the statical moment of balance from any causal reference to a particular past or a particular future,"[87] distills Bjørn Ekeberg, a philosopher of science, the static dogma of contemporary physics.

CHANGE IN LANDSCAPE

Customarily, it is believed that $E = mc^2$ is an equation by Einstein, but it is a special case of the general formula for kinetic energy, mv^2, when velocity, v, is the speed of light in a vacuum, c. Gottfried Leibniz and Johann Bernoulli knew of this relation, which the Dutch mathematician and natural philosopher Willem 's Gravesande confirmed by experiments in the early 18th century. Moreover, Émilie du Châtelet made it widely known through her books and translations.[88] It was only later that the factor ½, familiar from school books, was placed in front of the formula to make it solvable but then flawed.

Undeniably, as understood, a cart running down from a hilltop has kinetic energy. However, as the cart moves, the landscape also moves. Initially, the cart, as part of the landscape, is on the hilltop, and finally, it is at the valley bottom. So the landscape has changed. That change is contained in the original equation of balancing motion, whereas the textbook formula, $½mv^2$, is only about the condition of balance.

The original equation, in which the landscape also moves, is still used in astrophysics. It is known as the virial theorem.[89] An example of reality it describes is the movement of stars in a cluster and galaxies in a group where bodies' own motion changes their mutual gravitational field. Nonetheless, the universe, and its gravitational field due to expansion, tend to be missing from most calculations.[90,91]

The Newtonian worldview is perceived as deterministic, but it is we who have defined it as such. The familiar formula, $F = m$a, is about the equilibrium because we have dropped the change in mass, $d_t m$, from Newton's original law of motion, $F = d_t$p $= m$a $+ vd_t m$, where the force equals the change, d_t, in momentum, p $= m$v. The change in mass means energy flow into the system from the environment or vice versa.[90] For example, atomic fission generates heat from matter by decreasing its mass. Also, chemical reactions release heat per Newton's original equation. Thus, there is no change without a change in energy. Nonetheless, in theories of physics, the desire to predict excludes the fundamentally unpredictable aspects of reality.

UNIQUENESS OF HISTORY

Although we cannot calculate chains of events precisely, even the earliest events in the universe are not altogether beyond our range of experience as we feel the cold of the night sky on our skin. The coldness is the relic of the universe's hot creation.[92] So, we live amidst all the history that exists. Evolution as a unique process renders the universe asymmetrical in its details.[93] Nevertheless, the constraints of symmetry, i.e., stationarity, rather than comprehension of evolving reality, have guided theoretical physics since the mid-18th century. This is both the source of the success and the cause of the crisis physics finds itself in today.

At first glance, one might suppose that if one only knew a system's initial state exactly, the future could also be worked out precisely. However, as is familiar from everyday life, the start of a series of events does not dictate its outcome because an event will alter the driving forces, which in turn will change the course, and so forth. The equation of time can be written down but not solved. Both will change as the quanta move from the environment to the system and vice versa.[49] History is unique. The American baseball legend Yogi Berra's words strike home: "The future ain't what it used to be."[94]

Every ecologist knows that when a herd of animals grows, its consumption of environmental resources holds back further growth. Every economist knows that demand affects supply, whose change will then affect the demand, and so on. Nevertheless, few physicists acknowledge these irrefutable causes and consequences, privileging computable rather than realistic equations. I, too, recollect fancying those equations that I could solve, hardly bothering to understand what the calculation was all about.

The profession often impresses a stamp upon the professional, but apparently on our backs, because we find it difficult to see it on ourselves. An engineer considers their equations to be models of reality, whereas a physicist imagines that reality is a model of their equations. A mathematician pays no heed to reality. Unless scientists are aware of this reckless mentality of mathematicians, who are otherwise known to be meticulous, there is a risk that even the purest mathematics is interpreted as physical reality.

A prediction is excellent when the consequences have only a subtle effect on the ensuing causes. Even then, close but not good enough. It is easy to envisage how a stone rolls from a hilltop down to a valley bottom, but hard to notice that, as stones keep on rolling down, the hill levels out, and the valley fills up. The landscape, too, is in motion as the height difference, the cause of the rolling, decreases *along with* a stone's downward motion. Failing to note this, one will be led astray. Although the error is small for one stone, a catastrophic discrepancy between the calculation and reality will accumulate with time. When a mountain has eroded into a plateau, stones roll no longer.

The ideal of a study is that when one thing changes, other things remain intact. However, this *ceteris paribus* principle, a background-dependent model, does not hold, as Stephen J. Gould, an American paleontologist, biologist, and historian of science, stressed in his gigantic final compendium, *The Structure of Evolutionary Theory* (2002). Nothing can be changed without altering something else. Consequences give rise to new causes because the flow of quanta from one system to another changes both and their positions relative to other systems.

The exact sciences approximate: surroundings are taken as fixed. But a background-dependent model, an *effective theory*,[45] obscures why things happen. We cannot understand why a stone falls or why our motion continues at a sudden braking unless we understand that the surrounding void is thereby on its way toward balance. Nor can we grasp the evolution of the biosphere unless we understand that matter on Earth is thereby on its way toward balance with sunlight. Nor can we fathom embryonic development unless we acknowledge the role of surroundings. And our own actions – yes, motions, too – are influenced by the way we sense and see forces affecting us.

We may approximate, but we had better understand exactly why rather than how things happen. Even when taking the greatest pains, we cannot make repetitions of experiments wholly identical, for no system is entirely impervious and isolated from its surroundings. Nowhere in the universe could temperature be lowered to absolute zero and gravity could be switched off.

It is also an error to take the average as a representative case. The average event never took place; it was only calculated. We rely on statistics without knowing the essence of a state. We imagine that distributions are random and, therefore, symmetrical as the normal distribution. But distributions are skewed due to causes and consequences. Surprises await in those long tails.

Repetition disguises the historical character of time. The smaller the variation in data, the lesser the meaning of time. When nothing happens, certainty is absolute. Pursuing certainty as an ideal of science, we narrow our view of reality. Even when we can calculate the properties of a system in a state of balance, where time circulates on and on, we do not discern how and why the balance came about. Mathematical physics describes circulation as phase evolution[95] but does not reveal the quanta circulating. Circulation, clockwise or counterclockwise evolution of phase, is no true evolution, as it leads nowhere.

IN A THERMOS FLASK

The unpredictable nature of time was not clear to me at first. When I derived the equation of evolution, I tried to solve it – the reflex response of a physicist – but failed. So I began discussing it with colleagues. Dr. Tiia Grönholm, an atmospheric physicist, assured me that she would find a solution if the differential equation were solvable. It soon became apparent to us that no solution would be forthcoming. There is not even a numerical approximation because variables cannot be separated from each other. Everything hinges on everything else through flows of quanta. That is why the future is genuinely open yet not entirely arbitrary. Although the evolutionary equation cannot be solved, it can be analyzed. Soon we learned that evolution results in skewed distributions and S-shaped growth curves.[96] The cause of *Grand Regularity* began to be unveiled.

Later, in the spring of 2006, I gave a talk at the seminar series of the Helsinki Institute of Physics to get feedback on this new way of thinking. I certainly got it: Professor Kari Enqvist scorned the crucial characteristic of an open, evolving system, namely that light streams in. The distinguished cosmologist said that one should not think like that. I wondered cheerfully: "Why not? After all, the Sun is shining; it makes a lot of things happen."

Undoubtedly, textbooks on statistical physics are limited in scope to closed and balanced systems, but this restriction is contrived. Nothing happens within a perfect thermos flask, but our planet is open to sunlight. It is our star that has powered the genesis and evolution of life. True, many other factors, along with light, were needed to get this far. All the ingredients of history, quantum by quantum, must be included in the equation of time.

In 2016, I spoke to the faculty anew about this old principle, although my mood was very different from a decade earlier. At the end of the lecture, Keijo Kajantie, an emeritus professor, challenged essentially the same aspect that Enqvist had disparaged earlier: why does the system's probability include light in addition to matter? The mathematical notation, as such, was familiar to the attendees, but, at least for some of them, the power of sunlight was obscure. Of course, we know light through our own experiences but not in the form of an equation because modern physics has specialized in modeling steady-state systems. When nothing happens, measurements are precise. This is nice but out of touch with the real world.

Thomas Neil Neubert dissects the practical but unnatural dogma of invariance in his book *A Critique of Pure Physics* (2009): "... many physical phenomena from the elementary particle to the galactic level of organization can often be usefully understood within the context of the closed system assumption. This is analogous to the extremely useful (i.e., valid to great experimental precision) of the 'absolute space' or the 'absolute vacuum' assumptions."[97] When we want to calculate, we may assume, but we had better understand what we assume. Otherwise, we are prone to taking our assumptions as facts.

WHAT IS THE MEANING OF TIME?

For ages, the vexed question of time has preoccupied scientists and philosophers. The idea that time is quantum's property, like energy, might be surprising in its simplicity and concreteness. However, we would not talk about time if it had no substance at all. And we would not talk about the arrow of time if the substance had no sense of direction as the photon has. Tim Maudlin, an American philosopher of science, defends this simple, unfashionable view of time with a built-in arrow.[98] He thinks our direct impressions of the world are a better guide to reality than we have been led to believe.

We are used to associating energy with substance. Now we ought to relate time with substance as well. A few have already done this. Frank Wilczek, an American Nobel Laureate in physics, likened the periodicity of time to the symmetry of a crystal.[99] When we know the paths of quanta, we know the characteristics of a system. Elementary particles are sufficiently simple for us to test the theory of time.

Understanding is based on the understood. Time comprises periods as a trek comprises legs. To say that the photon's period of time is time itself is a mere trifle. It is only logical to take the period of a quantum as the fundamental element of time. By contrast, it would be confusing if this period and time were different concepts, having the same unit of measure. Paraphrasing Leibniz: if we cannot distinguish between two things, we must regard them as identical.[100]

Despite this evident logic, someone might still insist that time is not a physical property but only an abstract concept. After all, the explanation of the arrow of time, as the flow of quanta, does not seem to invalidate the quantitative results of modern physics. However, our object is not to contest mathematical modeling but to explain time.

Even if calculations were to remain as they are, the worldview does change when time is understood as concretely as energy to be the photon's property. Similarly, the Copernican model did not immediately make it easier to calculate the orbits of planets compared with the Ptolemaic system, but the belief system was nonetheless revised.

Even so, there might be those who are reluctant to reconsider their beliefs. Why should one think any differently until measurements force one to do so? Perhaps this could already be the case. No one seems to be able to get a grip on dark energy or dark matter. Is it not only so that the equations of contemporary cosmology lacking the essence of time require additional parameters to match up the astronomical data extending far back in time? These up-to-date models thus seem to already be behind the times.

It is difficult to break the habit of thinking that time is not a dimension. Still, there is no universal axis to organize all events because events occur in relation to an observer. Time is relative: the passage of time that I experience matters to me and the one you sense matters to you. Greenwich Mean Time (GMT) serves to synchronize events globally, but it is just a local convention in the universe. For example, what took place on our neighboring star, Proxima Centauri, about four years ago, is visible only here today. Time is not just what can be timed, so to speak, operational comparison. The running of a clock is also a series of events; the flow of time is a flow of quanta.

PERCEPTIBLY PROFOUND

Throughout the ages, people have understood the world more accurately as they have understood it more concretely. For instance, the physicians Alexandre Yersin and Shibasaburō Kitasato got a grip on the bubonic plague in 1894 when they were able to see the disease-causing bacterium under the microscope.[101] Molecules were comprehended as compounds in 1833 when Marc Antoine Auguste Gaudin, a pioneer in chemistry and photography, penciled atoms with bonds.[102] Next, the atom was portrayed as protons, neutrons, and electrons. Nowadays, the proton is pictured as three elementary entities. But what *are* these quarks? They and other elementary particles are said to be quantum fields. But, from our new perspective, we are bound to ask: what *is* the substance of these fields? Would it not be worth studying what concrete and comprehensible follow from positing the quantum as the fundamental element of everything?

Then again, why are time and energy aspects of substance? Why do energy differences diminish and time move on? Even though thermodynamics offers no answers to such axiomatic queries, the worldview that has now been opened is worth exploring.

FACTS AND FICTIONS

After having found the equation of time, I reckoned that perhaps contemporary science was missing something else besides time. So I began to attend seminars on theoretical physics at my university as well as meetings of the Finnish Society for Natural Philosophy. The blend was just right. The ideas in the two communities contrasted sharply.

I recall one time, after a meeting of the Society at the House of Science and Letters, when the physician Jyrki Tyrkkö said to me that the physicists always talk about energy, demanding an explanation of what energy *is*. I was stumped. Now I can say that energy is an attribute of the quantum, like time, but nothing more than that. A particle has its characteristic energy and period. The whole universe, too, has its energy and its age.

Wheeler recognized Einstein's source of insight by asking: "Why did Einstein discuss with people you and I call outsiders? Did not he feel that the amateur gives a fresh perspective that a specialist does not see from his narrow viewpoint?"[103] David Hilbert, a famous mathematician, also acknowledged Einstein's original thinking: "Every boy in the streets of Göttingen understands more about four-dimensional geometry than Einstein. Yet, in spite of that, Einstein did the work and not the mathematicians."[104] Einstein mathematized his imagination of free fall into general relativity, Newton his sight of a falling apple into the law of gravitation, Galileo his timing of running water into the equation of motion.

There are numerous examples in the history of science of how a simple question has pointed to an eye-opening insight.[26b,105] When I set off as a professor of biophysics, I asked myself what evolution *is*. But I did not understand that, in essence, I was asking what time is.

The first few studies with my colleagues Vivek Sharma, Salla Jaakkola, Ville Kaila, Sedeer El-Showk, and Peter Würtz showed how the universal law makes sense of time in physics,[77] chemistry,[72] biochemistry,[106] molecular biology,[107] and ecology.[108] Then, we had to understand what time is at the heart of substance. Otherwise, the theory would remain without a firm foundation. As Blaise Pascal, a French mathematician and physicist, put it: "I hold it equally impossible to know the parts without knowing the whole, and to know the whole without knowing the parts in detail."

The paper *In the Light of Time*, with Petri Tuisku and Tuomas Pernu in 2009, by and large, put forth new perspectives on the nature of substance and the grandness of the cosmos.[49] Quanta circulate steadily in a state of balance, whereas an imbalance forces them to either flow into the system from energy-rich surroundings or out from the system to energy-poor surroundings. We did not have any specific aim when getting acquainted with elementary particles, but the general impetus was to understand stability and change with the precision of a single quantum. Had we been unable to describe elementary particles in terms of quanta and their reactions as flows of quanta, the thermodynamics of time would not have been a viable theory.

It is not enough to know that all animate beings share the same stable metabolites, which in turn comprise atoms, and so on. We must ultimately identify the indivisible basic building block where no further division is possible. Nor could we justify the theory of time if we did not understand the evolution of everything, the expansion of the universe that displays *Grand Regularity* in astronomical data. Even today, I am still striving to find at least one phenomenon that cannot be explained by the thermodynamics of time.

My fervor in trying to falsify the thermodynamic theory in various contexts might give the impression of a menagerie. However, it is not that a professor of biophysics had wandered into fields well beyond his expertise. All studies follow from the axiom that everything comprises light quanta. This might be hard to appreciate by those who do not regard *Grand Regularity* as a sign of the unity of Nature or by those who do not cherish humankind's everlasting quest for a united view of the order of things. Then again, those who know some history of science expect that revolutionary results will come unpredictably.

According to the principle of sufficient reason, science is searching for the truth, the things that could not be otherwise. While the truth is not fully known, the method of testing theses is: logical reasoning and mathematical analysis must connect the conclusions to the foundations of a theory consistent with the data. If more than that is needed, the tenet is no longer valid. And if there is a possibility of a 'truth' being otherwise, it is not strong enough to bear the weight of the worldview. So expressly, the fundamental questions weigh on the foundations of science, testing their resilience.

THE FAILURE IN SUCCESS

Many physicists see time primarily as a philosophical question. Lee Smolin, on the other hand, anticipates that its explanation will resolve many other problems: "I believe there is something basic we are all missing, some wrong assumption, we are all making... What could that wrong assumption be? My guess is that it involves two things: the foundations of quantum mechanics and the nature of time... But I strongly suspect that the key is time. More and more, I have the feeling that quantum theory and general relativity are deeply wrong about the nature of time."[109] Jim Baggott, in turn, recaps the incongruence of modern physics: "With a very few exceptions, [physics] explains every observation we have ever made and every experiment we have ever devised. But those few exceptions happen to be very big ones... We know that the current version of reality can't be right."[21b]

A hundred years ago, this revision of equating time with the period of quantum would have been a welcome revival of realism. But today, after the howling success of modern physics, physicists are reluctant to reconsider; how could they have obtained the right results if the grounding of modern physics were wrong? Yet the question ought to rather be: have researchers got the right results? General relativity does not agree with the observations without dark matter and dark energy; quantum mechanics cannot handle the event of observation.

Doubting a well-established theory is a method of searching for the truth. However, searching for the truth is no longer the goal of every researcher, as other priorities prevail. John von Neumann, a pioneer in information technology, foresaw this trend more than half a century ago: "The sciences do not try to explain, they hardly even try to interpret, they mainly make models."[110] Vannevar Bush, an American science administrator, wrote in 1945 in his famous article, "As We May Think": "If scientific reasoning were limited to the logical processes of arithmetic, we should not get far in our understanding of the physical world."[111] With rich technical means and methods, we are poor in purpose. "Shut up and calculate!" summarized David Mermin on the rationality – or irrationality – of modern physics in 1989.[112] The physics professor at Cornell University warned us that scientific purism, susceptible to 'in-house' censure, prevents progress. Instrumentalism is not pragmatism, "what works is true," for it doesn't work to discover what is true, real, existing. Instead, instrumentalism flies in the face of the essential ethos of science. It just generates the right numbers without any intention of explaining the phenomenon. This bearing might well be the cause of our present problems.

Time occupied the minds of both Newton and Einstein. Now the issue is neither absolute nor relative time but tangible time – the quantum is the matter of time. Maupertuis inferred that everything complies with his principle of least action. Could not the very least action, the quantum of action, be the ultimate basis of *Grand Regularity*? Questions and answers intertwine. Einstein summed up the power of a worldview: it is the theory that decides what we can observe.[113]

Let us illuminate reality from another angle to see what lies in the shadows. Let us look at the whole in terms of details and the details in terms of the whole. Let us ask what the thermodynamic theory of time does and does not explain. The aim is not to justify the tenet but to find out whether we understand what we see.

Key points

- Time comprises periods of quanta.
- Time flows with the flows of quanta.
- Imbalance is a cause that makes things happen, quanta flow.

3. What is Everything Made of?

Elementary particles make up matter.
At rock bottom, photons make up everything.

As surprising as it may seem, data of various kinds are alike; only the names given to things are different. For example, the shape of a mollusk's spiral shell is akin to that of a spiral galaxy; biota and business grow similarly along S-shaped curves. Various distributions also look alike. For instance, microbes are much more abundant than big animals, and likewise, local stores vastly outnumber shopping malls. Flows of air, avalanches in semiconductors, and stock trading all exhibit chaos resembling our chaotic brain activity. Since this similarity, the *Grand Regularity* ranges from atoms to galaxies and from retail to world trade,[1] could it be that everything in the universe is fundamentally composed of one and the same substance?

Indeed, so thought Parmenides (ca. 510 BC). The philosopher envisioned that both matter and space are made of the same eternal elements,[2] which he called *atomos*, meaning non-divisible. While the atom was found to be divisible into elementary particles in the 19th century, and later, the particles were further transmutable in nuclear reactions into the vacuum, and vice versa, these and transformations of other kinds entail that everything is, in essence, the same substance (Chapter 1), and therefore everything happens by the same law (Chapter 2). Although this atomistic proposition is logical, it has taken considerable time to concretize that everything is composed of quanta of light.

(Chapter 1), (Chapter 2)

Ancient atomos, contemporary quantum

Where did Parmenides get the idea of *atomos*? Did he reason that nothing could come out of nothingness or that a thing might only transform into another? Perhaps Parmenides just generalized the common knowledge that the sugar in grapes turns into the alcohol of wine. How could this or any other change take place if everything does not ultimately comprise the same basic elements?

Parmenides was not alone in these thoughts. The earlier philosophers, Anaximander and Thales, had talked about a profound substance.[3] Later, Galileo reasoned that it could be those small, infinitely light and high-speed elements that carry heat from the Sun.[4a] While this comprehensive worldview has been around for a long time, at last, we have enough resolving power to bring it into focus.

At school, I considered the early theories about matter, which our history teacher taught us, to be precisely antique. I assumed that the ultimate composition of particles would be exposed using particle accelerators. I believed that a mathematical model that matches the data would give an explanation; however, I was later to realize that a good fit, as such, does not explain anything. Instead, we must learn to envisage elementary particles in the way we picture atoms to grasp their properties. The need to comprehend by visualizations dates back to Leonardo da Vinci, a paragon of the Renaissance Man.[4b]

Although the ancient idea of a fundamental element may seem amorphous, it is, in fact, a strict axiom. The logic of the early thinkers is ironclad. The universe cannot function unless it consists of only one type of elemental constituent.[5] This atomistic tenet is either true or false. There are no parameters that can be adjusted or added. When everything is composed of indivisible and permanent elements, nothing can emerge from nothingness, and nothing can vanish into nothingness, but everything depends on everything else. Therefore, no locus is identical to any other.[6] This relates to the ontological principle of the identity of indiscernibles.[7] It states that objects having all their properties in common cannot be separated from one another but are one and the same thing.

Despite its consistency, the old unified understanding seems not to be of much help to us because, in those days, not even the four fundamental forces of nature were known. Even so, Parmenides' thesis implies that gravity, electromagnetism, and the weak and strong nuclear force, too, are made of the same fundamental elements. Maybe this is exactly the intuition that we need. We know that photons mediate electromagnetism. Already Empedocles, a physician and philosopher (c. 492-432 BC), knew magnetism as a flow between a magnet and a piece of iron.[8] We should, therefore, ask whether this insight, the photon being the fundamental force carrier, holds for the other three forces too.

These thoughts should not be held as an exaltation of the foregone philosophers but rather as an invitation to consider what we might learn from their admirably sound reasoning. The early thinkers aimed to uncover general principles that set everyday experiences into a logical order. Now we seek to discover a universal law that sets both scientific experiments and everyday experiences into a consistent form.

Comprehensive thinking is inspiring, but is it physics? Along these lines, Newton was challenged to justify how force transmits through the void, although his law of gravity had already been confirmed through observations. The English physicist Michael Faraday was asked how forces could exist outside the body, although these electromagnetic effects were readily observable. Einstein, in turn, theorized the cosmos to be such a miraculous 'fabric' of space and time that it is no wonder general relativity was mistrusted for a long time. Some question relativity theory even today, although it aligns with the data. Furthermore, the Danish physicist Niels Bohr's dual image of wave and particle seemed bizarre from the start, although it, too, is backed up by the results of countless experiments. All in all, it has not been uncommon for a new theory to solve old problems in a way that feels as if nothing has actually been explained.[9] In this manner, the history of science advises us against deciding what physics is and what it is not if we hope to see continued progress.

Like Parmenides, Newton sought a unified worldview, as this query in his book of *Opticks* reveals: "Are not gross bodies and light convertible into one another?"[10] Presumably, Newton had reasoned that plant leaves bind a fraction of the sunlight they receive. After all, he saw how the sunlight decomposes into colors when passing through a prism and noticed twinkles of light escaping from substances he employed in alchemy experiments. These undertakings of the great genius are not to be judged as mysticism, for he simply lacked knowledge and means of how to transform one element into another.

Newton considered light and matter as essentially the same substance and similarly granular. This stance may have been one of the reasons for his disapproval of the conception of light held by the Dutch mastermind Christiaan Huygens. Light could not possibly propagate in a vacuum as sound does in air. Likewise, space could not be some sort of matter because even if it were supremely sparse, the celestial bodies would grind to a halt in the long run. Newton probably wondered whether a corpuscle could, after all, be a wave-like entity but failed thereby, too, to form a unified image of reality.

A hundred years later, the British polymath Thomas Young conducted his famous experiment demonstrating light's wave-like character. Nevertheless, the fundamental nature of matter, and that of the vacuum, remained mysterious.

Today, it is posited that the plethora of particles and their interactions are all manifestations of quantum fields.[11] In this framework, both particles and forces are viewed as consisting only of fields within fields.[12] Even so, shouldn't we strive to visualize them to see what determines the particle properties and what mediates gravity – if anything?[13]

WHAT IS LIGHT?

There is nothing more fundamental than light.

The ancient atomistic idea was put into practice in the early 1800s when John Dalton grasped that chemical substances are compounds.[14] Then, the enormous multitude of molecules could be organized in terms of relatively few atoms. However, one hundred years later, the physicists Hans Geiger and Ernest Marsden, led by Ernest Rutherford, discovered the atom itself consisting of a nucleus and electrons. So, the classic question arose again: What is the indivisible elemental entity? Is there one?

In the mid-1920s, the American chemist and physicist Gilbert Newton Lewis coined the word photon to put forth anew Newton's inkling that the particle of light could be the fundamental element.[15] He had good reasons to suspect the photon's primacy. Namely, the massless particle does not decay,[16] and the spectrum of light reveals an invariant entity, Planck's constant. Not odd but expected, the value of this natural constant is tiny ($h \approx 6.626 \cdot 10^{-36}$ Js), for atoms are small as well. Still, it was difficult to grasp what the constant meant.

Einstein's original idea in 1905 was that a ray *is* quanta of light.[17] The photon has its length, wavelength. It is like a tiny piece of wire. The constant, h, is, therefore, not just a mathematical gimmick by which Planck melded two equations into one that covers the whole spectrum of light, but a measure of the photon. The unit of the quantum of action (Js) tells us that energy (joule) and time (seconds) are the properties of the photon. Should we not surmise there to be a profound reason for Planck's constant to show up in so many equations that span scales of substance from quanta to black holes?

Even before Lewis' proposal of the light quantum as the basic element, the German-born mathematician Emmy Noether had written her famous theorem about quanta.[18] While stating conservation laws, it goes beyond just the conservation of energy and momentum, relating a system's energy and time to its amount of quanta. And since the universe is a system, we can calculate from its total energy and age that it contains on the order of 10^{120} quanta (Appendix G).[19] This number, with its 120 zeros, is unimaginably huge yet finite.

In this way, in theory, it was understood that the photon is the fundamental element of everything, but in practice, the thesis was not seen as convincing. The total number of photons did not seem to be constant. Specifically, an atom may go from an excited state to the ground state either directly by emitting only one photon or via intermediate states by emitting one photon at each stage. Since the number of detected photons depends on the route taken, it looked as if the photons appeared out of nothing and disappeared into nothingness. So it became a truth of physics that the photon could not be the indissoluble element.

Despite all its theoretical sophistication, physics is an empirical science. If a hypothesis goes against observations, it must be rejected. But of course, the thesis should and eventually will be re-evaluated if it turns out that something has gone unnoticed. Then again, the theory itself predisposes how we comprehend the phenomenon. Comprehension, in turn, determines our understanding of what we need to measure. The Hungarian-born philosopher Imre Lakatos even contended that no observation would suffice to prove a theory wrong.[20] Although a theory may appear to be in conflict with the data, we cannot conclude what causes the disparity without some theorizing.

Our cognition often exhibits blatant ambivalence and adamant bias. This is familiar from the image representing the silhouette of either a vase or two juxtaposed profiles. When we are acutely aware of our inclinations, we can be more confident in our conclusions. If we were to choose between alternatives, Einstein's sage advice was: the best theory is the simplest and the most comprehensive.[21]

A vase or two faces? We see what we have chosen to see. We find meaning in that which we understand. Can we think differently and understand different thinking as being meaningful?

The outcomes of our logic are no better than their premises. Therefore, we had better build on things that could not be otherwise – on truths. Is the quantum, the indivisible and eternal element, sufficient reason[22] for the law of cause and effect?

WHAT IS THE VACUUM?

The void is not all empty but embodies photons in pairs.

Would we not eventually come across the indivisible basic building block if we were only to mince matter into smaller and smaller bits? First, compounds divide into atoms. Each atom, in turn, breaks into a nucleus and electrons. Next, the nucleus splits into protons and neutrons. Each of them consists of three entities. They are called quarks.

Oddly, the quarks cannot be detached from each other. Instead of the quarks breaking apart, more quark-antiquark pairs will materialize the harder the particles are smashed into each other. These matter-antimatter particles are mirror images of each other, like the left and right hands. Where do these pairs come from? It looks as if they pop out of nothing. In that case, the bookkeeping of substance seems like worst-class fraud.

In truth, particles do not come from mere nothingness but from the vacuum. While there is no matter in the void, it is not just empty space either: a cubic meter of vacuum contains the energy of about one billionth of a joule (10^{-9} J/m^3).[23] This amount of energy is equivalent to that of a few hydrogen atoms. But in the void, there is no matter at all. Everything that can be removed has been removed. So what is left? What *is* the vacuum?

It is the big question of physics. Our ignorance is immense because the void is all-pervasive. But we do have a clue, even if we have not recognized it as such. Feynman wondered why there is as much energy in all of space, free of matter, as there is energy in all of matter.[24] The galaxies are scattered so far apart that the energy bound to matter in the vast universe is, on average, just one billionth of a joule per cubic meter. Is this correspondence just a coincidence?

Hardly. How could such a mutual energetic balance between the void and matter have come about and lasted over the evolution of the universe unless matter and vacuum are able to transform from one to the other in some circumstances? Therefore, they must comprise the same essence, namely, quanta.

It is no coincidence that Earth's gravity is in balance with its mass. The law of gravity states that equivalence. Should we likewise consider free space to be in balance with the total mass of the universe? Then the void must be the gravitational field of all matter.

The gravity of a distant galaxy affects us here, albeit only very weakly. However, the overall effect is not negligible because the number of galaxies in the universe is colossal, at least one hundred billion.[25]

The gravity of the entire universe is not a distant backdrop but a real hindrance we sense every day. When you step on the gas pedal, you feel you are being pushed to the back of the car seat, although this sensation is not particularly impressive in the rattletrap I drive. Conversely, you will continue in motion in the event of a sudden stop unless the seatbelt restrains you. Isn't it amazing that the whole universe reacts against you when you make a change? That's a solid excuse when you do not feel like getting out of bed.

The visionary physicists Ernst Mach[26] and Dennis Sciama[27] thought, like Newton, that distant stars are what causes inertia.[28] Their conclusion is easy to absorb. As gravitational potential decreases inversely with distance, $1/r$, and the number of galaxies increases with the distance squared, r^2, the overall effect *grows* with distance, r. Due to the numerous far-flung galaxies, universal gravity is very stable.

But how can the distant universe resist the change in motion *instantaneously* here on Earth? This bothered Newton too. Feynman was already wondering about the same thing as a kid. When he tugged a wagon, he noticed that a ball on it rolled and banged on the tailgate. And when he halted the cart, the ball bumped up against the front edge. So the young maverick asked his dad why this happened. "That, nobody knows," declared the elder Feynman. "The general principle is that things which are moving tend to keep on moving, and things which are standing still tend to stand still unless you push them hard. This tendency is called 'inertia', but nobody knows why it's true."[29] Decades later, Feynman praised his father for this lesson. The name of the phenomenon does not matter, but its cause does.

The cause of inertia is still unknown today, but you and I are making headway here, for the all-embracing void is the gravitational field of the universe. It is all around us. That is why it affects us in real time. This is, nevertheless, not enough for an explanation. We must also know what the vacuum *is*. We have to understand how the fundamental elements make it up, just as Parmenides argued. And we still need to comprehend how the vacuum couples to matter. Only then will we genuinely know what inertia is.

The leading idea from the 17th century onward was that the vacuum must be a medium to transmit light waves in the same way as air conveys sound waves. But the American physicists Albert Michelson and Edward Morley found no evidence that the Earth treks through the postulated medium, known as the ether.[30] The experiment was one of the most renowned negative results in the history of science, as expectations were the opposite. How does gravity propagate without a carrier if the vacuum is no substance at all?

A good swing ride gives you a gut feeling for gravity and inertia that is more memorable than anything you get from a lecture or textbook. Isn't it amazing that all riders, even empty seats suspended from a carousel, align at the *same* angle? This is because both gravity and the centrifugal force, due to inertia, are proportional to mass.

Einstein understood that this equivalence is no coincidence, as the gravity of the whole universe is the source of inertial effects. General relativity was devised to incorporate this idea but, in fact, failed to do so. As the British cosmologist Sciama pointed out in his iconic paper On the Origin of Inertia (1953), Einstein himself emphasized this failure.[27] Even though Einstein's model of the vacuum gives the right numbers, curved space-time[iii] without any substance is an incongruous idea. After all, the vacuum embodies energy and has electromagnetic properties. How could there be content and properties unless there is some essence? In this book, we demand concrete thinking due to our conviction that the universe is not an abstract theoretical construct but an altogether tangible reality.

DARK LIGHT

The conventional interpretation of Michelson and Morley's experiment refutes the *light-mediating* ether hypothesis. However, it does not exclude the possibility that the vacuum *itself* is light. This idea seems strange at first sight because the void is dark and transparent. How could the void be quanta of light and yet invisible?

Photographers and birdwatchers know the answer by experience: when coated with a thin film, the lens of a camera or telescope reflects remarkably little light. Actually, light waves are reflected from both the lens and the coating film, with the photons from the two surfaces

---

iii It was Hermann Minkowski who put space and time on equal footing as a block universe, a stance at odds with our own experience.

combining so that the crest of one wave then fills precisely the trough of another one.[iv] And hey presto: we see no light as the waves at opposite phases nullify each other. The photons themselves do not disappear but make up a pair that does not interact with matter in the same way as a single photon does.[31] Such a zero of an optical field due to opposing phases is familiar, for example, from singular optics.

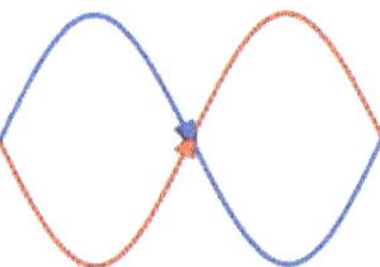

When two wavelets of light travel together so that one peaks exactly where the other bottoms out, we see no light. The out-of-phase-paired wavelets, embodying the transparent void, carry energy and time; that is why the void displays transport and wave phenomena.[v]

The out-of-phase pair is not discernable to the eye or detectable by a camera because the two photons, canceling each other's electromagnetic forces, do not displace electric charges. But the photons themselves have not vanished from existence. They are still moving at the speed of light. As such, could it be that these pairs are whizzing all over the universe and thereby *embodying* the vacuum? The idea is surprisingly simple. So has it not been proposed before?

Yes, it has. Faraday thought that the lines of force are themselves substance, like swirling fog, and hence saw no need for a substratum through which they transmit actions.[33a] In a highly praised paper published in 1865, Maxwell advanced this idea by mathematizing the electromagnetic field as a strained state of the void.[34] At that time, the essence of the void was a major problem, just as it had been since antiquity. Maxwell was aware of experiments by François Arago and Hippolyte Fizeau and of explanations by Augustin-Jean Fresnel and George Stokes that the ether could be a partially or entirely porous substance. Maxwell understood that the sparser the vacuum, the faster light goes. Technically speaking, the speed of light is set by the

---

[iv] The argument is more subtle. As single photons do not reflect back either but transmit through, the destructive interference involves waves of the vacuum.
[v] The massless photon pair of half-turn symmetry qualifies for the carrier of gravity, the graviton. The spin-2 boson breaks into two spin-1 bosons, photons.[32]

electromagnetic characteristics of free space. We also know that light moves faster in air than in water. That is why beams refract on the surface of water. Since Maxwell pictured light as undulations of the ether, the ether is, in his view, light.

The vacuum comprises paired quanta of light. These waves, without a net electromagnetic force, do not perturb electric charges yet couple to bodies with mass. Therefore, we experience the void as a reaction to acceleration and deceleration, as well as centrifugal and gravitational forces. We see as light, among the strings of paired quanta, the single quanta of the electro-magnetic field.

At the end of the 20th century, the British physicists John Poynting and Oliver Heaviside concluded from Maxwell's equations that the energy of the ether propagates at the speed of light.[35,36] However, they did not have the concept of the light quantum to account for the essence of the void. Planck's great discovery was still a decade off, and Lewis' interpretation of the photon as the primary constituent even further in the future. Before those revelations, the experiments by Michelson and Morley, as well as by Frederick Trouton and Henry Noble,[37] were interpreted as if the void was nothing. And as special relativity agreed with the observations, Einstein went on to formulate general relativity. Curved space-time is ether without substance.

From this angle, the age-old question of how forces transmit through space seems pointless. The mathematical model of gravitation works alright but still fails to explain why bodies move toward each other. The mathematical model of electromagnetism works fine as well but does not explain why charges move relative to one another. Explanations need substance.

In the wake of Einstein's principal works, quantum mechanics came to the fore in physics, and the void stepped into the background. The proposition that paired photons make up the void did not materialize. Had it done so, it could have unlocked a slew of problems.[38]

According to modern physics, the quanta of light are created out of the vacuum. But now we understand that they, in fact, are being *singled out* from the vacuum's paired quanta. Likewise, the modern tenet contends that the photons vanish into nothingness. In reality, we understand that they match pairwise to merge into the transparent vacuum. Experiments to date have not disclosed these 'hidden' photons to us because we have not thought about detecting them explicitly. As Lewis reasoned, we only have interpretations and, therefore, no proof against the proposition that the quantum of light is the permanent element. After all, the bookkeeping of quanta is perfect; no quantum has gone missing. Nevertheless, we need to consider whether this new interpretation leads to the old problems of the luminiferous ether hypothesis.

In a letter to his colleague Robert Boyle in 1679, Newton opined that the vacuum density relates to gravity.[39] Newton sought to grasp the balance of forces to fathom why the void does not slow down the planets and ultimately brings them to a standstill. But the energy concept, necessary for the answer, was formulated only after his death.

As a textbook illustration, space-time can be seen as an elastic fabric woven from numerous paired-photon fibers of varying lengths and periods.[40] From this perspective, single photons do not propagate through space; they are part of it, together with the far more abundant paired photons, i.e., weakly interacting photon pairs (WIPP).

Given that gravity is a much weaker force than electromagnetism, matter is almost transparent to the substance of space, as glass is to single photons. The strength of this very weak coupling is customarily called mass. Still, the void's quantum structure alone does not enable us to comprehend the meaning of mass. We need to know the quantum structure of particles, too. But, before that, we can already make sense of the effects of the vacuum, even if we do not see it directly.

THE DOUBLE-SLIT EXPERIMENT

Feynman graded the double-slit experiment as the only phenomenon whose examination calls for quantum mechanics, the remarkable theory of modern physics.[41] Yet, now that we know the substance of the vacuum, we can comprehend the experiment with common sense.

The setup is simple. Photons are sent off one by one toward a plate with two slits. Once many photons have gone through one or the other slit, we would expect to see two bright lines on a detector behind the two slits, wouldn't we? But we see not just two but many lines. The alternating bright and dark lines, crests and troughs, tops and bottoms, are familiar from ripples in water. In the bright band, the crests of the light waves add up. The waves in the same phase reinforce each other. In the dark band, the highs and lows of one wave cancel out those of the other. The out-of-phase waves silence each other.

From the detected pattern, we thus infer that one wave on its way interferes with another. But how could that be possible? If only one wavelet of light is on its way at a time, with what could it interfere?

To circumvent this quandary, physicists have resorted to quantum mechanics. This effective theory claims that one and the same photon goes through both slits *simultaneously* and interferes with *itself*. Weird, isn't it? Despite its eeriness, physicists tend to uphold the truth of this interpretation.

A similar pattern of interference appears when electrons are used instead. Again, it seems as if one and the same electron had passed through both slits *simultaneously* and interfered with *itself*. And not only so, but molecules, such as the Buckyball (a.k.a. fullerene) comprising 60 carbon atoms, seem to pass the slits at the same time.[42] Even bigger molecules exhibit the same behavior when hurled through the slits.[43]

Alas, quantum mechanics is not a theory complying with our experience of reality; its interpretation of the phenomenon bamboozles more than enlightens us. For example, how could a stone split in half and then rematerialize as a single solid? Surely, we can accept that the stone consists of grains, for we know the pebble could be ground into sand, and the sand could be pulverized all the way to atoms. But the double-slit experiment is not about smashing things into smithereens.

In quantum mechanics, the pebble is described as a cloud of probability. This mathematical concept, the so-called wave function, was needed in the early part of the last century to calculate the outcome of the double-slit experiment. In truth, modern physics does not even try to explain phenomena, only to model the results without interpreting them. Could this insane instrumentalism be the result merely of the inability to conceive of a realistic theory?

In other words, the double-slit experiment has not yet been understood. We have only a mathematical model to calculate what we detect. The model is impeccable. Calculations agree with data, but the problem is more profound than mere mathematics. Quantum theory fails to tell us what this thing called probability *is* – not mathematically but physically.

Smolin sees instrumentalism as a fatal flaw in modern physics. Physicists have mathematized Nature without substance, but physics cannot go on ignoring substance. Knowing the electron only as a wave function from the Dirac equation, do we really know the electron?[44]

THE UNDULATING VACUUM

Wielding our new conception of the vacuum, we can make sense of the double-slit experiment (Appendices C and H). As the particle moves, it perturbs the void, as a ship makes waves in water. When the particle passes through one of the slits, the waves of the void it generates pass simultaneously through the other slit. These waves then deflect the electron before it hits the detector. In other words, the particle experiences the vacuum waves to which its own motion gave rise. So the interference pattern forms.

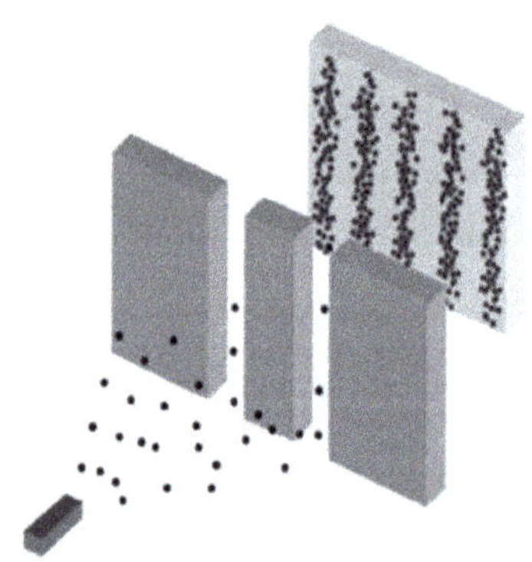

Particles are shot one by one through the two slits toward a detector screen, where we see a striped interference pattern rather than just two stripes.[45] When the particle passes through one of the slits, its waves in the vacuum also pass through the other. The interference pattern emerges because the vacuum waves deflect the particle on its way to the detector.

The same principle is at play on a boat's arrival in a harbor shielded by a breakwater. When the craft goes through one of the two entrances, it makes waves that also pass through the other. These waves, which came through the other entrance, rock the boat, i.e., interfere with it as it arrives at the pier.

When we consider the void as a wavy substance of photon pairs, we perceive the double-slit experiment in the same way as we do rippling water. There is no reason to believe that the particle miraculously goes through the two slits simultaneously, but all the reasons to think that the waves it raised in the void do go. This revelation is a great relief for our sanity. The reality of the way particles behave is no stranger than real life.

Let us not get ahead of ourselves. Our new interpretation *could be* true, but the case is not closed. Science would not be science if we casually embraced what is given. We must question positions. We should investigate whether the waves indeed arrive in time to affect the electron that gave rise to them. And is it possible to calculate where the particle ends up? Do the waves also strike other electrons, fired before or after this one? Does the physical vacuum around the photon also explain simply by the Fresnel-Arago laws the interference when a single photon goes as if simultaneously via two arms of an interferometer?[46] What substantiates the claim that photons do not just move in the vacuum but constitute it?

DELAYED CHOICE

In 1978, John Wheeler conceived an experiment that clashes with causality. First, a photon strikes a beam splitter. Then another splitter combines the split rays into an interference pattern. Next, the experiment is rerun by removing the second splitter while the split rays are on their way. This time a single speck, instead of a pattern, is detected. So, how can the photon delay its choice to behave either as a wave or a particle until the second beam splitter is in or out of place?

Consider the photon perturbing the vacuum. The interference pattern emerges as the vacuum waves recombine at the second beam splitter. Conversely, in the absence of the splitter, the waves cross without interfering. So, the true trouble is that a vacuum wave is not readily detectable when split away from its source, the photon, the quantum of the electromagnetic field. The paired photons are apart in the vacuum wave, but their pairwise electromagnetic effect is still zero.

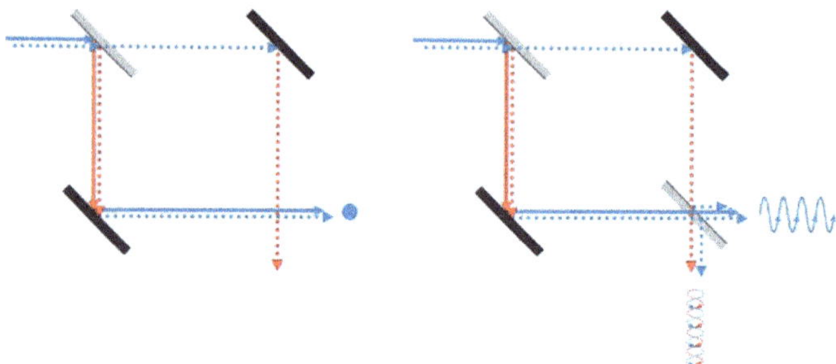

Left: First, a photon (solid blue arrow) either goes through a beam splitter or gets reflected (red). Then it goes on and gets reflected straight to one or the other detector port. Thus, Wheeler argued the photon had decided to behave as a particle (sphere). Right: When another beam splitter is put in front of the detectors while the photon is on its way, a destructive interference pattern develops on one port and a constructive on the other. Thus, Wheeler maintained that the photon had reversed its decision and behaved as a wave. In truth, the photon gives rise to vacuum waves (dotted arrows) that interfere with it when realigned parallel.

THE EXPLANATION IN SUBSTANCE

The nature of light is confounding. On the one hand, the English physicist Paul Dirac claimed in his textbook of quantum mechanics that "… each photon then interferes only with itself. Interference between different photons never occurs."[47a] On the other hand, Leonard Mandel, a pioneer of quantum optics, demonstrated with his colleagues that two laser beams do interfere and thereby aroused controversy in the 1960s.[48] About 20 years later, Mandel and his colleagues also showed that two photons always emerge parallel when arriving at the beam splitter in synchrony (Appendix C).[49] The field of a photon, i.e., the vacuum's quantized structure modeled by the photon wave function, extends far and displays itself in such correlations. Waves interfere when guided in parallel; when orthogonally, they do not. Mandel understood that there is no better knowledge than our own experience: "Concepts like these can seem remote until you explain with analogies taken from everyday life. But unfortunately, some ideas do not lend themselves to familiar analogies."[50]

The invisible vacuum becomes visible, for example, in the school class experiment where a ray of light enters a glass block so that it gets totally internally reflected back into the glass. At the point of reflection, the adjacent vacuum gets disturbed as the light causes electric charges in the glass to move to and fro. You can infer this evanescent wave existing next to the otherwise reflecting surface, for example, from making your fingerprints visible by pressing hard against a glass of water. Rather than explaining the evanescent wave in this way, textbooks say that the electromagnetic field, or wave function in technical jargon, cannot be discontinuous at the interface between the glass and the air. However, the mathematical condition of continuity does not tell us that the field is light-like substance.

The vacuum as the paired-photon substance does not, per se, question quantum mechanics, as we can frame the wave function as a mathematical image[51] of the wavy void around the particle rather than of the particle itself. For example, the photon produces vacuum waves of its own kind. The electron generates waves that are typical of the electron. Likewise, the waves of a boat are typical of a boat, and the waves of a ship are characteristic of a ship. Thus, our concepts are revised, not so much our calculations.

Of course, we must look for observations that might contradict the thesis of the paired-photon void. We need to examine the implications of the vacuum's physical being as opposed to comprising virtual photons or geometric space without substance. For instance, can we understand why the position and momentum of an electron cannot be quantified more precisely than a single quantum? This uncertainty is a central principle of modern physics.[52]

We must also ask why the void was not envisioned in this way already a long time ago. After all, it is textbook knowledge that two photons combined out-of-phase cannot be seen. Also, in line with relativity theory, the speed of light in the vacuum comprising light can hardly be other than the speed of light, and the passage of light in light can hardly be other than the shortest in time. And the photon-embodied substance permeating a dense medium just as constituting the sparse vacuum can hardly be other than in balance with matter. So, it may well be that this understanding prevailed until the end of the 19th century but died at the hands of the Michelson–Morley experiment. When evidence, however tenuous, is proclaimed over and over again without alternatives, one is disposed to believe it.[53]

UNFRIENDLY RECEPTION

As the electron goes through one of the two slits, we may try to see which way. However, the monitoring itself causes the void to ripple, and the ripples mask the electron-induced waves. Even though the interference is disturbed, the path can be determined statistically by probing photons gently near the detector.[54] Then, the photon passing through either one of the two slits has already experienced most of the waves that also went through the other slit. Accordingly, the interference is washed out if the photon is checked right after the slits.

Powerful lenses are fashioned out of dense glass, for the density of the medium influences the path of the light. In the same way, vacuum density affects the flight of an electron. To see this effect, the physicists Yakir Aharonov and David Bohm proposed a modified double-slit experiment,[55] where the electron goes briefly through a denser vacuum due to vector potential, and expected the interference pattern to alter. Likewise, it was recently demonstrated that whole atoms sense the gravitational potential of a body.[56]

Observations matched the calculations, but that was not a sufficient explanation for Bohm. He pondered what the void *is*. Bohm had already questioned the Copenhagen interpretation of quantum mechanics[vi] when working with Einstein. He sought a causal theory, unable to accept that results could be random, inexplicable, mysterious.

Bohm's original ideas got an unfriendly reception. His response to Nobel Prize winner Isidor Rabi is telling: "Exactly the same criticisms that you are making were made against the atomic theory – that nobody had seen the atoms, nobody knew what they were like, and the deduction about them was gotten from the perfect gas law, which was already known."[57] Moreover, the reply to the nuclear physicist Herbert Anderson exemplifies how difficult it is for dissent to be accepted. "All I wish to do is to obtain the same experimental results from this theory as they are obtained from the usual theories, that is, it is not necessary for me to reproduce every statement of the usual interpretation." By all accounts, the new understanding is expected to comply more with consensus than observations. Robert Oppenheimer's

---

[vi] According to Copenhagen interpretation a system does not have definite properties prior to being measured. Only the outcomes of a measurement can be known.

reaction to Max Dresden's scientific talk in 1952 is absolutely alarming: "If we cannot disprove Bohm, then we must agree to ignore him."[58] The guiding principles of science are not always uppermost in our minds, not even in the mind of the then director of the Institute for Advanced Study in Princeton, as the science writer Adam Becker points out with this quote in his book, *What is Real?* (2018).

Likewise, the claim that for the thermodynamics of imbalance, the theory of time, to be valid, should produce all of modern physics is slippery in its criteria of completeness and consistency. Instead, it is legitimate to demand an empirically falsifiable axiomatic theory rather than a malleable model fitting the data but not explaining it.

The Aharonov-Bohm effect would hardly be known by its name had not Harold Dodds, the rector of Princeton University, refused to renew Bohm's associate professorship in 1951. Bohm had just been cleared of allegations of communism raised when he had refrained from witnessing against his colleagues. So it seems Bohm was spurned not for questioning the political order but rather, and more problematically for us, for questioning the dominant scientific doctrine. The upright Bohm left to tour the world and became world-famous.

Quantum mechanics models an interference pattern that results from numerous events. Yet, it does not describe any one of those events. Based on the Copenhagen interpretation, the passage of particles through the slits is arbitrary. It is a sterling model of the vacuum quanta dashing in all directions at the speed of light. While convenient in computation, randomness is devoid of physical meaning. How could there be any effect without some cause?

To hold quantum mechanics as the truth rather than as a mathematical model, one has to imagine that Nature is random. Then one would be forced to believe in miracles. In fact, the three-slit experiment suggests the quantum mechanics explanation for the simultaneous passage of the particle through all the slits is not exactly correct.[59]

It is worth emphasizing that the physical nature of the vacuum does not imply determinism, as Einstein and Bohm are thought to have assumed. The future is not predetermined, but neither is it random. The vacuum affects the electron and the electron affects the vacuum. Likewise, our environment acts on us and we act on it. When everything depends on everything else, the equation for flows of quanta can be formulated but not solved.[60] Although the course of any one event is inherently unpredictable, a regular pattern emerges from numerous unique events. This is how a certain rule, the natural law of causality, can lead to an uncertain outcome. Laughlin found this to be an important and interesting issue.[61a]

CRESCENT WRENCH

At the end of the 1980s, I did my doctoral thesis at the Low Temperature Laboratory at the Helsinki University of Technology. Ultra-low record temperatures had been achieved a good many years before I got involved as a rookie, but the basic research continued to thrive. In 1987 we conducted an astonishing experiment. First, we cooled atomic nuclei of silver down to a few tens of a billionth of a degree above absolute zero. Then we heated the nuclei for a short period, but soon they cooled down again as if all by themselves! Where did the heat go? What was retaining the cold?

I knew from the theory of solid-state physics that the interaction reservoir between the hot nuclei would still be cold. "Yes, yes, but what *is* it?" asked Kai Nummila. The then newly graduated superfluid helium researcher from the neighboring group was after something tangible substance, the reservoir of cold photons that warmed up as the nuclei cooled down. After all, often, we had a wrench in our hands to do something practical than a pencil to do something theoretical.

We cannot imagine something nonexistent having properties. Yet this is how quantum mechanics portrays the vacuum. It is a pool of virtual particles or undetectable quantum fluctuations. General relativity, in turn, wraps the void in space-time geometry. However, space has electromagnetic properties as materials have electrical resistance and impedance. The whole void houses as much energy as matter in the entire universe.[23,24] The vacuum is real, for sure.

A scoop of water is a measure. The vacuum also has a measure. Textbooks of physics mention it but only as a mathematical condition.[62] In 1867, the Danish physicist Ludvig Lorenz used that measure to solve the motion of the electromagnetic standing wave. Likewise, water keeps moving back and forth unless it spills out of the scoop.

The same continuity condition also holds for a sand pile leveling out. The sand does not disappear but flows to the side. A categorically concrete worldview is solid.

We sense the vacuum with our own body as inertia, but a device is needed to detect the vacuum waves. I grew familiar with the wavy void when doing spectroscopy by setting up an electromagnetic wave bouncing back and forth between two atomic nuclei. This remarkable condition was discovered by the American physicists Erwin Hahn and Sven Hartmann in 1950. Likewise, you can set up water to slosh end to end in a bathtub when you keep moving at the right pace.

The works of the pioneers seem long gone, even when they are not. I realized this in 2001 when Hahn himself stepped onto the stage and held an entertaining after-dinner speech at a conference in Orlando, Florida. Hahn began by saying a young student had recently asked him what the above-mentioned Hartmann-Hahn condition is. Erwin had answered: "I don't quite know what Sven's condition is today, but I have a sore back."

Undoubtedly, the vacuum is a remarkable substance, but substance it is. Already Aristotle reasoned that Nature abhors emptiness. Today, we can calculate the vacuum's electromagnetic properties assuming fleeting quantum fluctuations and gravitational forces in terms of non-Euclidean geometry, as if the vacuum were not a substance at all. But it is. Using electron holography, Akira Tonomura recorded tiny vacuum vortices moving on the surface of a superconductor. At a memorable public lecture at The Royal Institution, the Japanese physicist showed how the oppositely spinning vortices encountered and extinguished each other.[63] In the same manner, a particle and its antiparticle annihilate each other. By all accounts, the tension between realism and the instrumentalism of modern physics is worth resolving.

WHERE DO THE PHOTONS COME FROM?

Photons are forever.

Faraday speculated shaking the lines of force would produce light.[33c] But first, a few years ago, researchers at Chalmers managed to shake visible photons out of the vacuum two by two at a time.[64] Also, researchers at the Max Planck Institute showed in 2019 that light quanta emerge in pairs out of the vacuum when electrons flow from a needle to a conductor surface.[65] By all appearances, the wavelets of light, both single and paired, embody the vacuum.

The waves of vacuum spread far from a particle, just as waves of water disperse far from a boat. But, on the other hand, the particle itself impinges on the detector just as a boat crashes on the rocks. The German Nobelist Werner Heisenberg reasoned along these lines that no object could be at the same time both a particle, a small speck of substance, and a wave, a widespread field.[66] Indeed, fluctuating quantum fields model the rippling vacuum around the particle rather than the particle itself.[67] In the same way, an electric field spreads from an electric charge and a magnetic field curls around a bar magnet.

The quanta of light inside matter, distinct from the atomic nucleus and the electrons, is akin to the surrounding free space but denser and richer in structure, as revealed in intricate spectra. A vibrating atom, just like a propagating photon, generates waves in this Coulomb field, which in turn perturb other atoms.[68] You can witness this effect, too, by your own senses, for example, as sound waves of a percussion drill echoing through the framework of a tower block.

Quantum mechanics works well, but it does not explain what's going on. Physicists are the first to acknowledge the need for an explanation, a causal chain of events. So they delineate events in Feynman diagrams, where virtual photons or other virtual particles crop up from the vacuum and mediate interactions from one particle to another. Having done their job, the virtual particles drop back into the vacuum. Yet, this nifty narrative leaves us wondering how causes and effects emerge from nothing and disappear back into nothingness. We might think telling stories would do no harm as long as we only illustrate rather than explain the events. True enough, but any fiction tends to become the gospel truth when the truth is unknown.

Where do the photons come from when a lamp lights up? This is what Feynman said his father had asked him.[29] Being unable to explain bothered Feynman a lot. We now understand that the photons do not emerge from nothingness but instead come out of the vacuum, where they already exist in the invisible form of paired quanta ([Appendix H](#)).[69] Since an electric charge splits the pairs apart, the photons become visible as an electromagnetic field. The photons pair up again upon neutralization, and the field vanishes.

In contrast to this tangible thinking about quanta, Niels Bohr's aide Aage Petersen summarized the master's view this way: "There is no quantum world. There is only an abstract quantum physical description. It is wrong to think that the task of physics is to find out how nature is. Physics concerns what we can say about nature…"[70]

We indeed perceive only properties, not the substance itself. We experience the void as inertia, not the void itself. Likewise, we see the color of light, not the photons themselves. Newton, too, saw color as an attribute of light. But Bohr's view is poles apart. For him, the quantum of light is an abstraction, the virtual photon. He maintains it would be absurd to think that there is something to interpret, let alone explain, behind a mathematical model. A model is a model.

Many a physicist, including myself back in the day, has adopted Bohr's anti-realistic stance without ever understanding what the calculations are all about. Many do not even consider there to be anything besides the modeling itself to be understood. Is this science?

THE STANDARD TRUTH

"One of the hardest lessons to learn in academic life – and for me, the most disconcerting – is the speed with which a radical insurgency can become orthodoxy,"[71] says Lee Smolin in his book *Einstein's Unfinished Revolution* (2019), referring to the ascent of quantum mechanics in just one decade from 1920. By now, many professional physicists think quantum mechanics is beyond doubt, not to say dogma. They might look upon a researcher interested in the ultimate nature of light, matter, and space as an eccentric or, worse, a philosopher. A seeker of truth may even be branded as a denier of modern physics.

Since questioning its grounds has long been deemed to be of no avail, physics is bound to become ever more mired in sophistication, perhaps never returning to reality. The abstruse doctrine proves hard to challenge not only by those who are not versed in its intricacies but also by those who are root-and-branch indoctrinated.

When we defend the standard tenet by saying, "everyone knows that...," then everyone knows our logic does not hold. We make this classical reasoning mistake most easily when defending the fundamentals, those 'facts' that everybody knows. And so, throughout history, those who have pondered the self-evident truths have made us see Nature more realistically.[72]

THE PILOT WAVE MODEL

The results of quantum mechanics are precise, but the theory does not make sense of the phenomena. Nevertheless, this insidious instrumentalism is good enough for many professionals. Its critique is not fruitful, as malleable mathematical models are not falsifiable.

The pilot wave model is one of the most convincing counterpoints to the Copenhagen interpretation. Louis de Broglie presented it at the legendary fifth Solvay conference in 1927. The French physicist argued that a field of some substance envelopes a particle. Even at the meeting, Wolfgang Pauli torpedoed de Broglie's drafts. According to the Austrian-born physicist, the idea did not agree with observations. Yet, we now know that de Broglie's idea was plausible after all. Bohm showed in 1952 that the physical vacuum could absorb the energy that is released in events, say, reactions.[73]

The image of reality given by quantum mechanics is incomprehensible. Consequences cannot appear as if coincidences. So, could some cause or another be hiding? This was explored in the early 20th century. John von Neumann, a Hungarian-born genius, proved in 1932 that no hidden variable theory could be correct.[74] Only three years later, the mathematician Grete Hermann, at one time Emmy Noether's student, disproved von Neumann's proof. However, her illuminating work was consigned to the darkness. Quantum mechanics established itself. Eventually, John Bell found Hermann's paper and showed that hidden variable theories and quantum mechanics are irreconcilable.[75] Many assumed this to have brought closure to the topic. But, now it is time to re-evaluate not the hidden parameters but the overt proposition that the photon pairs embody the void.

The paired-photon void is not just an idea. In 2007, Jung-Tsung Shen and Shanhui Fan showed that photons propagating in parallel correlate strongly.[76] More recently, Claudius Riek and Alfred Leitenstorfer from the University of Konstanz, together with their colleagues, hit the void with a laser pulse and measured its vibrations.[77] The void is pretty solid stuff. It takes the electromagnetic force, which is about a 10^{43} times stronger force than gravity, to tear apart the paired photons (Appendix G). This strength of photon pairing is evident from a magnet that sticks to a refrigerator door, even though the whole globe is pulling it off.

In 2006, Yves Couder and his colleagues from Diderot University in Paris illustrated in a YouTube video how a particle makes the vacuum undulate by showing how an oil droplet makes ripples on the surface of oil. The research team also demonstrated the double-slit experiment.[78] In the contested experiment[79] a droplet went through one slit while waves went through both. John Bush, a professor of mathematics at MIT, and his team showed that droplet motions range from straight to chaotic and from circular to coiling.[80] Based on the universality of patterns, we should expect quanta to form similar shapes even as elementary particles.

Quantum fields

Our experience tells us that fields envelop bodies, say, a gravitational field, the Earth. By lucid logic, a field is around a particle. In contrast, the standard doctrine claims the particle itself is an excited state of the quantum field,[81] say, an electron is an excitation of the electron field.

But what do these words mean? We cannot relate the idiosyncratic concepts of modern physics to our experience of reality. That has led us to a hornet's nest of problems. As Popper noted: "But science and rationality have very little to do with specialization and the appeal to expert authority."[82]

Thinking clearly, is not the excited state the vacuum state around the particle rather than the particle itself? Is not the vacuum thereby in balance with the particle structure? What does history have to say about this?

First Faraday, then Maxwell viewed the vacuum as some kind of substance,[34] but Michelson and Morley uncovered no ether.[30] Notwithstanding the conclusion, Heaviside found it impossible that the void should be nothing.[36] In this regard, Sagnac's 1913 experiment was perplexing, just as it still is today.[83] If there is no ether, why does it take longer for light to travel in one direction than the other on a rotating optical track? If the vacuum is nothing, why does it seem to affect movement? What causes inertia, if not the vacuum?

In the face of these unresolved dilemmas, de Broglie jettisoned his idea of the wavy vacuum engulfing the particle and went along with Schrödinger thinking the particle is a field, an excited state of the void. The pioneers were not unanimous about what the vacuum might *be*.

The history of science reminds us of turning points when a seemingly irrefutable tenet turned out to be incorrect. Scientists, indoctrinated by their profession, are often oblivious to this possibility. The vacuum of virtual particles is the truth of our time. But we should not take one theory or another on faith; rather, we should reason how things must be. Confidence in one's wits and senses is a prerequisite for acquiring knowledge.[21]

Next, we need to find out what the particles look like. Logically they must also consist of quanta. If not by exchange, how else could there be a local and universal energetic balance between the void and matter?[23,24] Particles emerge from the vacuum and submerge into it, as has been vindicated on many occasions.[84]

The exchange maintains balance. For example, room temperature remains stable when heat is released from objects, such as radiators, and bound into them. Similarly, national economies evolve toward balance as goods, capital, and labor move from one country to another. This is not a metaphor for particles but the same principle on another scale. The velocity distribution of gas molecules in a room is skewed, just as the distribution of wealth in the national economy is skewed. Comparisons help us to comprehend the motions of particles just as goods. Quanta flow everywhere toward energetically more favorable distributions. That's what the economy is all about too.

THE PURPOSE OF SCIENCE

This comprehensive thinking and logical reasoning may appear like crackbrained pseudo-philosophical claptrap to a scientist who is not familiar with the atomistic tenet of the ancient thinkers,[2] shared by the natural philosophers of the Enlightenment,[10,14] and modern physicists.[15,21,85,86] For strait-laced specialists, *Grand Regularity* does not mean anything; hence, they are not after its explanation either. General conclusions appear to them as a peculiar hodgepodge of different phenomena. But is it not scientists themselves who have pixelated the unity of Nature into a columbarium of disciplines?

An unsubstantiated solitary sentiment can be challenged by evidence, whereas the refutation of a view squarely based on a theory requires revocation of that theory. Even so, it must be possible. After all, we just challenged quantum mechanics' interpretation of the double-slit and delayed choice experiments. So perhaps reality is not bizarre after all. Such a change in our way of thinking may be dramatic, but dramatizing or dodging it is pointless; nonetheless, that happens.

Universities commit to science by their values. The University of Helsinki, for instance, declares that the pursuit of truth and new knowledge is the prerequisite of research and education and states that the essential characteristic of a researcher, teacher, or student is a critical mindset. Presumably, Newton, too, taught physics by commenting on Aristotle's doctrine critically.

In light of this, I was astounded to discover that not everyone at the university shares the ideals of science. In the spring of 2015, Professor Hannu Koskinen, director of the Department of Physics at the University of Helsinki, called me to his office. He needed to respond to a student with the question, "Is it not unethical that Professor Annila said, when referring to the current paradigm of virtual photons, that the vacuum could well be comprehended as actual photons in pairs?" I provided a reply that it would have been wrong if I had not informed the students about my concrete understanding of the vacuum, which explains rather than models electromagnetism and inertia.

I had explained to the students attending my lecture series on nuclear magnetic spectroscopy why the electron, proton, and neutron are magnetic and what a magnetic field is. Since publishing my papers,[60,87,88] the paired-photon vacuum was not a new idea to the director. Besides, the university magazine had written about my research promoting open discussion. My faith in the hall of knowledge, adhering to its values and its commitment to seeking truth and new knowledge, was tested as the department head let me understand that I should leave the vacuum and particles for others to study and teach. Yet, he did not explain in any way how my thinking, teaching, or research results would disagree with observations or measurements.

We returned to the same theme a year later. The head told me quite frankly that he found my conclusions unbelievable. But he would not say what was wrong with them. Koskinen only reasserted that all the experts simply could not be mistaken. In physics, telling crank from expert ain't easy.[89] It is possible that dark energy, dark matter, wormholes, multiverse, and entanglement signal that the whole field has descended into crankiness. Even physicists do not grasp these irrational concepts. So, it seems that the ideas have a grip on physicists rather than physicists having a grip on ideas.

You may wonder why there is so much ado about almost nothing, that is, about the vacuum. When I reify the vacuum in terms of photon pairs, instead of just swallowing a 'bubbling soup' of virtual particles,[90] I question the worldview of modern physics. It is neither pointless nor preposterous.

In the history of science, changes in the worldview inexorably produce confrontations. Therefore, the research results should not be judged only from the dominant perspective because that very tenet is being explicitly re-evaluated. These cases are rare but, for all intents and purposes, manageable. It only takes courage to stand up for the principles of science and adhere to academic values.[vii] As Aristotle said: "Courage is the first of human qualities because it is the quality which guarantees the others."

It takes people to have a debate. Even so, the focus should be on the matter under dispute rather than the disputants. Einstein and Bohr disagreed on how to interpret observations yet respected each other.[91] Defending one's position while understanding different views is critical in finding out what reality is. Building a worldview is a complicated process. History highlights the harm of condemning hastily in the heat of the moment. Our aversion to the pain of admitting error too often

---

[vii] The International Center for Academic Integrity recognizes courage as indispensable in enacting the fundamental values, honesty, trust, fairness, respect, and responsibility, especially under pressure to do otherwise.

overrides our desire for truth and justice. "In questions of science," said Galileo, "the authority of a thousand is not worth the humble reasoning of a single individual."[92]

WHAT IS MATTER?

Quanta make up particles.

"Most of the fundamental ideas of science are essentially simple, and may, as a rule, be expressed in a language comprehensible to everyone."[93] Einstein's opinion seems overly optimistic. Is not modern physics hard to grasp? Maybe Einstein meant that the laws of nature are simple; only mathematical models of reality are complicated.

Newton's vision of light and matter as the same substance[10] suggests that particles are quanta of light. Faraday concluded similarly that everything, matter and the void, must be one and the same substance. After all, the lines of electromagnetic and gravitational force in space, curving and whirling around matter, emerge from matter and end in matter.[33] Let us, therefore, ask how photons *could* make up elementary particles.

THE PITH OF THE MATTER

When a ray of light originating in water encounters air, it bends sharply. That is why an upright reed seems to bend sharply at the surface of water. A beam of light also bends when passing through layers of air hovering over a hot desert, giving rise to Fata Morgana. Rays bend likewise when hurtling through the gravitational field of a star.[94] When a ray refracts, its element, the photon, also bends. The light quantum is like a small string. It stretches, shrinks, and twists, but no force can cut it into pieces.

This portrayal of light points to string theory, a model of particle physics, where elementary particles are thought to be strings that vibrate. The pitch of a guitar's string sound is analogous to a property of a particle, such as its mass.

For all its merits, string theory is problematic.[95a] Science requires verifiability, but there is too much freedom in string theory to be testable. The tenet lacking in solid axioms churns out an incredible number of possible worlds. It is also awkward that the geometry of space does not follow from the theory but must be inserted into it. Moreover, it is strange that there are supposedly more dimensions than the three we know. The extra dimensions required by the 11-dimensional supergravity theory must be curled up because we do not discern them. This oddity can be illustrated by a garden hose. It looks like a one-dimensional cord from a distance but is exposed as a hollow tube on a closer look. We get the idea, but we still do not grasp 11-dimensional space, as we have no experience with super-dimensionality. Mathematically consistent is not necessarily realistic.

Brian Greene, a string theorist himself, identifies the missing essence in his book *The Elegant Universe* (1999): "As we look to the next stage in the development of string theory, finding its 'principle of inevitability' – that underlying idea from which the whole theory necessary springs forth – is of the highest priority."[96] So, what is this basic bit of string, this elemental piece of wire? We need to find the truth, as it could distinguish a valid theory of our world, as it truly is, from the innumerable models of imaginable worlds.

String theory has been twisted toward being the theory of everything. But it has refused to bend into a form compatible with both quantum mechanics and general relativity. Should we not seek a theory that matches reality rather than one that matches modern physics? Smolin summarizes: "The real question is not why we have expended so much energy on string theory, but why we haven't expended nearly enough on alternative approaches."[95c] Indeed, we do not have to hang ourselves in the loops of string theory. Let us free ourselves to think differently.

When we consider how bits of strings, the quanta of light, could make an elementary particle, the facts leave little room for the imagination but still plenty for another viewpoint. A great deal of solid information is already available in the standard theory of particle physics, known as the Standard Model.[97] So, let us use it since we are not aiming to overthrow that theory; we just want to make sense of it.

A particle as a coil of wire, a string of photons, may seem like an amateur idea. However, a few years ago, it was noticed that closed curves of light are also solutions to the Maxwell equations.[98] A train of photons could curl up and close up as a particle. A one-dimensional string could fill three-dimensional space the same way a thread fills a ball of wool.

Geometry is concrete. It is mathematics in its most comprehensible form. The properties of particles could reflect the characteristics of quantized strings in much the same way as in string theory.[88] If we only knew the particle structures, we would understand how quanta translocate in reactions from one particle to another, just as we visualize how atoms transfer in chemical reactions from one compound to another. The particles and the void as wireframe models might provide insight into modern physics. Perhaps we would also understand why quantum mechanics and general relativity are pretty good models of the void. Wheeler anticipated the vision to be thereby gained: "Surely the magic central idea is so compelling that when we see it, we will all say to each other: Oh, how simple, how beautiful! How could it have been otherwise? How could we have been so stupid so long?"[99]

The vision from a viewpoint

Quarks are constituents of elementary particles but are not immutable. So, do even more fundamental elements make up the quarks in the same way atoms make up molecules?

This was behind the reasoning of Jogesh Pati and Abdus Salam in 1974. These pre-eminent physicists promoted a particle called the preon as the basic element of the quark.[85] But their profound idea was not entirely cogent. The point-like preon implies enormous energy, but there is no evidence of such. The preon is not really needed either, as it does not resolve the issues burdening the Standard Model.[100] This well-established theory organizes the electromagnetic, weak, and strong forces and classifies known elementary particles but explains neither the forces nor the particles.

Quarks cannot be the basic building blocks because one quark converts into another in nuclear reactions. Moreover, it appears that quarks and photons have some common structure.[101] If the point-like preon is not the fundamental element, could it be the photon?

As the standard theory assumes particles are quantum fields, we see the data from that angle. And when we presume particles are made of photons, we see the same data from that angle. Although no data is free from interpretation, we might find some data that a theory does not explain, or we may realize that another theory explains something not in the data. We may also evaluate how consistent, simple, concrete, broad, and fruitful one theory is compared to another.

NEUTRINO

Is there a particle comprising only one photon? What is the simplest shape you could loop from the elemental piece of wire?

The *Grand Regularity* of Nature implies that we may relate what we see to what we cannot see. For example, a paddle stroke makes waves of water the same way an accelerating electron makes light waves visible in the vacuum. Often, the paddle stroke also creates a small swirl. Occasionally, an ocean wave rolls over to form a tunnel-like breaker, to surfers' delight. Waves and vortices are optimal flow patterns. That is why these geodetic lines are ubiquitous, perhaps even manifesting themselves as elementary particles. Let us imagine the photon curling into a loop. Is the loop any one of the known elementary particles?

To think about elementary particles like this may again seem amateurish. But science begins with a root idea. Einstein came up with curved space-time while musing on how meridians curve toward the Earth's poles.[102] First, he only had a vision of how mass relates to curvature. Then he endowed the idea with a mathematical form and compared calculations to measurements. So inspiration became a testable theory. The original intuition nourished the whole world of physics. The imagination of the then-patent officer changed our worldview. In a similar fashion, why don't we try to envisage what particles could look like?

Images such as those of Einstein or others may well be denounced as metaphysics. This branch of philosophy seeks answers to the questions: What *is*? What is it *like*? Nevertheless, many a physicist despises metaphysics rather than drawing inspiration from it. This conception of what science *is* influences what the scientist regards as a credible research topic.[4h] Thus, the examination of Nature is not open but framed for scientific inquiry by the impressive technical skill base (*tekhnè*) of modern physics. By contrast, for Aristotle, physics was a form of philosophy, a way of reasoning about Nature.

Let us be honest. Modern physics has abandoned the guiding principle of modern science: liberating humankind from the medieval magical-hermetic tradition and opening reality to everyone's comprehension.[4a] We must, therefore, trudge our way back to reality. As unstructured as it might be, our experience is the most factual foundation for a theory.

The neutrino is the tiniest particle we know. It is hard to catch, but Wolfgang Pauli figured in 1930 that it must exist. Indeed later, tracks of neutron decay into the proton and electron revealed the neutrino, the little neutron as Edoardo Amaldi, an Italian particle physicist, jestingly baptized it.[103] The name stuck after the world-famous Enrico Fermi used it in his papers.

The neutrino's angular momentum equals Planck's constant per orbital period, just as the photon's momentum equals Planck's constant per wavelength.[104] In other words, the neutrino spins about itself. Could the neutrino be a circling quantum?

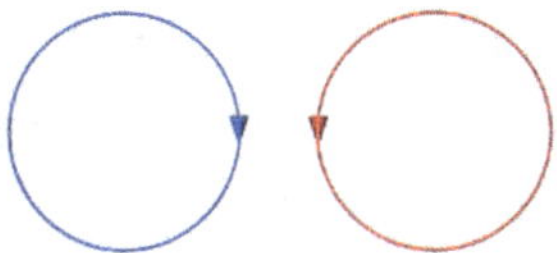

The neutrino is one quantum loop. The spin direction depends only on whether we view the loop from the front or rear. Thus, the neutrino and antineutrino are the same particle.

In the 1930s, the physicists Pascual Jordan and Ralph Kronig theorized how the photon and neutrino relate to each other.[105] Now, we can test the idea that the neutrino is a loop of quantum. The loop is a mirror image of itself, just as a wave is a mirror image of itself. This strikes a chord; the neutrino is thought to be its own antiparticle, just as the photon is known to be its own antiparticle. As the photon winds into a circle, the wavelet's electromagnetic crest and trough cancel each other out. This makes sense: the neutrino does not evince electromagnetic properties. For that reason, the neutrino makes its way through just about anything. It literally breaks through the rock on its way from the European Organization for Nuclear Research (CERN) near Geneva to the Gran Sasso laboratory in Abruzzo.[viii]

Neutrinos form in nuclear reactions as it takes a strong force to coil a quantum of light into a small spinning loop. Every second, about a million billion (10^{15}) neutrinos hailing from the Sun pass through each of us.[107] We do not sense them. For comparison, we easily see the same number of photons that a laser pointer shines. On a pleasant summer day, ten-million-fold more, some 10^{22} photons pour on us every second.

THE CASE OF CURVATURE

A loop curves differently than a wavelet. The difference in curvature shows itself as the mass of the neutrino,[97] for mass means curvature. The Earth is curved compared with the surrounding space. The curvature of a particle, too, is relative to the minute curvature of the vast void.

The photon is massless because its curvature is identical to the void's paired photons. That is why the whole of space weighs nothing, even though it holds as much energy as all matter in the universe.

The neutrino is completely planar. On the contrary, the cosmos is not absolutely flat. Although immense, it is slightly curved by containing matter, albeit, on average, very little. The mean density is about $0.6 \cdot 10^{-26}$ kg/m^3.[23] Thus, when passing by the completely planar neutrino, the photons of the vacuum slightly straighten their trajectories, manifesting as the neutrino's mass. Therefore, the neutrino does not move entirely freely, precisely at the speed of light in the vacuum. However, this coupling between the void and the neutrino is too tiny to be measured with state-of-art instruments.

The early 18th-century Swiss-born math genius, Leonhard Euler, calculated the curvatures of waves, loops, and other curves. Later, Einstein geometrized the void about a body as curved space-time, but he did not say anything about the geometry of particles. In 1964 Peter

---

Higgs,[108] François Englert and Robert Brout,[109] as well as Gerald Guralnik, Carl Hagen, and Tom Kibble,[110] explained masses of certain particles by their coupling to a surrounding field, known as the Higgs field.

However, the Standard Model of particle physics does not explain why a neutrino has mass. How could it? The Standard Model only classifies all the known elementary particles without clarifying their structures. This is one reason why physicists do not consider the Standard Model entirely satisfactory. In his book, *Seven Brief Lessons on Physics* (2014),[111] Carlo Rovelli, when addressing issues of contemporary physics, reminds us that theories tend to be temporary. More than that! The Italian theoretician thinks we have yet to look at the particles from the right perspective, the one that reveals their true simplicity.

We are after that elegance. A particle itself could be simple, whereas the field of vacuum quanta around it can be complicated. Similarly, a boat, simple in shape, is surrounded by complicated waves that hardly reveal its elegance.

Let us consider mass as a measure of how strongly the vacuum couples with the particle. Then the mass depends not only on the particle's structure but also on the density of the photon pairs around it. For example, electrons move exceptionally fast in graphene, layers of graphite, familiar from the lead of a pencil, almost like a massless particle, as physicists Andre Geim and Philip Kim demonstrated this phenomenon during the first years of the 21st century.[112] Now, we understand this strange result. In thin graphene layers, the electron itself is as usual, but the surrounding field of photons is different from that of the vacuum. We likewise understand the claim by special relativity that mass increases with speed because, for a particle in motion, the surrounding vacuum seems denser than for a particle at a standstill. Similarly, it sure feels quite solid when you dive into water at high speed.

Against the backdrop of history, the multitude of mass concepts (relativistic, effective, bare, inertial, and gravitational mass) is telling.[113] In the early 19th century, the spectrum of electricity and magnetism was still a mess. Nonetheless, many a contemporary did not complain about this, having grown up in scientific obscurity.

WHAT DOES THE ELECTRON LOOK LIKE?

The electron is divisible as it breaks in nuclear reactions.

Let us suppose that the electron is made up of photons. What would it look like? Where does the electron get those properties that the photon lacks? These are its electric charge, magnetic moment, and mass. Or let us ask more precisely: how can we predict the structure of the whole electron from its constituent photons?

This question reflects common assumptions about the nature of a scientific explanation. However, have we any proof, or do we merely assume that the whole could be predicted from its parts? It is not the same thing to predict the cascade of events by which photons assemble into the electron as it is to figure out the structure of the electron. Not only the pieces but also the circumstances shape what happens.

At the beginning of the 19th century, André-Marie Ampère found that an electric current in a copper wire gave rise to a magnetic field. From this observation, the self-taught physicist and mathematician inferred that there must be a constituent of matter with combined electric and magnetic properties. This electrodynamic molecule was to be known as the electron.

In 1823 Ampère suggested that the electron is a serpentine-like circular coil, for he knew by experience that electric charge relates to the number of windings of copper wire. He also knew that the magnetic moment was proportional to the coil's cross-sectional area. You can easily make a wireframe model of Ampère's electron. Just strip off the helical spring from the back of a notepad, or take a Slinky and bend it into a full circle. When the ends meet, the coil makes a seamless toroidal ring.

But how many loops are there in the electron torus? Ampère did not know that. After his death, it took a half-century to obtain the measurement from which the number of loops could be deduced.[88]

The natural constant α, known as the fine-structure constant,[97] unveils the electron's structure. The value of $1/\alpha$ is 137 plus. It denotes the ratio of the total length of the torus to the length of a single quantum, the neutrino loop. In other words, the constant hints at how many loops there are in the overall helical coil. Likewise, Pi (π) is the ratio of the rim of a circle to the diameter. The helical coil rises like a corkscrew. Due to the rise of the thread, known as the pitch, one quantum trails the other in a slightly retarded position along the helix. Throughout the 137 loops, the lag amounts to precisely one quantum. Thus the electron torus has exactly 138 quanta (Appendix B).

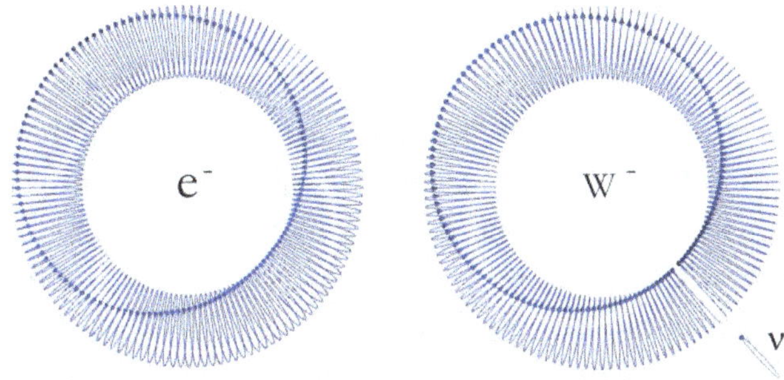

The electron is a closed helical coil of 138 quanta (left). The electric charge accumulates from all the windings, and the magnetic moment relates to the torus' enclosed area. The electron's mass is small because the loops on the opposite sides cancel out each other's effects on the vacuum except for the pitch. By contrast, the vacuum couples strongly with an open torus where a quantum loop, the neutrino, has been excised (ν) (right). The W⁻ boson is heavier than an iron atom. Note that the drawing proportions differ some-what from the value dictated by the fine-structure constant to enable easier perception of the form (Appendix B).

The first time you confront the claim that the electron comprises 138 photons, it may seem bewildering. On second thought, there is nothing extraordinary about it. For instance, we are used to the ben-zene ring having six carbon atoms. Yet two hundred years ago, when Joseph Proust realized that chemical compounds comprise a fixed number of atoms, his French colleagues thought he could not be in his right mind. The revelation was nonetheless soon endorsed by John Dalton, an architect of modern atomic theory.[114]

Neither are we boggled at the football-like molecule fullerene, which is made of sixty carbon atoms. The stable C_{60} molecule has ex-quisite symmetry, as does the electron torus. Why would the electron, a compound of 138 quanta, be any more exotic? Strange is the thing we do not know yet. At its best, science reveals the unseen.

Of course, we can ask why it is 138, not some other number. In the same way, the science fiction author Douglas Adams was asked why 42. In his most famous work, *The Hitchhiker's Guide to the Galaxy* (1978), a computer named Deep Thought gives this number as an inexplicable "answer to the Ultimate Question of Life, the Universe, and Everything". Well, the proof is in the pudding; we can explain at least the structure and properties of the electron from the fine-structure constant (Appendix B).

The pitch of the torus forces apart the surrounding paired photons, manifesting as the electron's electric field. Likewise, the curl of the torus causes vorticity in the surrounding photons, manifesting as the electron's magnetic field. The electron's magnetic moment is a little larger ($\alpha/2\pi \approx 0.00116$) than that of a current loop of the same radius, known as the Bohr magneton,[52] because the torus area is slightly larger than the area of a torus the current loop.

The torus also explains the mass of the electron. Since the loops on opposite sides of the torus curve in almost opposite directions, the photons of the vacuum, when inbound, bend in one direction, and when outbound, in the opposite direction. As the net effect is small, the electron's mass is small.[115]

The electron torus complies with the Standard Model. Its corkscrew geometry matches the symmetry group of the electroweak force, known as SU(2).[97] This chiral symmetry means that the electron and its mirror image, the positron, are different particles, in the same way as the left hand differs from the right. The Chinese-American physicists Tsung-Dao Lee and Chen-Ning Yang anticipated this,[116] and Chien-Shiung Wu proved it in 1956.[117] The result was a sensation. Wolfgang Pauli had jested, "I cannot believe God is a weak left-hander."[118] He soon went back on his solemn words but blessed and cursed the riddle of the electron structure.

The torus also makes sense of the electron decay to the W^- boson. The loss of one quantum neutrino loop introduces a narrow gap in the torus. The electron is indeed divisible into its elements.

THE FINE CONSTANT

Wolfgang Pauli brooded over the electron structure to the last. In 1958, the incurably sick virtuoso asked his assistant Charles Enz, who came for a visit to the hospital, "Did you see the room number? [137]."[119] Pauli was on the right track. The meaning of the fine-structure constant is the key to the riddle of the electron structure.

Pauli is best known for his exclusion principle, a textbook maxim about two electrons that cannot be in the same quantum state. From the electron and vacuum structures, we understand what the rule means. The vacuum vorticity around two electrons is lowest when the two tori orient antiparallel. Thus, the void forces the electrons to follow the Fermi-Dirac statistics. Likewise, the void forces photons and other particles of the same quality to distribute according to the Bose-Einstein statistics (Appendix H). The general distribution law of open systems applies to the states of imbalance as well.[120]

Pauli was one of Arnold Sommerfeld's many students who became Nobel Laureates. The German physicist Sommerfeld had studied the natural sciences deeply and broadly, and mathematics had been his major for a long time. In 1916 Sommerfeld reasoned that the fine-structure constant relates the speed of the electron orbiting the nucleus to the speed of light in the vacuum. There is a grain of truth in this, for the geodetic line is not the atomic orbit but the electron torus itself. The quanta circulate in the compact torus very fast but not as fast as in the sparse vacuum. The internal motion is a standing wave, as Schrödinger concluded from the Dirac equation in 1930.[121] Kronig had already considered this electron "trembling" (Zitterbewegung in German) in 1925, but, at that time, Pauli deemed the idea too unrealistic for publication.

Eddington shared our goal for an all-inclusive theory: "Our challenge is to find such a coherent theory of charged particles and light in which the electrostatic effect and the quantum action are derived from their sources."[122] The riddle of the fine-structure constant irked Feynman too: "It's one of the greatest damn mysteries of physics: a magic number that comes to us with no understanding by man."[123] However, David Hestenes, an architect of geometric algebra, expressed the solution of the Dirac equation as a closed helical path of photons.[124] He reasons geometry could classify elementary particles because mass relates to curvature.

Were you to take the textbook on trust and believe that photons appear from nothing and disappear into nothingness, then Planck's constant would be a mere number to you. You would not understand it as the measure of the quantum, not even though it comes with a unit of measure (Js). As your point of view delimits your field of view, you would not get either that the fine-structure constant is the ratio of the electron torus to the quantum loop.

Let us recall that Nature is unproblematic. None of the outlined problems inhere in Nature but instead in the domain of thought called Physics. Their persistence tells us that we misunderstand something. There shall hardly be progress as long as clearing up confusion is looked upon as a problem in itself.

THE JEWEL OF PHYSICS

Quantum electrodynamics[125] is said to be the jewel of physics.[123] To wit, there is no gainsaying its exquisite power: the model gives the electron's magnetic moment down to the tenth decimal place in agreement with data. But, it does not explain the value itself. So, no matter how thin you slice it, it's still baloney. The long series of mathematical terms sums up not the magnetic moment itself but its effect on the vacuum quanta. Likewise, a series of epicycles summed up the planet's orbital motion as seen from the Earth rather than from the Sun.

The textbook claims that the electron is a point-like particle.[47b] But, how could a dimensionless pinprick possibly have properties like charge, magnetic moment, and mass? Moreover, the point-like electron leads to an infinite reaction of the electron on itself.

The quantized torus is consistent with the data, as researchers concluded from their measurements: "If the electron is a composite particle, its constituents are strongly bound, giving the electron the observed point-like quality at experimentally accessible energies."[126] Indeed, the constituents, the quanta in their tight coil, are strongly bound to one another like the strongly bound quarks.

The electron's size cannot be determined with light more precisely than within the Compton wavelength. The American physicist, Arthur

Compton, thought that the electron could be a torus, as the English physicist and chemist Alfred Parson had suggested in 1915.[127] Originally, Compton aimed to determine the electron structure by scattering light off the electron. In the process, he realized the photon's particle-like character, obtaining the Nobel Prize for this revelation.[128]

QUANTIZED VORTICES

If we did not know the structures of both the vacuum and the particle but only the corresponding equations, we would be bewildered, as most physicists today, that it takes *two* revolutions for the electron to return to the initial state. Undoubtedly, one spin would bring the electron to its initial position in the emptiness but not in the vacuum. The second round unwinds the twists of the vacuum quanta that the first round introduced to the electron-coupled field.

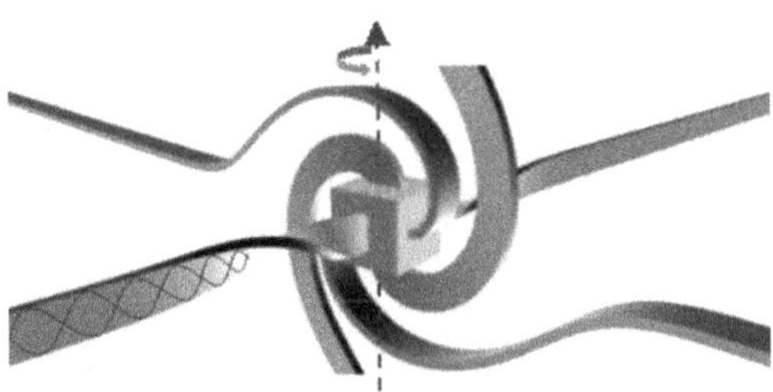

The illustration shows how a particle (represented by the cube), for instance, an electron, couples with the vacuum. The particle must revolve twice before the vacuum's threads of quantum pairs unwind (shown as the curving strips).[129] (Screenshot of YouTube video by Jason Hise.)

You can demonstrate the unwinding of a winding by holding a book on your palm and swinging your arm once below and once again above your shoulder, as Alexander Unzicker, a clear-thinking German physicist, shows in his video *Quaternions and Fundamental Physics* (2020).

The vacuum and matter are tightly coupled. This became apparent in a 1961 experiment by the American physicists Bascom Deaver and William Fairbank.[130] Electrons circulating in a superconductor make the surrounding vacuum vortices periodic, magnetic flux quantized.

About twenty years later, the German physicist Klaus von Klitzing discovered that the electrical resistance across a thin layer of semiconductor increases in steps as the magnetic field strengthens. This quantum Hall effect discloses that quantum vorticity matches the periodicity of electron motion. The vacuum's quest for a balanced structure is also seen as a force between two parallel uncharged conducting plates. The phenomenon is known as the Casimir effect.[131]

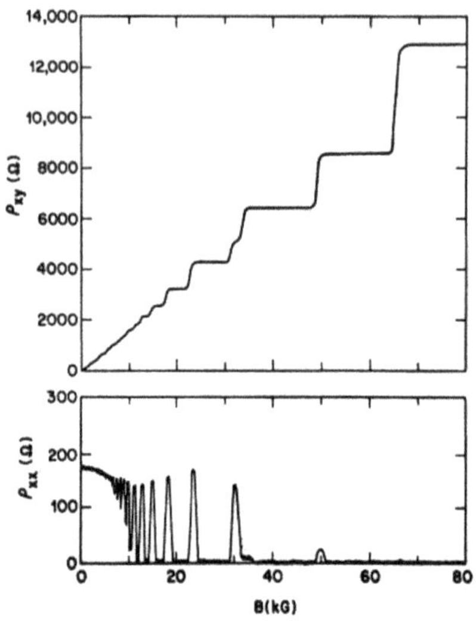

As the magnetic field B increases, the transverse electrical resistance at the boundary layer $\rho_{xy} = h/ne^2 = 2\phi_0/ne$ of the semiconductor grows stepwise (n), and the longitudinal resistance ρ_{xx} decreases in jumps.[132] The resistance remains stable when the layer fills up with one quantized whirl after another. When the layer is jam-packed, the resistance jumps as the vortices and electrons rearrange themselves. New space is cleared for more vortices as the area A of the flux quantum ϕ_0 decreases and the field strength B increases so that $\phi_0 = BA$ remains constant. When one or more electrons circumscribe two or more vortices on their least-time orbits, the resistance also spikes at fractional ratios of the flux quanta to the electrons.

Today, physicists wonder why the electrical resistance in layers of various materials exhibits the same power-law dependence on temperature. This universality points to the universal structure of the vacuum; the same substance absorbs heat from different surfaces.[133] Confusions clear up from the holistic perspective.

Positron

The electron torus leads us to the structures of other elementary particles. The positron is the mirror image of the electron, an antimatter particle. Antimatter is not strange, merely rare, for matter is the universal standard of substance.[134]

Electrons circulating a positive nucleus, instead of positrons orbiting a negative nucleus, is merely a norm, like the handedness of traffic. Many countries enforce driving on the right, while the traffic is on the left-hand side in the British Isles. The reason for such a rule, one way or another, is plain; it makes the system work, whether it is the universe or a smaller empire. Of course, it does not preclude the appearance of a wrong-way driver; there are occasional positrons to contend with. A standard will be adopted when it helps to gain balance and abandoned when it hinders the system from attaining balance.

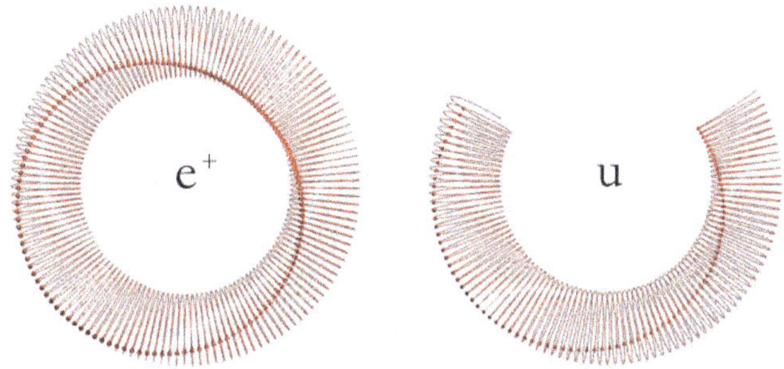

The positron (e^+) (left) is the mirror-image particle of the electron. The positron's electric charge and magnetic moment are the opposite of the electron because the handedness of the coil is reversed. The positron's mass is the same as the electron's mass because the curvature is the same. While the vacuum's paired quanta couple only weakly with the positron, they strongly couple with the particle that spans $2/3$ of the full torus (right). The mass of this up quark (u) is big because half of its coils are without opposite coils to cancel out their effect on the vacuum.

When a particle meets its antiparticle, they annihilate each other.[97] This reaction is useful; for example, positron emission tomography delivers images of the central nervous system after a transient ischemic attack. In the annihilation, however, the matter and antimatter do not vanish into nothingness, *ad nihil*, but their quanta escape into the vacuum. The electron and the positron structures suggest that the tori open and unwind so that the quanta discharge pairwise into the void. We do not see these photon pairs bereft of the electromagnetic force. We witness only two out-of-phase photons propagating in opposite directions that stem from the unwinding of the two opposite electric charges.

The whole making more than the sum of its parts is branded emergence.[135] For instance, atoms in a molecule make more than free atoms. Likewise, quanta of the electron and the positron make more than when part of the vacuum. We are puzzled about emergence when not considering all the ingredients of a transformation. Namely, photons are either released from matter to the void or bound from the void to matter.[136]

THE REALM OF A THEORY

In the quest for balance, quanta move from one particle to another, ultimately subsuming into the vacuum. This *rational mechanics* of quanta may seem like waking up to the naked truth, and so it is.

While the thermodynamic theory, like other theories, can be hard to falsify within its realm, its flaws may show up in other ways. The tenet would be invalidated by discovering something unexplainable, such as some substance not made of light quanta. The theory would also be overthrown by explaining something nonexisting. Questioning strengthens trust. So let us keep on asking the basic questions.

WHAT DO THE PROTON AND NEUTRON LOOK LIKE?

Particles' structures explain their properties.

The nucleus of every element divides into protons and neutrons. Hints of even deeper divisibility came in the 1950s when short-lived particles were discovered one after another using particle accelerators. Willis Lamb joked about those rewarding times in 1955: "I have heard it said that the finder of a new elementary particle used to be rewarded by the Nobel Prize, but such a discovery now ought to be punished by a $10,000 fine."[137]

Long ago, the periodicity of elements suggested atoms were divisible. So it was not too surprising that the atom could be split into protons, neutrons, and electrons.[138] Also, particles display periodicity in their properties. In 1964, physicists Murray Gell-Mann at Stanford and George Zweig at the European Particle Physics Research Center (CERN) realized that the proton and neutron are composed of three elementary particles, quarks. The neutron's magnetic moment also implies the existence of these electrically charged constituents.

Gell-Mann was awarded the Nobel Prize for discovering quarks. However, things turned out differently for Zweig. His insights were looked upon as humbug at CERN, where Leon Van Hove, the leader of the theoretical division, did not entitle Zweig to publish or give talks.[139] Apparently, science exhibits the full spectrum of human nature. Today, peers acknowledge Zweig's work, although not been published in a peer-reviewed journal.

The proton consists of two up quarks and one down quark. The neutron consists of two down quarks and one up quark. But what are the quarks made of?

Because the charge of the up quark is $^2/_3$ of the positron charge, it *could be* a $^2/_3$ arc of the positively charged torus. Because the charge of the down quark is $^1/_3$ of the electron charge, it *could be* an arc of $^1/_3$ of the negatively charged torus. This straightforward inference of the quark structures agrees with observations; when a proton and an electron transform into a neutron, the charges cancel out.

The idea that the quarks are fractional structures of the electron and positron is not new. In 1974, Howard Georgi and Sheldon Glashow reasoned that the proton ought to break into the positron and the pion. The pion, also known as the Pi meson, is a particle that comprises a quark and an antiquark.[140] However, as there has not been any sign of spontaneous proton decay to this day, the theory is no longer of much interest. Nonetheless, the fundamental idea of unity is. "All elementary particle forces (strong, weak, and electromagnetic) are different manifestations of the same fundamental interaction involving a single coupling strength, the fine-structure constant," Georgi and Glashow wrote, continuing, "Our hypotheses may be wrong and our speculations idle, but the uniqueness and simplicity of our scheme are reasons enough that it be taken seriously."[140] In 2009, Glashow harked back to these times: "Although our attempt at grand unification has been ruled out, many theorists are convinced that the underlying idea is correct."[141]

As the torus comprises 138 quanta, the down quark is made of $^1/_3 \times 138 = 46$ quanta, and the up quark of $^2/_3 \times 138 = 92$ quanta. The toroidal helix winds along the arc so that the short-wavelength photon, known as the gluon, connects one quark to the next at a dihedral angle of 60 degrees. Thus, in a particle composed of three quarks, i.e., a baryon, such as a proton or a neutron, the three quarks cannot but reside on the three faces of an equilateral pyramid. This tetrahedral symmetry is consistent with the rule found by Gell-Mann and Zweig, known as the strong interaction gauge group SU(3).

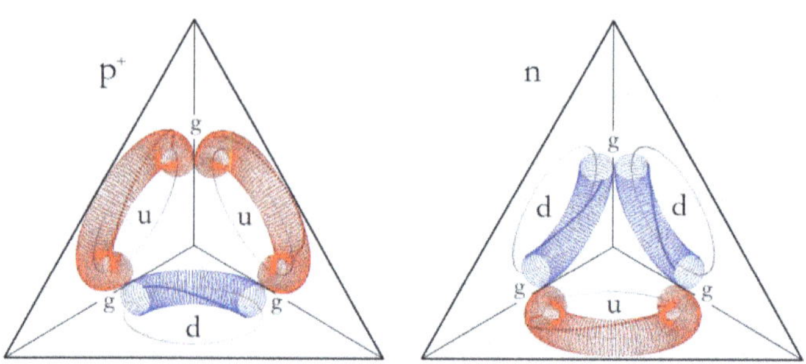

The proton (p⁺) consists of two up quarks (u, red) and one down quark (d, blue). The neutron (n) comprises two down quarks and one up quark. The three quarks bonded via short-wavelength photons, gluons (g), lie on three faces of an equilateral pyramid. Note that the tetrahedron's vacant face points toward the viewer.

The proposed proton and neutron structures, while surprisingly simple, explain the properties of the particles. The total number of windings gives rise to the electric charge. In the proton, the two up quarks ($^{+2}/_3 + {}^{+2}/_3$) containing positive windings and the one down quark ($^{-1}/_3$) with negative windings total a charge of $^2/_3 + {}^2/_3 - {}^1/_3 = +1$. The neutron is neutral because there are as many right-handed as left-handed threads: $^2/_3 - {}^1/_3 - {}^1/_3 = 0$. The magnetic moments of the quantized curves can be calculated like the moment of a current loop (Appendix B). The proton moment is bigger than the neutron moment because the proton quanta enclose a much larger area than the neutron quanta. The results agree with the measurements.[88,115]

A molecule is not merely the arithmetic sum of its atoms but a compound of its own kind. Neither is a particle solely the sum of its parts but a system of its kind. As a result, the internal orbital motion of the proton, the spin, is not a simple sum of the momenta of its quarks and gluons, as the physicists prefigured, but more.[142]

The vacuum couples to the proton and neutron much more strongly than to the electron because the arcs of quarks are imperfect without the opposite curvature, unlike the full torus. So the proton and neutron are much heavier than the electron. The masses calculated from the wireframe models comply with the measurements.[88,115]

However, the simple calculation of curvature does not explain why the neutron is a tad heavier than the proton. The minor mass difference suggests slight structural differences.

FIELDING QUESTIONS

Quantum field theory[143] dominates the current understanding of subatomic particles. Although it is an excellent mathematical model, measurements may imply that particles are not quantum fields. For instance, physicists are perplexed as to why the proton seems a bit smaller in a hydrogen atom having a muon instead of an electron.[144]

Carl Anderson and Seth Neddermyer found the electron-like muon in cosmic radiation in 1936.[145] The discovery caught the physicists by surprise. Isidor Rabi quipped, "Who ordered that?"[146] As the common constituents of matter were already known, the muon sure seemed a superfluous supplement to the proton, neutron, and electron.

When a muon, instead of an electron, circles around the proton, the proton itself is still the same. However, the observed field around the proton is different because the muon mass is 200 times the electron mass. Technically speaking, the field is denser in energy. So, correspondence between the proton's charge radius for electronic and muonic hydrogen is not trivial.

From the proton and neutron structures, we also grasp why some atomic nuclei are magnetic while others are not. For example, the most common carbon isotope, ^{12}C, is not magnetic because its six protons pair up in an antiparallel fashion, as do its six neutrons. Likewise, bar magnets pair up in opposite orientations canceling each other's magnetic fields. On the other hand, ^{13}C is like a compass needle, as one neutron is without a pair. The most common nitrogen isotope, ^{14}N, is also magnetic since one proton and one neutron are unpaired. In this manner, we can put together models of other nuclei from the models of the proton and neutron.[115]

Physicists ponder why the partition of positive and negative charge inside the neutron, i.e., the electric dipole moment, seems vanishingly small compared to what the neutron's sizable magnetic moment suggests.[147] Failing to measure the effect lies in physicists erroneously expecting that the neutron's spinning rate rather than alignment depends on the applied electrical field. Temperature does not affect the spinning rate, only the degree of alignment under an applied magnetic field. In short, a particle spins in the vacuum vortex and aligns along the vacuum gradient.

TRUE COLORS

I have stood at my front door a few times, trying out keys to find the right one. I eventually gathered that it helps turn the middle of my three keys upside-down, whereby I can distinguish between the two right-side-up keys. Of course, the keys can also be labeled with different colors, for a label does not change the key itself.

As with three keys on a chain, in the neutron, one down quark is before and another after an up quark. However, in quantum chromodynamics, a type of quantum field theory, the orientation of one quark relative to another is not specified, as the particle structures are unknown. Instead, the quarks are stamped with a label known as the color charge. In reality, the quark does not carry a color charge, implying symmetry breaking,[148] of which there is no sign. So let us face it: the Standard Model is imperfect with imaginary features.

The standard theory brands gluons, too, with a color label. In substance, the gluon is a short-wavelength photon between two quarks. Thus, by tailing one quark and leading the other, it has its position as part of the particle without a color label.

As we see from the wireframe model, the quarks can pivot about the gluons quite freely, but they cannot be pulled apart, for the gluon, as a photon, is uncuttable. Smolin describes the strong force precisely in this way: "Experiments show that when the two quarks are very close to each other, they seem to move almost freely as if the force between them is not very strong. But if an attempt is made to separate the two quarks, the force holding them together rises to a constant value which does not fall off, no matter how far they are pulled."9ª Thomas Neil Neubert, an insightful critic of modern physics, sees the strong force as one-dimensional.[150] The gluon is like a piece of wire.

The American physicists David Politzer, Frank Wilczek, and David Gross formulated the strong force as the theory of asymptotic freedom in 1973 and were awarded the Nobel Prize in 2004. On that trip to Stockholm, MIT professor Wilczek also visited the Department of Physics at the University of Helsinki. From his talk, I recall watching an animation of the dynamic vacuum. The ether is Wilczek's favorite substance.[90] I could not imagine that only a few years later, I would see in those animations the photon pairs sloshing around and the single photons as the biggest splashes. At that time, I still thought that Parmenides was thoroughly outdated with his atomistic idea.

Regarding the gluon as a short-wavelength photon, asymptotic freedom applies to all quanta, not just the quarks. While the idea of no free photons at all may seem surprising, the spectrum of free space discloses that the photons are not truly free.[38] They do not distribute hither and thither but rather according to the Bose-Einstein statistics. When a photon breaks free from a particle, it binds to the vacuum's strings of paired quanta (Appendix H). The quanta stick together – the universe is one, mathematically speaking, connected. Since the paired photon rays cannot be scissored sharply but switch smoothly, fields extend way beyond bodies.

NATURAL IMPRECISION

The proton and neutron alike are small particles, with a size of about one femtometer (10^{-15} m). The electron has been measured to be even smaller.[126] But it is not clear what exactly has been measured, for the measurement itself modifies its target. To detect an object, we must extract at least one quantum from it, and this loss changes the object. Conversely, an observer may prevent a system from decaying by watching it, thereby supplying quanta in place of those lost.[151] This observer effect is no paradox but common sense. For example, a steady flux of sunshine renews an ecosystem.

If we were to see the electron truly, we would have to pull at least one quantum out of it rather than one out of its surrounding vacuum field of quanta. That extraction would convert the electron to the W^- boson. Pascual Jordan put it vividly in 1934: "Observations not only disturb what is to be measured, they produce it."[152] He was dead on. The measurement entails a flow of quanta from which particles may materialize. In this way, Nobel Laureate Patrick Blackett created electrons and positrons in 1933.[153]

Nothing can be measured more precisely than to a single quantum. Heisenberg presented this profound principle of quantum mechanics in 1927.[66] This natural imprecision[52] means that the more precisely an energy difference is determined, the longer it takes. The relation follows from the quantum's, $h = Et$, diametrically opposed properties, energy, E, and time, t. Momentum and position pair likewise, $h = px$. However, position, x, along a dimension, is an abstract notion of wavelength, λ, just as time is an abstraction of the period. Coherently, the meter is defined as the path length traveled by light in a given time.

Physics involves questions of epistemology: what we can know, what kind of information is true, and whether knowledge is possible at all. Without further speculation, let us follow the guiding principle of this book: the abstract is abstruse, the concrete is comprehensible.

THE SIGNIFICANCE OF SCIENCE

At the dawn of humankind, humans gazed at the world with naked eyes. Then, a knife cut brought the body's organs to light. Much later, a microscope revealed the secrets of cells and their organelles. And not too long ago, X-rays exposed molecules and atoms, and particle accelerators enabled the spotting of nuclei and quarks. And now, from the fundamental element on, the photon, we see the whole.

As we cannot position contemporary science without perspective, we should once again step back and see where we stand. We should use both logic and ontology. "The universe would not know how to deal logically with more than one substance," Marcel-Marie LeBel wrote in an essay issued by the Foundational Questions Institute in 2009.[5] With this, he concurred with the reasoning of philosophers, like Baruch Spinoza, and called for concreteness.

What is everything made of? The question has been around for ages. It has also been answered. Parmenides spoke about *atomos*, Lewis about the photon, Pati and Salam about the preon. These ideas were critiqued according to the state of understanding of the time. The ancient philosophers Leucippus and Democritus questioned the atomic nature of the void because they could not fathom how such a substance could possibly change its shape. The idea of the photon as the fundamental element was discarded in the early part of the last century. It was not understood how the photon could appear from the vacuum and disappear back into it. The notion of the preon, in turn, did not match observations because its point-like character entailed properties of infinite magnitude.

Everyone who has taken a step toward unraveling the fundamental constituents of Nature has met mistrust. While there is no reason to take any proposition for granted, there is none to reject any straight off either. The axiom of everything comprising quanta is simple, perhaps unbelievably so. But neither simplicity nor mistrust makes it wrong; neither complexity nor trust would make it right.

The earlier revisions to the worldview have followed from uncomplicated conclusions. Aristarchus of Samos figured out the size of the Moon simply from the shadow of the Earth thereon. The ancient astronomer proposed the Sun-centered model, having measured the Sun to be far greater than the Earth. He also understood that stars are bodies akin to the Sun but incredibly far away, as suggested by their seemingly stationary positions despite the rolling of the year. Similarly, Giordano Bruno, a visionary cosmologist, highlighted that the stars are other Suns, each with its planets – for this, he was burned at the stake in 1600. We reproach past intolerance of truth as if it were different from present intolerance of dissent.

An explanation is not only about what we want to know but also about what is already known but fails to cohere. In his book, *The Trouble with Physics* (2006), Smolin sums up the five main goals of physics:[95b]

1. *Combine general relativity and quantum theory into a single theory that can claim to be the complete theory of Nature.*

2. *Resolve the problems in the foundations of quantum mechanics, either by making sense of the theory as it stands or by inventing a new theory that does make sense.*

3. *Determine whether or not the various particles and forces can be unified in a theory that explains them all as manifestations of a single, fundamental entity.*

4. *Explain how the values of the free constants in the Standard Model of particle physics are chosen in Nature.*

5. *Explain dark matter and dark energy. Or, if they don't exist, determine how and why gravity is modified on large scales.*

These tasks may not seem like the most urgent ones, but many problems may well be caused by our notions not measuring up against reality. We should, therefore, derive a mathematical theory from reality, as Galileo did, rather than defining reality by mathematical theory.

EXPECTATIONS

"Where are all your equations?" asked Miklos Långvik after I had requested his comments on a manuscript of mine. The young scientist, who was caught up in quantum gravity, verbalized what many senior physicists stayed silent on: namely, my concrete results do not meet the expectations of contemporary physics. This I am aware of to the point of distress. The thing is, I try to guide my thinking through observations rather than expectations. Thus, when a few equations suffice to state the structures and reactions of elementary particles, I see no need to obfuscate matters.

Why do physicists admire theories with copious equations while they claim to aspire to parsimony? After all, is not the idea of physics to describe as much as possible with as little as possible?[21]

There are far more equations in my PhD thesis than in my papers on the holistic worldview. The thesis was about magnetic fields around atomic nuclei as the nuclei reoriented upon being agitated. The fields changed, whereas the nuclei themselves remained intact. Similarly, it is hard to calculate the dynamic effect of a particle on the plenitude of vacuum quanta to the ends of the world, whereas easy to present a particle as a localized string of its quanta. Thus, simplicity and complexity are not at odds with each other. This plain view of wave-particle duality[149b] is surprising in our byzantine era, where it is radical to put effort into unsophisticated understanding that elementary particles are indeed elementary.

Feynman said: "You can recognize truth by its beauty and simplicity. When you get it right, it is obvious that it is right — at least if you have any experience – because usually, what happens is that more comes out than goes in."[154] At the onset of this chapter, we only took Planck's constant as the measure of the quantum and the fine-structure constant as the characteristic of the electron, and we got the structures of the elementary particles.

WHAT HAPPENS IN A NUCLEAR REACTION?

In all processes, quanta move from one form to another.

Sunlight is a prerequisite for life. The vital photons emanate as the heavenly furnace transforms hydrogen into helium. We can grasp the reaction now that we know the structures of the reactants.

The nuclear reaction starts in a proton's vicinity when an electron torus opens up by losing one of its neutrino loops, becoming the W^- boson.[97] The highly reactive particle annihilates with the adjacent up quark. The quanta uncoil pairwise into the vacuum. The up quark discharges completely. The gluon left behind latches onto the ends of the remaining arc of the W^- boson, a down quark of the new structure, the neutron. The nucleon conversion can also be described tangibly as an exchange of a meson, quark-antiquark particle.

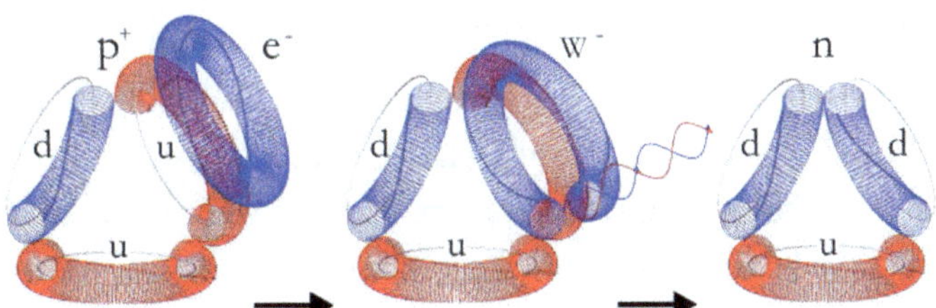

Stages of a reaction where an atomic nucleus captures an electron. From left to right: an electron (e⁻) next to a proton (p⁺) opens up by losing a single quantum, a neutrino. The quanta of the resulting W^- boson and those of the up quark (u) discharge as pairs (red and blue wavy arrows) into the vacuum. The up quark dissolves totally, while the remaining arc of the W^- boson becomes a down quark (d) to form a neutron (n).

LOST IN A MAZE

Surely you know the feeling of finding yourself in a familiar location after being lost in a maze of streets. Isn't it surprising how close you were and yet couldn't recognize the familiar neighborhood? Research is also a sort of wandering. If only we had someone to lead us straight to the goal.

I once had a guide like that. As an exchange student in the State of Washington, I often went riding with a wise old horse, Dandy, up into the mountain forests. I did not have to pay much attention to where we ended up. When it was time to return, I just turned the horse around. The mare was happy to take the fastest way back home. She joyously galloped through thickets, and I ducked to avoid being smacked by branches. Soon we came to the stables, occasionally from a different direction than I had expected. On those occasions, it would have been hard for me to find the way back on my own.

Once we have reached the goal of a scientific endeavor, the most straightforward path to it is evident in hindsight. When we realize that Planck's constant means the photon, the string-like fundamental element, we will stumble upon the torus, even without knowing Ampère's model of the electron, and invariably end also up with the wireframe models of other particles and the void. We can verify by simple calculations that the models are consistent with the characteristics of the particles (Appendix B) and, therefore, also in harmony with the Standard Model.

From this transparent perspective, it seems that the weird lens of quantum mechanics has distorted our view of reality for decades. Philip Ball, a science writer, worded this labyrinth of theses as a joke about a tourist who, lost in rural Ireland, asks a passer-by how to get to Dublin. "I wouldn't start from here," comes the reply.[155]

Now that we grasp the quantum in tangible terms, we also understand previous interpretations. For de Broglie and Bohm, the particle and the field are distinct but invariably inseparable since the vacuum envelopes everything. On the other hand, the Copenhagen interpretation does not discern the particle but only models its effects on the vacuum with the wave function.[156] Newton's view of particles is not that erroneous, only short of the truth that the fundamental element is like a piece of wire rather than a corpuscle.

When we grasp substance in corporeal terms, we comprehend what we can and cannot do with it. We need realism about our options for manufacturing goods and energy production. It is a tough enough job to sustain both a living Earth and a high standard of living, considering the daunting challenges we face today.

FROM THE ATOM TO THE QUANTUM

While the Standard Model works well, it falls short of explaining. For instance, physicists wonder why the proton, neutron, and other baryons consist of either three quarks or three antiquarks but never mixed forms of matter and antimatter. Paul Hoyer, a professor of elementary particle physics, raised this question, too, when I joined him for a discussion over coffee at a break during the symposium of the Finnish Society of Sciences and Letters in the fall of 2013.[157] As you are by now aware, three quarks or three antiquarks can be bound together by gluons to form a stable structure. In contrast, quark-antiquark combinations cannot but remain open. Quark-antiquark pairs, known as mesons, are unstable for the same reason.[158] A similar phenomenon is manifest on the molecular scale; chemical compounds remain reactive as long as there are open bonds.

Physicists figure out what the original particle is from decay products. Similarly, chemists deduce the starting compound in a reaction from reaction products. In this way, it is possible to find out, for example, what substance caused poisoning. Likewise, it is possible to reason that no particle could mediate a proton's decay into a positron and pion, as Georgi and Glashow outlined.[140] Neither is there a reaction in which the neutrino and antineutrino annihilate each other into nothingness,[159] for quanta are permanent.

Two hundred years ago, chemistry was an abstract, nonfigurative subject like elementary particle physics today. Then Dalton realized that compounds react in integer ratios because they are composed of atoms. Now you realize that elementary particles react in integer ratios because they consist of quanta.

About one hundred years ago, scientists worked out how atoms bond together. Today, kids learn how to assemble molecular models from models of atoms, even invent new compounds. Now, similar possibilities are at hand with particles. We can construct models of elementary particles from models of the quarks, even predicting exotic particles.[115,160] Quanta move from one particle to another in nuclear reactions as atoms move from one compound to another in chemical reactions. Moreover, in cyclic reactions, particles convert from one form to another over and over again in the same way as metabolites circulate in organisms and nutrients cycle in ecosystems. Such circulations are also manifestations of *Grand Regularity*.

WHAT IS THE HIGGS BOSON?

A particle, irrespective of its role, is still a particle.

When an observation meets expectations, it will hardly shift the worldview, whereas an unexpected result will be upsetting. Likewise, an interim report will hardly shift the market price when a company's returns are well-anticipated, whereas a shocking release will. What if the postulated dark matter is not found? At least the Higgs particle, theorized in 1964, was duly discovered.

The Higgs boson decays, among other things, into two electrons and two positrons.[161] This tells us that the particle has a certain symmetry. An open torus on each face of a tetrahedron explains the particle's properties: charge neutrality, mass, spin, and its symmetry (even parity),[115] as well as the Higgs boson being its own antiparticle.

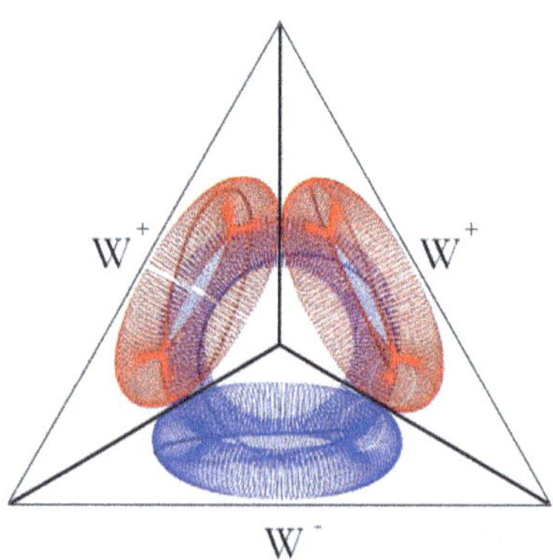

The Higgs particle consists of two Z bosons, each comprising a W⁻ (blue) and a W⁺ boson (red). The particle's mass is large because the vacuum's quantum pairs couple strongly to the narrow gaps of these open tori. At two gaps, the bottom left edge and the front right edge, a gluon links the W⁺ and W⁻ bosons of each Z boson.

The mass of the Higgs boson is enormous, being made up of two linked W^- and W^+, or equivalently, two Z bosons, akin to the ZZ diboson found in 2008, mediating the weak nuclear force. These particles are massive because the strings of vacuum quanta couple tightly to narrow, single-neutrino-wide slots of the open tori. This is similar to how a fork gets caught up in strands of spaghetti. The masses of many exotic elementary particles derive from such structural details, which can be seen as imperfections when considering the electron's pure circularity. Against this backdrop of tangible thinking, the paired-photon vacuum rather than the postulated Higgs field endows the elementary particles with their masses.

ASSORTED FAMILIES

Elementary particles can be grouped into three families; nobody knows why.[148] Each family has the same members. Their characteristics are very alike: only the particle masses of the second family are higher than those of the first, and those of the third are higher still.[97]

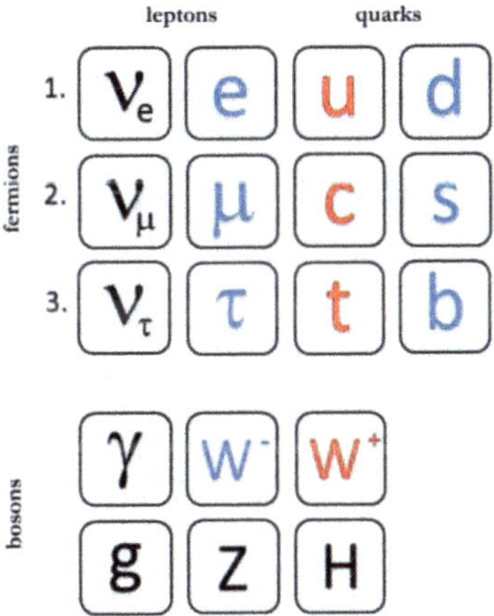

In the Standard Model of particle physics, elementary particles are grouped into fermions and bosons. The ordinary substance is made of the first (1.) family ingredients. In the atomic nucleus, up (u) and down (d) quarks bind together through gluons (g), forming protons and neutrons. Photons (γ), in turn, couple the nucleus with electrons (e). Bosons (W, Z) mediate nuclear reactions involving neutrinos (ν). The Higgs boson (H) is surmised to give these intermediate bosons their masses. The particles in the second (2.) and third (3.) family have similar properties as the corresponding particles in the first family, except for being heavier. No particle carries gravity.

Two hundred years ago, the logic of the periodic table became clear by understanding that atoms are divisible into nuclei and electrons and that the nuclei divide further into protons and neutrons. The logic of the Standard Model is alike: the particles divide into quarks and gluons and quarks further into the quanta of light.

The similarity of particles across the three families is akin to chemically similar elements belonging to the same group in the periodic table. Copper, silver, and gold are very alike but differ by mass. Likewise, the electron, muon, and tau have the same charge and almost the same magnetic moments. Only their masses differ. Independent of the family background, the electron, the heavier muon, and the even heavier tau react in a universal manner. Small differences display themselves in the fine-structure constants, which are not exactly equal, and in the lepton universality,[162] which may not be perfect[163] either.

The family patterns suggest that the fundamental element, the quantum, adopts the same geometric shape in each member of a given particle family. The elementary planar loop, the electron neutrino, characterizes the first family. The second and third family geometry is unknown in detail. Still, we may rely on the *Grand Regularity*, envisioning that in surroundings of increasing energy, the planar neutrino twists around itself like a rubber band twists. Twisted light is, in itself, a well-known phenomenon. For instance, its polarization rotates when light goes through a helically wound optical fiber.[164] The number of twists of the basic element might be the hallmark defining each family. Perhaps there are more than three families, despite no trace of any fourth family member.

When the neutrino goes through space and matter, its mass varies. As Bruno Pontecorvo[ix] indicated in 1957, this oscillation tells us that the electron neutrino twists into the muon neutrino and further into the tau neutrino to gain balance with the stuff it goes through.[165] Also, a chemical compound, such as a macrocycle, twists from one conformation to another, depending on conditions. So, a biologist would say that on its way, the neutrino adapts to its environment – a physicist could say the same. Clearly, the photon adapts to the gravitational field

---

ix Having served on the Manhattan project, the Italian physicist's defection to the Soviet Union in 1950 evoked a sensation.

by shifting its period.[166] Unlike a closed loop, the open wavelet shortens or lengthens continuously to gain balance with its surroundings.

Since the particle structures are unknown to the Standard Model, transformations between conformations are not described tangibly. Instead, mixing between quarks is modeled similarly to the neutrino oscillation by a matrix of rotations. However, the rotations about the three axes do not correspond to the twists in the particle shape.

Since the electron and its heavier counterparts, the muon and the tau, transform into each other, not directly, but via the W^- boson, a mathematician would conclude that the topologies of the three particles differ. By contrast, a doughnut can be morphed into a coffee cup because the topology is the same. Thus, it is possible to mold soft clay from a doughnut shape into a coffee cup so that the hole in the ring becomes the cup's handle.

THE ESSENCE OF INERTIA

The discovery of the Higgs particle was taken as evidence of the Higgs field, proposed in the mid-1960s. A few scientists explained that the weak force carriers, the W^+, W^-, and Z bosons, have mass due to coupling with the universal field. One of them, François Englert, a Belgian physicist, visited Helsinki in 2010. The soon-to-be Nobel Laureate began his lecture by referring to Galileo. So I knew to expect a vision of what inertia *is* rather than getting introduced to a model of inertia.

Whether the universal field permeates all of space as the Higgs field or as the paired-photon void depends on how we construe the observations. Already, Euler understood mass as the measure of inertia.[167] Using his formula for curvature, the particle-to-vacuum coupling can be calculated from the particle and vacuum structures.[88,115]

After Englert had finished his talk, I asked him about the connection between curvature and mass. After all, these concepts are at the gist of general relativity. Even years later, Masud Chaichian, who hosted the visit, upheld the significance of my question. As a professor of high-energy physics, he was keen on understanding particles as strings of quanta and the vacuum as the quantum field around the particles. I felt that Masud's broad-mindedness was quite exceptional.

Inertia is a world-class mystery, yet everyone has first-hand knowledge of it. When your motion changes, you sense inertia like a gust of wind. That experience has eluded scientific rationalization. We need a theory most acutely when we grasp hardly anything about the subject. But a false line of thought leads us astray. Using Galileo's method to mathematize our own experience of time as the flow of quanta seems amateurish next to the mathematics of Higgs' mechanism. But still, it paved the way to concrete and consistent comprehension of the void, particles, and their reactions.

WHICH VIEW ON REALITY IS RIGHT?

Wolfgang Pauli demanded the verifiability of science with his legendary retort, "It is not even wrong."[168] Unless a theory is clearly stated, it cannot even be critiqued. Abstract thoughts are hard to check against reality, whereas the tangible idea that everything is quanta can be proven wrong. For one thing, the tenet would trivially turn out false if a quantum were to split, and for another, if energy were found to stay constant in an event. Imre Lakatos demanded such a principled attitude: "Intellectual honesty does not consist in trying to entrench, or establish one's position by proving (or 'probabilifying') it – intellectual honesty consists rather in specifying precisely the conditions under which one is willing to give up one's position."[169]

SPOOKY ACTION AT A DISTANCE

Einstein shunned quantum mechanics.[170] He regarded action at a distance as an impossible fable.[171] How could a measurement of one particle possibly betray the attributes of another in less time than light takes to cover the distance between them? Instantaneous violates causality; nonlocal is noncausal.

This odd notion of action at a distance concerns an experiment where two photons are emitted in opposite directions from the same source at the same time, each ending up at its own detector. When the phase of the electromagnetic field of one photon is detected, the phase of the other is immediately known to be the opposite.

That corollary is not itself odd because the radiated photons were mirror images of each other as they departed. So until and including the event of detection, the phases cannot be anything other than the opposite. But by quantum mechanics, the photon phase is indeterminate until detected. So only when measuring the photon phase at one detector is the other photon assumed to take the opposite phase at the other detector. This creates the extraordinary impression of information transmitting from one detector to the other faster than the photons themselves move.

Einstein did not approve of this spooky action at a distance. He tried but failed to refute the unnatural notion that photons are 'entangled.'[171] So it is that still today, theories are expected to comply with quantum mechanics. But does that compliance make a theory tenable?

The action at a distance experiment has been performed numerous times. Yet, claiming the photons are entangled does not make the outcome comprehensible. We cannot fathom the entanglement and collapse of a wave function at detection through our own experience. An explanation without cause is an explanation without sense.

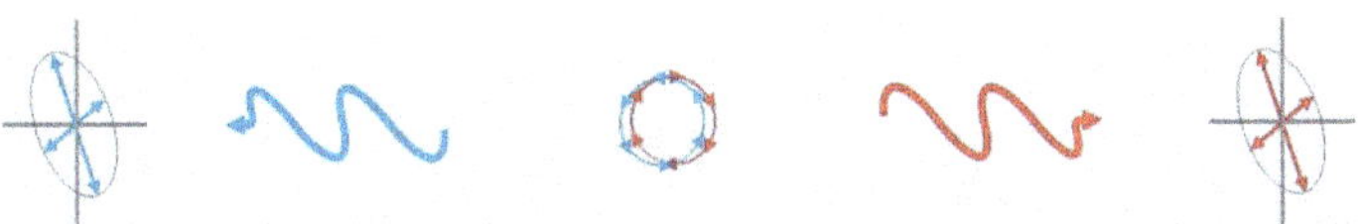

Two photons (blue and red waves) discharge from the same event (center) in opposite directions so that their phases, polarizations, and vibrations mirror one another. So, when the phase of one photon is registered with one detector (left), it is known right away that the phase of the second photon is the reverse. The conclusion is consolidated with another detector (right). This outcome does not mean that information propagates faster than light from one detector to the other. It only means that the correlation between the photons' phases survives until detection.

By common sense, understanding the experiment cannot even be called a challenge. So long as the photons maintain their phases on their way to the detectors, we know at once from the phase of one that the phase of the other must be the opposite. For example, suppose the phasor of one photon happens to point to '6 o'clock' relative to that of the detector. In that case, we immediately know that the other points in the opposite direction, at '12 o'clock.'

When explained in this way, nothing is spooky about the experiment. We just do not know the photon phase before it is measured. In fact, the photon has a defined phase only relative to the reference phase, say, of the detector. If the detector were pivoted upside-down, the photon phase would be the opposite. Likewise, a clock dial face without numbers cannot by itself tell us what time it is. From this operationalist perspective, the hands do not indicate any time until compared, for example, to the vertical direction. Quantum mechanics translates this ignorance to the absurdity that Einstein memorably dubbed "is-the-moon-there-if-nobody-looks."[172] Sure is. The vacuum embraces everything and nothing escapes interaction, say, detection.

The photon phasors are the opposite of one another but unrelated to the detector phasors until detected. The outcome is thus contingent on the detector phases, so to say, on the background. This is a self-evident and essential point, for quantum mechanics is a background-dependent theory.[6a]

Perhaps there has been confusion about the correlated phases because the photon phases are, in fact, not detected. Instead, the photons that made it through the polarizers to the receivers were counted. Specifically, when the photon phase happens to be 45 °, the probability of the photon wavelet entering either one of the two channels is proportional to the phase projection, the cosine of the angle, i.e., $\cos(45°) \approx 0.71$. Thus, although it is equally likely for either channel of a polarizer cube to catch the photon, the probability of the photon wavelet entering one or the other is not 0.50 but 0.71.

Does the probability exceed 1 when both channels catch the photon with the probability of 0.71? This concern reminds me of a legendary phrase by a Finnish ski jumper: "It is fifty-sixty how it will go." In the experiment, we have the same situation as in real life. Only one of the alternatives will happen. The probability of catching the photon with the 45 ° phase is 0.71, although the photon number registered through either one of the two channels is likely to be almost equal.

In the same way, as the photon wavelet goes through a polarizer, you can think of yourself as going through a doorway. The probability of getting straight through is proportional to the width of the opening in front of you. When you see the doorway at an angle of 45 °, approximately 70% of the width is visible. So when two doorways on either side of a corner are equally visible, the overall width of view is $0.7 + 0.7 = 1.4$. Even so, you do not go through both openings but through one or the other. When you face one of the openings straight on, it is visible to you by its full width, whereas the second opening is not visible at all. Then the total view is $1 + 0 = 1$.

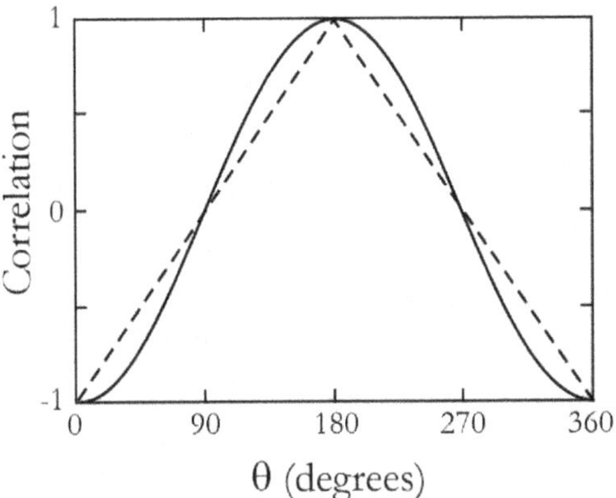

Two photons of opposite phases are registered, each by its own detector. When one of the detectors is pivoted a full circle from 0 to 360 °, the correlation follows the cosine curve (solid) because the probability of the photon wavelet entering one or the other channel is proportional to the projection of its phase. When the detector phases are parallel ($\theta = 0$ °), the correlation is $-\cos(0$ °$) = -1$ because the photon phases are opposite. Conversely, the orthogonal phases are completely independent, uncorrelated, $\cos(90$ °$) = 0$. When $\theta = 45$ °, the correlation is $-\cos(45$ °$) \approx -0.71$. As rotations in general, the detector rotation produces trigonometric (solid) rather than the erroneously assumed triangular function (dashed). Specifically, when the photon phase is 45 ° relative to the receiver, the correlation is not -0.50 but -0.71.

The difference between the total views of the two detectors is at its largest $1.4 - 1.0 = 0.4$ when the phases of the two receivers are at 45 ° relative to each other, and least, zero when the receiver phases are parallel. The passage through the phase-sensitive entry depends on the photon's phase, whereas the registration at the counter depends on the photon's energy. If one erroneously expects the photon detection to be independent of the detector phase, one sees a discrepancy between the expectation and reality. The difference must be explained. The adventurous explanation of entangled photons leads to the illusion of the spooky action at a distance.

The explanation is as simple as it is real. The photons start off and remain in opposite phases. The measurement merely reveals the phase difference between the detectors, known already at the onset (Appendix D). So too, the entanglement of macroscopic oscillators can be understood as classical correlation. As the study demonstrated, the oscillators are independent of each other because the position of one and the momentum of the other were measured more precisely than Heisenberg's uncertainty limit. Thus, the entanglement belongs among those historical concepts of science, once vehemently defended and experimentally verified but came to nothing.[4]

PUT INTO PERSPECTIVE

Quantum mechanics is a tool but also a taboo. If we believe reality is fundamentally incomprehensible as presented by modern physics, we will be prone to accept other authorized but unwarranted judgments.

Einstein did not approve of the incomprehensible. Action at a distance was not in itself new to him, having thought about how gravity could act over great distances and yet seem like an immediate inertial effect. "Matter cannot act where it is not."[174] Einstein knew the adage but not the substance of space. So, time and energy without substance feature profusely in the equations when space-time, an elastic cosmic fabric, 'informs' a body of the masses in the rest of the universe.[40] But, we have no experience and no evidence of such vacuity.

The problem of entanglement is different from that of inertia. Entanglement is nothing but classical correlation, whereas Einstein was truly puzzled by how the interaction could relay through space in no time. Yet, Einstein knew the purpose of physics: it should explain, not mystify, the world.

Today, quantum mechanics is upheld by referring to Bell's theorem, which says that quantum mechanics is incompatible with hidden variable theories.[75] Indeed, but that is not even the issue. The real issue is the erroneous thought that the photon phase would be a number, whereas it is a vector. As a result, even seemingly spurious entanglement between photons that never coexisted has not set alarm bells ringing.[175] In reality, the photons are not entangled but correlated, showing the phase difference between the two detectors.

Someone might think that such esoteric physics experiments do not carry real-world weight. Unfortunately, not everyone takes such a reasonable stance toward science. Some scientists trust, even defend, a doctrine they do not comprehend. Such an attitude is incomprehensible, given that a measured value does not mean anything by itself, no matter how precise. The meaning follows from the interpretation. This profound uncertainty is way beyond the experimental error bars.

THE PRINCIPLES OF SCIENCE

"Perhaps we can't make sense of [quantum mechanics] simply because it isn't true. It is instead likely to be an approximation to a deeper theory that will be easier to make sense of...,"[6b] as Smolin thinks.

At one time, the geocentric model was seen as precise, yet it had to give way to the heliocentric worldview. Knocking modern physics from its pedestal, demoting it from a worldview to a <u>mathematical</u> model, would not be scandalous, only the latest evolution in our time-honored intellectual lineage. An unambiguous connection between concepts and reality is the prerequisite of a valid theory encompassing the universe, including our clocks, our instruments, and ourselves.

Newton aimed at all-inclusive comprehension. He reasoned that the gravity of the universe curves the water surface of a spinning bucket. For the same reason, the rotating Earth must not be perfectly round but flattened. This he pointed out to Leibniz, even though he could not pinpoint what mediates gravity.[176]

Maupertuis measured Earth's form by triangulation on his 1736–1737 expedition to Tornio River Valley, Lapland. The mission's return to Paris with invaluable data settled the quarrel that had raged at the French Academy of Sciences about the shape of the Earth. The permanent secretary, Bernard le Bovier de Fontenelle, handled the dispute in an exemplary manner by respecting the principles of science. Despite his strong Cartesian prejudices, he insisted that the institution would promote "no general system [at all], out of fear of falling into the disadvantage of [promoting] rash systems,"[177] as Mary Terrall, professor of history at UCLA, notes in her biography of Maupertuis.

Later, although Einstein's space-time model gave numerical values that matched astronomical data, Nikola Tesla judged the curved space-time concept to be unreasonable. The daredevil inventor and startling showman insisted on geometry corresponding to some substance. His 1931 complaint echoes ever truer all these decades later: "Today's scientists have substituted mathematics for experiments, and they wander off through equation after equation and eventually build a structure which has no relation to reality."[178]

At about the same time, Husserl saw scientific thinking becoming ever more a technical methodology that drifts away from explaining Nature.[179] That is why a return to Galileo's method is fruitful. We can relate our experience of time flowing to the flow equation of quanta. When gaining or losing one quantum, the present transforms into the past and the future into the present. This thermodynamic stance is in stark contrast with theorizing the universe as a four-dimensional space-time block, where the present moment holds equal status with any past or future moment.

The world – heedless of our viewing it through this frame called Physics – is what it is, in its causality and entirety.

ONE MORE CHANCE

In a sense, the Michelson–Morley experiment[30] was a conjuring trick. Since the void was not seen as ether, the objectives of physics became defocused. However, already at the time, there were also clues to a better understanding. Maxwell had explicitly stated that light is the wave of the void.[34] He envisaged that we could sense the vacuum in ways other than light, too. That is true. We feel inertia with our own bodies. So why did the bright idea of the void as light fade away from the scientific consciousness?

Instrumentalism took over when relativity theory inhibited us from perceiving the vacuum as a physical substance. We got wave functions, pseudo-particles, and virtual photons, but we did not get explanations. This split view of the void as abstract on the one hand, and concrete on the other, was echoed in the dialog between Einstein and the Nobel Laureate Dutch physicist Hendrik Lorentz.[180] In 1917, Lorentz reasoned, "It is always risky to close a path of research completely and perhaps it is good, considering everything together, to grant the ether one more chance. Conceivably a time will come when speculations over its structure, from which we now abstain, become fruitful and effective."[33b]

Dirac thought about a vacuum consisting of quanta propagating at the speed of light.[181] Robert B. Laughlin, known for his contentious but substantiated arguments, thinks likewise. The physics professor at Stanford University deems it deplorable that theoretical physicists are ill-disposed toward the ether. Relativity does not in itself exclude its existence, provided that the stuff complies with the symmetry of space-time. In his book *A Different Universe* (2005), Laughlin continues:

> About the time relativity was becoming accepted, studies of radioactivity began showing that the empty vacuum of space had a spectroscopic structure similar to that of ordinary quantum solids and fluids. Subsequent studies with large particle accelerators have now led us to understand that space is more like a piece of window glass than ideal Newtonian emptiness. It is filled with 'stuff' that is normally transparent but can be made visible by hitting it sufficiently hard to knock out a part. The modern concept of the vacuum of space, confirmed every day by experiment, is a relativistic ether. But we do not call it this because it is taboo.[61b]

The taboo silences. And so, we fail to deal with the problem. While experiments have irrefutably shown particles materializing from the vacuum, others have disproven its substance, the light-carrying ether. This dilemma accentuates the divisive character of modern physics. On the one hand, a particle seems too real to be only a transient quantum field; on the other, the luminiferous ether seems unrealistic. How could it be superbly mobile and, at the same time, a stiffer and more elastic substance than steel to transmit light? So those who deny the ether are illogical, while those who hold onto it are irrational.

Rather than choosing sides, we should seek a solution to the problem. Advocating quantum mechanics reproducing data, and the whole point is missed. Why are the results correct, even if the ontology is empty? Thus, we should be aware of off-the-point arguments and other common flaws in reasoning.

Since antiquity, the human mind has been intrigued by how forces are mediated through space. While effects can now be calculated, causality is as unclear as it was in Newton's time. Yet, we can ask:

* Does not the vacuum itself comprise photons instead of just being a medium for them?
* Are not paired photons the substance with transparent, penetrating, and fluid-like characteristics?
* Would not light be undulations of this paired-photon vacuum, as Maxwell reasoned?
* Would it not be the relativistic ether, as Laughlin inferred?
* Would it not be the quantized medium, as Newton assumed?
* Would it not be sensed as inertia when perturbed?
* Would it not be sensed as gravity when out of balance?

When the answer is at hand, its simplicity is astonishing. As Galileo Galilei remarked, "All truths are easy to understand once they are discovered; the point is to discover them."[182]

Changes in our worldview are like swings of a pendulum.[9] Einstein's acausal theory contains ingredients from the period prior to Newton. All that ultimately matters is what can and cannot be explained by the paired-photon void. When there are alternative ways of interpretation, Newton advised: "Truth is ever to be found in simplicity, and not in the multiplicity and confusion of things."[183]

KEY POINTS

* Everything is quanta of light, the fundamental elements of Nature.
* The vacuum comprises paired quanta of light. We cannot see them, but we can sense them as gravity and inertia.
* The neutrino, a quantum loop, is matter in its most elementary form.
* The electron is a torus of 138 quanta.
* Quarks are fractional arcs of the electron and positron tori.
* Mass is the measure of the void's coupling with matter

4. Why Does the Universe Expand?

The universe expands
as matter morphs into space.

The universe is not just a faraway starry sky. It is all around; it is all that there is. The universe is so naturally present that we hardly pay any attention to it. Yet, we should comprehend no less than the whole. Otherwise, our view of reality remains incomplete, even incorrect.

The ubiquitous patterns in data are visible in the sky, just as on Earth: spirals, S-curves, skewed distributions, and power laws characterize stars, galaxies, and voids. These patterns, free of scale, also known as fractals, recur from the largest to the smallest structures, as the astrophysicists Pekka Teerikorpi and Yurij Baryshev elucidate in their book *Discovery of Cosmic Fractals* (2002). Teerikorpi, an associate professor of the University of Turku and a researcher at the Tuorla Observatory, impressed me in 2011 with his talk at the House of Science and Letters in Helsinki. He lavished the attentive audience with insights about the expanding universe, drawn from ancient astronomy to contemporary cosmology.

In their most recent work, the authors remind us that the state-of-the-art understanding of any given time should be treated with caution.[1] As we cannot do experiments with the whole cosmos but gauge everything from only one place, here on Earth, our assumptions and our vantage point unavoidably influence our interpretations of observations.

Cosmology is literally the study of all that there is, from the bright birth to the dark demise. Cosmological questions are among the most significant in contemporary physics. What is dark energy? What is dark matter? Why do the constants of nature have the values that they do? Why do we exist?

Answers are fervently sought. Distant objects are cataloged and postulated particles are probed to find evidence for or against the present paradigm. The hottest issues in cosmology are not only about hypothetical cold dark matter but also about the whole worldview. Although most of what we know seems sound, not everything is.

When the *Grand Regularity* of the data does not depend on the subject or scale (Chapter 1), could it be that the endless least-time quest for balance (Chapter 2), rather than the sudden Big Bang, is what dictates the cosmic evolution? The expansion of the universe resulting from all events where quanta break free from matter and integrate into the void addresses many cosmological problems. But before delving into them, let us summon up how contemporary puzzles came about and what they are all about.

Perplexing observations

In 1912 at the Lowell Observatory in Arizona, the American astronomer Vesto Slipher spotted signs of the expanding universe.[2] Light coming from a nebula was slightly redder than anticipated. Light waves originating from many other nebulae had also lengthened,[3] suggesting that all those hazy objects in the sky were receding. Similarly, the screech of an ambulance siren lowers in pitch when it passes by. Just as we can infer the vehicle's speed from the falling pitch, we can calculate how fast a celestial object is receding from the redshift.[4]

Slipher was amazed. As a rule, the nebulae are moving away at high speed, even over a thousand kilometers per second. The Andromeda Galaxy is an exception. The big spiral is coming straight at us, as are a couple of smaller nebulae. Nonetheless, Slipher did not claim right away that the universe was expanding, as he did not know what the receding objects were. But a few years later, he surmised that the nebulae are galaxies like the Milky Way.

Edwin Hubble attested that the nebulae lie outside our home galaxy. The pioneer of extragalactic astronomy found a rule: the farther away a galaxy is, the faster it is moving away.[5] Even so, Hubble, too, refrained from declaring the universe is expanding. He had back-calculated from the recession speeds the time the galaxies had flung apart, but then the universe appeared as though it were younger than the Earth. The cosmic yardstick was defective, but recalibration corroborated the expansion. The worldview changed. Honoring Hubble's work, the legendary space telescope was named after him.

General relativity came to be the mathematical model of the expanding universe, however, not in a straightforward way. First, Einstein added the so-called cosmological constant to his equations to make them compatible with a steady-state universe.[6] This belief was still the consensus in 1917. Ten years later, when cosmic expansion had been ascertained, Einstein reckoned, so they say, this fiddling with the theory to be his worst blunder.[7]

We would not have that anecdote to tell had Einstein accepted Alexander Friedmann's conclusions without delay. The prominent Russian physicist derived the formula for the expanding universe from Einstein's equations years before Hubble's publication.[8,9] Sadly, Friedmann's early death deprived him of the recognition to come.

Friedmann's unheard-of conclusion about the expanding cosmos was not even considered before Hubble's observations in 1929. Likewise, Georges Lemaître, a Belgian priest studying astrophysics, was ignored in predicting Hubble's law in 1927.[10] Bypassing insightful thinking often predates revolutions in the worldview. In retrospect, the new view seems the only viable option.[11]

CRITICAL ISSUES

Physicists measure and model expansion but hardly ask why the universe is blowing up. As a mathematical model of the cosmos, general relativity is not about causes and effects. So, the expansion is seen as an inborn and inexplicable attribute of the Big Bang, not a consequence of a still-prevailing cause.[12]

The reason for expansion is worth reconsidering because astronomical data on stars, gas clouds, galaxies, and voids are no different from any other data displaying the *Grand Regularity*, the universal patterns.[13] The evolution of the universe cannot be a process in its own right but sums up all events. We should thus be able to infer something about cosmic evolution from any event. So let us do that.

The shining Sun transforms matter into photons. As explained in the previous chapter, we see the bright light but not those photons that break out from matter in pairs. Space consists mainly of these dark, weakly interacting photon pairs (WIPP), as stars produce much less other stuff, including neutrinos.[14] From the structures and reactions of particles, we understand that space does not emerge from nothingness; it unfolds out of the stars, where matter transforms into the void (Appendix H). What does this result mean? Which observations would support it; which oppose it?

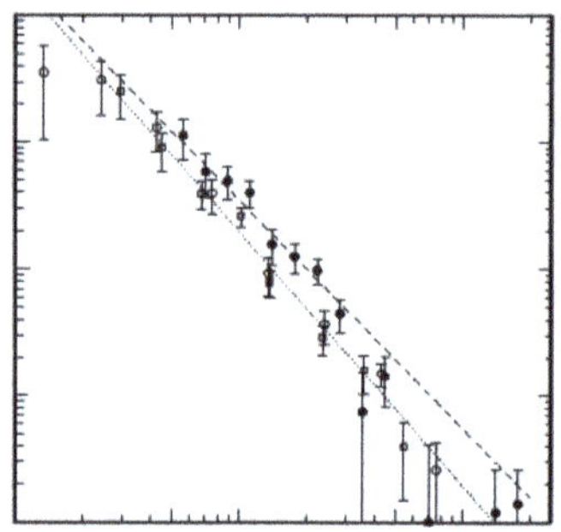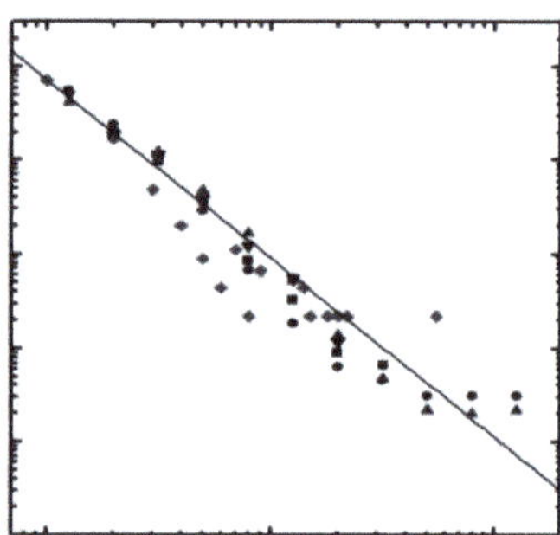

Galaxies are typically in groups, and the groups belong to massive clusters. There are plenty of short distances between the neighboring groups while relatively few very long distances between the outlying groups. The distance distribution closely follows a straight line on the logarithm-logarithm scale (left).[15] The degree distribution of yeast proteins also follows the power law (right).[16] The same form suggests the same law.

At first glance, it may seem a far-fetched idea that matter is transmuting into the void, especially if discrediting the light quanta of being the indivisible and eternal elemental constituents of everything. But, likewise, as late as in the 1850s, physicists still had reservations about there being such things as atoms, while chemists had already been confident about the atomic nature of matter for a hundred years. This issue was finally settled in 1905 when Einstein explained that the characteristic motion of dust particles, known as Brownian motion, arises from molecules bumping into the fine specks of dust.

Assuming the void emerges from the quanta of matter, the most powerful sources, such as quasars, should be receding the fastest. Indeed, most quasars are about 10 billion light years away, and some even further at 13 billion light years.[17] A quasar beam, thousands of times brighter than the entire Milky Way, may rapidly vary in brightness. The variation over hours to months is characteristic of a dense object undergoing a chain reaction.

As the amount of matter diminishes, the expansion slows down, and eventually, the void will be everything there is. The geometry of a universe expanding in this way is said to be flat, and its density is called critical.[12,18] Naturally enough, the density of the universe is precisely critical. Since the imbalance between matter and the void decreases in the least time, the universe is expanding everywhere in every direction.

By the same least-time principle, the more massive a star, the more it radiates.[18] Thus, the stellar radiation displays the same form as the power radiated from a human or an insect consuming food or any other thermodynamic machine consuming fuel, say, a vacuum cleaner or an airplane. Sure, the Sun is devouring matter and radiating more efficiently than any of us are converting food into heat. Yet, the principle is the same; only the mechanisms are different.

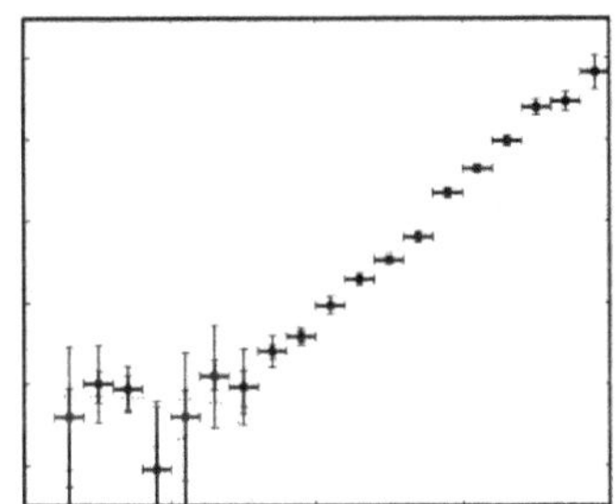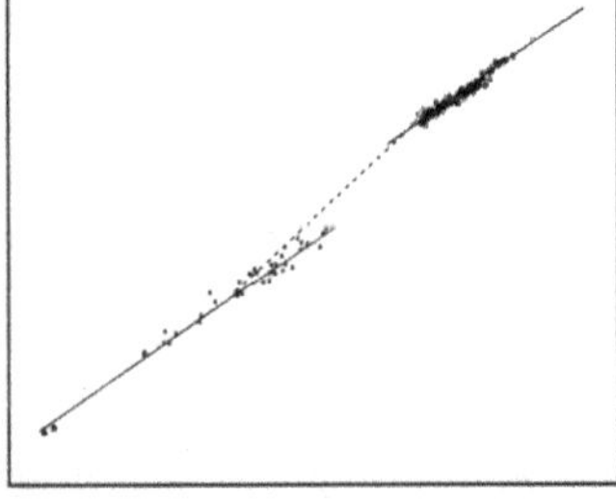

The luminosity of galaxies, galaxy groups, and clusters increases with increasing mass (left).[19] Likewise, the metabolic power of organisms grows with increasing mass in a power-law manner (right).[20] The same form implies the same principle.

In contrast to the thermodynamic theory of time, the standard model of cosmology, the ΛCDM model (Lambda Cold Dark Matter), does not explain why the density is critical because the model is fitted to the observed flatness.[21] If we did not know that space emerges from matter, the flatness would seem like a lucky strike – or be mistaken as something designed.[22] But neither Fortune nor a Designer has a finger on the scale. Flatness is not a genuine problem. Kinetics cannot but balance potential when space stems from matter. The bookkeeping of everything is correct when quanta do not come out of nothingness or vanish into nothingness.[23,24] This core principle, *ex nihilo nihil fit*, follows from the principle of sufficient reason.[25]

THE END OF TIME

More than a thousand years ago, the polymaths Alhazen (Arab. al-Haytham) and Avicenna (Pers. Ibn Sina) inferred that the speed of light is finite. The natural philosophers Galileo, Boyle, and Hook shared the same thought. Then, having gotten acquainted with the measurements of the Danish astronomer Ole Rømer, Newton concluded that light from the Sun arrives at the Earth in 7 to 8 minutes.

Later, Maxwell realized that the speed of light is not exactly a natural constant; it is a characteristic of the vacuum.[26] In the same way as light goes more slowly in glass than in air, it propagated more slowly in the early dense universe than in the sparse surroundings of our era. As space becomes increasingly thin by expansion, the speed of light will only increase. Although the speed of light is not a constant, it serves as a universal measure. The wavelength of light divided by its period is a constant for all photons, irrespective of density.

The expansion cannot exceed the speed of light, for the universe is made of photons. Through all events, the universe has enlarged enormously. As the photon periods have lengthened, spanning about 14 billion years from the past to the present, the average energy per cubic meter is only about a billionth of a joule (10^{-9} J/m^3).[18]

When space stems from matter, we may abandon the assumption that the universe could decelerate, start to contract, and eventually collapse.[27] Evolution is irreversible. No force can turn the course of all events around back toward the Big Crunch. We may also disregard the possibility of the universe billowing out ever more rapidly by dark energy. There is no fuel to power ever-faster expansion. Distant galaxies will not recede beyond the range of light. The galaxies will cease to exist in due course when all the quanta bound in matter have been released into the vacuum. We can also discard the hypothesis that the vacuum might break apart.[28] There is no force to rip apart the indivisible and eternal photons. Indeed, Parmenides' idea of the primary element of everything limits interpretations of the data on the universe's evolution more sharply than many a model of the cosmos.

Fundamentally, *the* imbalance between matter and the void makes everything that happens happen. It is the final cause of which Aristotle wrote. From this perspective, cosmic evolution in all its richness is merely an austere process: the quanta of matter become the quanta of the void. When all energy differences, forces, the causes of events have at last vanished, everything that can happen has happened. This ultimate fate, the end of time, is called heat death.[29] Nothing could be colder than that void of ever-lengthening photons.

WHAT IS GRAVITY?

Bodies move along with the void in motion.

Gravity is the most prosaic of phenomena. Bodies fall. Why? – The cause is unknown. Our worldview is incomplete.

From the 16th century, the cause of gravity was speculated time after time until Einstein came up with an unparalleled perspective. There is no reason whatsoever for bodies to fall. Gravity is mere geometry.[x] Bodies and light alike move along the optimal paths, most favorable trajectories, geodetic lines, unless disturbed or obstructed. Raindrops fall straight down to the ground unless the wind is blowing.

Despite its towering success, general relativity cannot be the complete theory of gravitation because space cannot be a continuum but must consist of quanta as matter and light do. As expounded in the previous chapter, the quanta are real, whereas space-time without substance does not explain but models the motions of bodies. Similarly, the geocentric celestial orbs of age-old cosmology did not explain the true motions of planets and stars but modeled their apparent motions. An explanation craves some essence.

Gravitons can be theorized to mediate gravity in the same way as photons are known to mediate electromagnetic force. In this picture, the gravitons emerge from mere emptiness, when necessary, and

---

[x] The tenet that gravity *is* geometry originates from Weyl, whereas Einstein employed space-time as a geometrical concept only to calculate gravitational effects.[30]

disappear back into nothingness, when no longer needed, like the virtual photons. So, the virtual particle is a way of calculating, not a way of explaining. The model works fine when forces are weak but fails, for instance, in the atomic nucleus, where the forces are strong.[31]

A quantum theory of gravity is yet to be formulated. But, experiments provide only minimal guidance.[32] The technology at CERN or elsewhere is not powerful enough to create circumstances where physicists expect the quantized nature of gravity to expose itself.

THE VOID IN MOTION

Gravity still an open question may simply imply that the track we are on does not lead to the goal. Since we cannot see the way forward, let us look back at what Newton supposedly saw.

What is actually happening when an apple falls? Why a distant galaxy is receding and a nearby one is approaching? From these observations, we should recognize a single law for all motions, just as Newton discerned the same law in the falling of an apple and in the orbiting of the Moon.

The space swells through numerous processes as quanta break out from dense matter into sparse space. So, distant galaxies, coupled with the outspreading void, are receding. Conversely, our neighboring galaxy, Andromeda, is moving toward us because fewer quanta are entering between it and us than are streaming out into the greater universe. The apple falls earthward for the same reason. The fruit's speed increases with decreasing distance as quanta carry more and more energy away from the closing space between it and the ground.[33,34]

We may compare the void's motion to that of water. When water flows between floating bodies, they move apart; when flowing out from the gap between the bodies, they approach each other. Analogously, Maxwell used hydrodynamics when deriving his field equations for the void.[35b]

From this hydrodynamic perspective, bodies neither attract nor repel one another but move because they couple with the flows of the void. This understanding of the vacuum as a substance is in line with our own experience. When the driver suddenly slams on the brakes, our bodies continue in motion because we are coupled to the surrounding space. We sense the coupling in the opposite direction when the driver steps on the gas. That is inertia.

We also experience the grip of the vacuum as centrifugal force. Physics textbooks refer to it as a virtual force,[36] but the force feels very real, for example, on a carousel. In the same manner, a hammer thrower gets, with a few spins, the ball out of balance with the vacuum, and when letting loose, it shoots off to regain balance.

Similarly, for the Moon to take off from its orbit as a stone shoots off from a sling, more and more quanta would have to enter between it and the Earth. Conversely, for the Moon to fall to Earth, more and more quanta would have to exit into the rest of the universe. When the forces are in balance, there are neither causes nor effects. Centrifugal force is said to tally gravity when the paired quanta neither flow from the gravitational field out into the universe nor vice versa.

Even if we can model gravitational and inertial effects using virtual forces, it does not mean that we do, thereby, understand gravity and inertia. But now, enjoying a concrete view of the vacuum and particles, we finally grasp what we experience.

Like any other force, the gravitational force is an energy difference. The denser the gravitational field, the local vacuum, the higher the difference relative to the sparse gravitational field of the universe. For instance, an apple falls faster on Earth than on the Moon because the difference between universal and terrestrial gravities is greater than between universal and lunar gravities.

The thermodynamic theory maintains that quanta flow so that any imbalance diminishes in the least time. Why would gravity be an exception? The apple falls *straight* down and the universe expands in *every* direction because the void moves toward a balance in the least time.[37,38] Thus, astronomical data display the *Grand Regularity*, skewed distributions, spirals, S-shaped curves, straight lines on logarithmic-logarithmic scales, and even chaotic trajectories.[4,39,40]

The conclusion that gravitation is the manifestation of the void in motion is logical but not sufficient. We must examine the theory from many perspectives. We must ask: What exists but is not explained by it? What does not exist but is explained to exist?

THE OLD SCHOOL

"No machinery has ever been invented that 'explains' gravity without also predicting some other phenomenon that does not exist,"[36] said Richard Feynman about mechanistic explanations of gravitation. For example, Robert Hooke proposed in 1671 that a body causes other bodies to move toward it by emitting waves in every direction. Two hundred years later, James Clerk Maxwell pointed out that this could not be the case because the source would run out of energy in no time.

Newton's friend, the Swiss mathematician Nicolas Fatio de Duillier, and later the Genoese physicist Georges-Louis Le Sage proposed that bodies are attracted to each other because they shield each other from the ether in motion.[41] Maxwell and Poincaré concluded that this could not be the case; the celestial bodies would heat up horrendously if the ether particles were impinging on them all the time. With Newton's law of gravity, Euler reasoned that the ether's density decreases with increasing distance from the body.[42]

Bernhard Riemann's theory from 1853 is also noteworthy. The German mathematician imagined the gravitational field as a fluid whose sources and sinks are bodies.[43] Also, in general relativity, energy in matter is the source of gravitation. Likewise, the Russian engineer, Ivan Yarkovsky, suggested in 1888 that the ether converts into matter when being absorbed into a celestial body.[44] Today, we know by observations that particles, such as electrons and positrons, may emerge from the vacuum and transform into it.[45] Oliver Heaviside thought in 1893 that bodies do not attract each other but rather vacuum density differences push them toward each other.[46] The self-taught English engineer, mathematician, and physicist, understood that gravitation, like electromagnetism, is not the property of bodies but the void.

Although there is a lot of sense in the old explanations, they are deficient in concreteness. Euler did not say what the ether substance is; Riemann and Yarkovsky did not say what matter is, and neither did they explain how matter converts into the void and vice versa. Heaviside admitted that his analysis was weak. Mathematical equations, even those of the greatest interest, do not illuminate in the slightest the ultimate nature of gravitational energy.[46] Theories have come and gone, but the nature of the phenomenon has remained elusive.

THE SUBSTANCE OF GRAVITY

Since Newton, it has been understood that gravity is a force. Although its substance remained ambiguous, the field concept, albeit unnamed, can already be recognized in the introduction to Kepler's main work. "If two stones were placed in any part of the world near each other, and beyond the sphere of influence of a third cognate body, these stones, like two magnetic needles, would come together in the intermediate point, each approaching the other by a space proportional to the comparative mass of the other."[47] However, such an action at a distance without a force carrier was impenetrable to the natural philosophers Descartes and Galileo. They saw the world as mechanistic: causes entail collisions.

Newton acknowledged that he did not know the cause of gravity when contemporaries recognized instrumentalism in his law of gravity.[48] Unless the force carrier is known, the theory of gravitation does not explain why bodies move. It remains only a model.

The Victorian physicists Michael Faraday and James Clerk Maxwell toiled strenuously to comprehend how electricity, magnetism, and gravity are conducted across space – or, more accurately, by the void. They perceived that the vacuum, although not appearing mechanical, must be physical to mediate the forces.[49] Maxwell finally understood: it is light that carries electromagnetism; photons propagate along Faraday's lines of force. These geodetic lines *are* the photons embodying space. However, the precise character of the carrier of gravitation, the graviton, remained obscure.

The concept of a field is Faraday's.[35a] The field lines outside a body mediate forces. Faraday, like Ampère, was mostly self-taught. This may well account for the tangible thinking that the industrious inventor and plainspoken lecturer espoused. His relentless search for the truth is evident from his twice turning down the offer to be the Head of the Royal Society of London.

References to bygone scientists may seem only a matter of curiosity. Still, present-day problems seem disconnected without a historical perspective, and so efforts to solve them are, at best, technical tinkering and, at worst, misleading. Today physicists struggle to unite electromagnetism and gravitation, while Faraday had already discovered them united in the substance of space.[35a] The question he confronted was only about what the substance *is*. Faraday thought that contiguous particles in species of opposite polarity constitute the conserved essence of everything.[50] When considering the paired quanta as the substance of space, his intuition was prescient: "... there must be something in gravity which would correspond to the dual or antithetical nature of the forms of force in electricity and magnetism."[51]

As is well-known, Faraday failed to validate the link between gravitation and electromagnetism. He did notice some effects that could have been interpreted as evidence thereof but demonstrated his superiority as a scientist by not allowing his assumptions to drive his interpretations, identifying the effects as artifacts of experimentation. It is perhaps even more impressive that Faraday did not jump to the opposite conclusion either but left his prophecy for us: "The results are negative. They do not shake my strong feeling of the existence of a relation between gravity and electricity, although they give no proof that such a relation exists."[51]

From this perspective, could it not be the paired photons that mediate gravity and the unpaired photons that carry electromagnetism? The paired quantum (WIPP) graviton would be in line with Laughlin's vision of an emergent particle.[52] The graviton would emerge when two photons combine in opposite phases. It is assumed to have no mass, and the quantum pair indeed lacks mass.[53] The spin of a paired photon is two, just as the spin of the carrier of gravitation is thought to be. This tangible line of reasoning does thus not overturn theories. It makes them comprehensible, maintaining that no effect is without cause and that no cause is without substance. So let us stick to Parmenides' idea of the fundamental elements making up everything.

History illuminates the problems of our time. Had Newton, Faraday, or Maxwell succeeded in explaining the void as the physical ether as they aimed, the spirit of our time and our aims would be different – concrete. So why did they fall short?

The characteristics of the void reflect the vastness of the universe, whereas we are used to the structures of our world. Newton was puzzled about the ether's extraordinary elasticity, density, and strength, vastly exceeding those of any substance he knew. Even so, Faraday and Maxwell thought that space was itself some substance. Only when no luminiferous ether was found in Michelson's and Morley's experiment was it assumed that the void is nothing. Lorentz thence took audacious steps toward instrumentalism devoid of substance. Einstein went further by geometrizing gravitation as non-extant space-time.

Decades later, when particles began to be generated from the vacuum by accelerators, there was no need for conceptual revision. It was presumed that the particles were fields like space itself. Then the transmutation from the vacuum to the particle could be theorized to be just a transition from the ground state to an excited state of the field. So, a concept was premised on the previous one. Over generations, long lines of reasoning built up into the contemporary doctrine, a paradoxical palimpsest. But have physicists modeled the vacuum in an excited state around the particle rather than the particle itself? Is this failure to distinguish between the quanta that make up the vacuum and those that comprise the particle the reason why the nature of gravitation confounds physicists?

THE POWER OF A DOCTRINE

In school, we are taught that when an apple falls, the gravitational energy converts into the apple's kinetic energy. Conservation of energy holds nearly but not exactly. When the apple is on the ground, the Earth is no longer the same as it was. So, its gravitational field is not the same as when the apple was still hanging from the tree. If we do not take this change into account, our bookkeeping of quanta will be inexact. In practice, the difference is negligible because the apple is tiny compared with the globe. However, for understanding gravity, the difference is significant. Neglecting it, we erroneously believe the gravitational field to be no substance.

The credo of physics keeps a tight grip on scholars, as one conversation at the annual meeting of the Finnish Physical Society in 2016, with the professor of cosmology Kimmo Kainulainen exemplified. This theoretician at the University of Jyväskylä acknowledged without reservation that kinetic energy is released as heat when the apple hits the ground. Still, he couldn't see that energy was released from the gravitational field *while* the apple was on its way down.

The orthodox physicist, for whom the gravitational field is nothing but curved geometry, reasons that something which is absolutely nothing cannot release anything. Because there is no essence in space-time, there are no events in it either. Thus, an expert on modern physics denies that the falling of an apple is an event, a change, even though acceleration expressly spells a change in motion. No wonder physics, going here and there against our own experience of reality, is inconsistent and incomprehensible.

Moreover, physicists take it for granted that gravity is an attractive force. On Earth, bodies fall vertically, earthward, but horizontally, where Earth's gravity is constant, we experience the gravitation of the whole universe. We sense it as inertia; the reaction of the universal gravitation to acceleration and deceleration are opposite. While contemporary physics makes a distinction between gravity and inertia, gravitational mass and inertial mass could not possibly be different from one another because both gravity and inertia display the body's coupling with the same vacuum.

Newton, too, understood gravitation as a universal phenomenon. It applies to an apple, the Moon, and anything else alike. He used Galileo's method to mathematize an apple's fall into the law of gravity. The thermodynamic theory gives this equation for acceleration and the well-known condition for a balance where the flow of quanta ceases.[4,38,40,54] In contrast, the truth of an effective theory, as a mathematical model of data, is not guaranteed to the same degree as the theory founded on our first-hand experience.

WHY ARE GALAXIES NOT EXPANDING?

The cosmos expands as galaxies coalesce.

On the one hand, matter aggregates into galaxies and neighboring galaxies merge; on the other, distant galaxies recede.[23,55] These opposing flows balance each other at the edge of a galaxy group, where the efflux of quanta to the universe and the influx from the universe tally. Based on astronomical observations, this boundary grouping the Milky Way, Andromeda, and their small neighboring galaxies lies about four million light years from us. As Hubble's law states, all objects beyond it are moving away from us[56] because the quanta from the rest of the universe are pouring in between. In turn, all objects within this realm are moving toward us because the quanta between them are departing into the rest of the universe.

As the universe expands, nearby galaxies ever further away will become increasingly closer to one another in relation to the ever-larger and ever-thinner universe. So they begin to move toward each other, and the network of galaxies prunes.

Cosmic rays

The redshifted light implies that faraway galaxies are receding, while the blue-shifted light indicates that nearby galaxies are approaching. Likewise, the down-shifted velocities of some cosmic particles indicate that their sources are receding, while the up-shifted velocities of others show that their sources are approaching. However, unlike light with a constant speed, the faster a galaxy is moving away, the faster a particle therefrom must depart for it to end up in our detector.[37]

Theodor Wulf discovered these cosmic particles in 1909, finding their radiation stronger at the top of the Eiffel Tower than on the ground.[57] People refused to believe this Jesuit priest, who had majored in physics, deeming it ludicrous that we could be continuously showered with particles of no apparent origin. Nevertheless, the journal *Physikalische Zeitschrift*, only two decades old then, did not yet have much prestige to lose by publishing the baffling news.

Three years later, Victor Hess, an Austrian-American physicist, ascended by balloon to a height of over five kilometers proving that the particles come from somewhere far out in space beyond the solar system. As he conducted the measurement during an eclipse, the Sun could not have been the culprit.[58]

Even though detection methods have improved enormously since then, it is still a matter of dispute from where the particles originate.

I got interested in cosmic rays for the same reason as I did in other findings exhibiting the *Grand Regularity*: could the thermodynamic theory explain their spectrum? Like many other phenomena, the flux follows a power law. It covers a staggering ten orders of magnitude from slow to high-speed particles, including a few bends like joints in a leg. What these characteristic bends tell us is subject to debate.

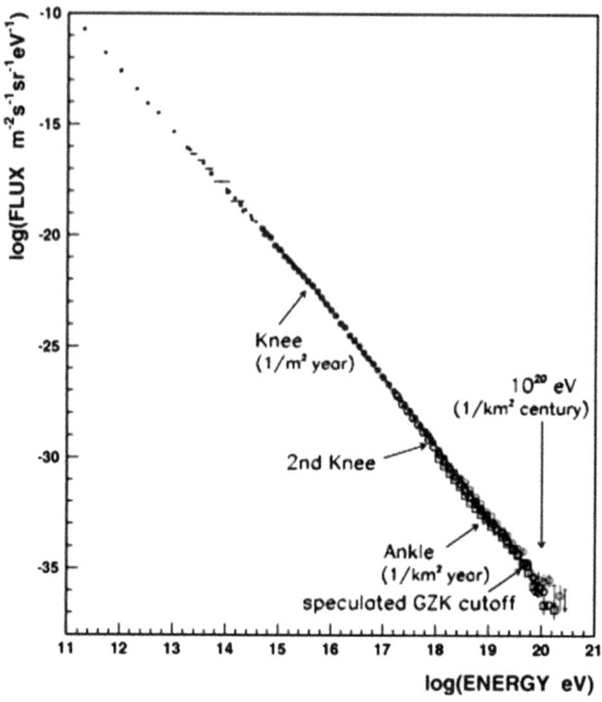

The number of particles coming from space decreases with increasing energy but not precisely in a straight line.[60] As a rule, the particles whose energy is lower than the prominent bend, the knee, come from far away in the universe. Conversely, the particles with higher energy arrive from the Milky Way and neighboring galaxies.

It is generally believed that the high-energy particles above the spectrum's most prominent bend, the knee, come from somewhere far away.[59] From time to time, a particle darts down from the sky so fast that it could smack you as hard as a baseball struck by Yogi Berra. Fortunately, you cannot be hurt because the atmosphere slows the particle down. Copious low-speed particles, in turn, are thought to stem from sources in the Milky Way.

Reality proved to be just the opposite.[37] The abundant low-energy particles come from the numerous receding galaxies, whereas the few high-energy particles originate from nearby sources, mainly the Milky Way and Andromeda. Analyzing the data according to the thermodynamic theory, I found the dividing line, the knee, to correspond to a distance of about four million light years. Beyond it, the expansion begins, as indicated by optical observations.[56] At balance, where forces $F = Mc^2/R = c^4/G$ tally, the size of the local realm, R_o, relates to the mass of the Local Group, M_o, as the size of the universe, R, to the total mass, M, consistently with space emerging from matter.

Most of the high-energy particles originating from nearby sources slow down on their way.[61] Their wakes in the vacuum give rise to electrons and positrons, i.e., quantized vortices. Even faster particles slow down by leaving behind mesons, quark-antiquark particles, i.e., quantized waves (Appendix B).[45] Due to these pair production processes, the flux of cosmic particles drops off steeply in the high-energy end. Thus, like the spectrum of light, the spectrum of particles discloses the universe in evolution.

On track

The incomplete image of gravitation as a solely attractive force arose long ago from the vagueness of the void. But Newton was on track. He thought that the ether might be granular. Small grains would fit next to a body, whereas large ones would have to lie further out. So, instead of action at a distance, leveling out these density differences would display itself as gravitation.[62] However, this brilliant insight was eclipsed by ether's ambiguous essence. Neither was the genius willing to guess it, declaring, "I contrive no hypotheses." (*Hypotheses non fingo*.)

Two hundred years later, Maxwell understood that the void is, in its essence, light. Energy is carried through space in the form of light.[26] Gravitation and electromagnetism only give the impression of action at a distance because the vacuum reacts to changes without delay, being present everywhere. Even so, maintaining balance takes time, as quanta propagate from here to far away and from far away to here.

Newton was a trailblazer lacking leads, whether right or wrong, relying on his own intuition. Einstein, too, opened a trail of his own. In the thrall of excitement, his devotees hardly thought about where they were heading. Thoughtless thinking is not inconsequential, for thinking changes the mind. Yogi Berra hits the home run in expressing this: "You've got to be very careful if you don't know where you are going because you might not get there."[63a] When trying to find a way out of a dead end, have physicists even ended up being out of their minds imagining wormholes, multiverse, action at a distance, and superluminal expansion?

Newton noticed, just as Faraday did over a century later, that gravitation and electromagnetism are alike, but only now do we understand that unity. Photons mediate both forces. Gravitation is about bodies coupled with the paired-photon flows, leveling off the vacuum density differences. Electromagnetism is about charges and magnets coupled with the unpaired-photon flows along the vacuum gradients and vortices. By all accounts, the vacuum is neither virtual particles nor space-time geometry but substance, the ether.

In the teaching of electromagnetism, as of mechanics, the vacuum is framed as implicit and inexplicable. There is scarcely a textbook not promoting the falsity that opposite electric charges attract each other. Namely, it is not the charges themselves but the surroundings that set the charges in motion. For instance, when table salt dissolves in water, the positive sodium cations move *away from* the negative chlorine anions as water comes between them. That is how the imbalance diminishes in this case. Conversely, opposite charges move toward each other in the vacuum because that is how the vacuum approaches the state of balance without electric fields. The vacuum thus causes bodies to be, by and large, net neutral.

Similarly, the opposite poles of bar magnets attract each other because the void minimizes the overall magnetic field, i.e., vorticity. And irregular celestial bodies tend to become round because the vacuum is thereafter as flat as possible. Thus, the void's pursuit of balance is the mother of motion.

Faraday's idea that a body generates forces in its vicinity was once deemed eerie. Nowadays, we are already comfortable with the field concept but have yet to get used to the fact that fields are quanta of light. It may seem unbelievable that the substances of the gravitational and electromagnetic fields would still, in this day and age, be unclear to physicists. Yet, they are.[64] Equating reality with the instrumentalism of modern physics merely impedes the long-overdue sobering up of the purview.

The profound problems of physics seem hard as granite. Many sharp minds have been blunted in scrutinizing them. So, it is high time to work things out differently, right from the beginning. There is no shame in simple reasoning, as long as the reasoning is correct. Then one is in good company. As Steven Weinberg, one of the most quoted physicists of our time, puts it: "Our job in physics is to see things simply, to understand a great many complicated phenomena in a unified way, in terms of a few simple principles."[65]

WHAT ARE THE FUNDAMENTAL FORCES?

Forces display the universe's structures.

Physicists and philosophers wonder why the four fundamental forces, gravity, electromagnetism, and the weak and strong nuclear forces, have the strengths they do. This puzzle is known as the hierarchy problem.

Let us inspect this matter through the lens of time. With the universe's expansion, energy differences diminish and eventually disappear altogether. At the end of time, there will be no matter left, as all quanta will have become part of the vacuum. Then gravity will be all gone, as there will be no matter to cause density differences in the void. Electromagnetism will likewise cease to exist, as there will be no electrical charges. There will be no weak force either, as there will be none of its carrier particles, W^-, W^+, and Z bosons, and there will be no strong force, as there will be no quarks for gluons to link together.

As the fundamental forces vanish with time, they may not have been fixed at the beginning of time, either, but display universal structures, from the largest to the smallest, rather than dictating them. The laws of nature are thus contingent on the law of causality. Laughlin awaits this unity of Nature: "I am increasingly persuaded that all physical law we know about has collective origins, not just some of it."[66a]

Paul Dirac is best known for discovering antimatter in his equations, but in the 1930s, the British physicist also contemplated the relationship between structures and forces. He thought that the immense ratio between the electromagnetic and gravitational force reflects the size ratio between the electron and the universe. But this large number hypothesis[67] was not taken seriously. The structures of particles were not known down to the accuracy of one quantum, the expansion of the universe was not understood as a process in which the sparse vacuum stems from dense matter, and the quantum was not recognized as the common denominator of everything. The large number hypothesis thus remained a hypothesis.

Eddington was also enthused about the ratios of natural constants. As the story goes, Samuel Goudsmit, having listened to Eddington's passionate lecture about the meaning of the fine-structure constant, asked his elderly colleague Hans Kramers: "Do all physicists go off on crazy tangents when they grow old? I am afraid." Kramers assured him, "No Sam, you don't have to be scared. A genius like Eddington may perhaps go nuts, but a fellow like you just gets dumber and dumber."[68] As a matter of fact, Goudsmit, together with George Uhlenbeck, both Dutch-American physicists, were well-known for showing that the electron has a spin. The Dutch theoretician Kramers was well-known, too, for his work with Niels Bohr.

Now we understand that gravitation is feeble compared with electromagnetism because the universe is huge and hence low in density compared with the tightly coiled electron. Similarly, the fine-structure constant, the ratio of electromagnetism to the strong force, expresses the ratio of the electron torus to the neutrino loop (Appendix G).[39] In general, curvature means force, as the famous mathematician Carl Friedrich Gauss had already figured out in the early 1800s.

NEITHER CHANCE NOR DESIGN

The fundamental forces are outcomes of the evolving universe rather than of chance or design: the speed of light relates to the structure of the vacuum, the fine-structure constant to the torus of the electron, and Planck's constant to the photon wavelet. None of the natural constants are thought to be true constants either, let alone fine-tuned quantities.[69] The anthropic principle, proposed by the Australian theorist Brandon Carter,[70] is only apparent. It is not so that we would not be observing the universe unless it had such properties that enabled us to evolve into existence. The data relating to us disclose the same patterns as any other data. This *Grand Regularity* speaks for the Copernican principle, whereas nothing speaks especially about us.

Jim Baggott hurls some well-deserved flak at modern physics. He believes that the strengths of the fundamental forces, the values of the natural constants, and the number of spatial dimensions are not fine-tuned.[71] We are just not aware of the physics that determines them.

At one time, the geocentric model contained an analogous mystery: why do the Sun's epicycles happen to be vanishingly small? Nowadays, it would be intellectual dishonesty to posit that the extremely low density of dark energy precisely dictates the universe to be as it is. How could precision to tens of decimal places be possible if the density could be something other than what it is?

By describing everything in the same way, thermodynamic theory accepts the cosmological challenge.[72] The precision in the fine-tuning is no coincidence, as there is nothing without reason, *nihil sine ratione*.[73] Every fundamental element is in the universe's bookkeeping. Hence the accuracy is over 120 decimal places ([Appendix G]).[33]

There is a consensus about the importance of the fine-tuning problem; the disagreement is about the answer.[74] As Laughlin anticipates, "The fact that it [the cosmological constant] is so small tells us that gravity and the relativistic matter [the vacuum] pervading the universe are fundamentally related in some mysterious way that is not yet understood...."[66c] we now understand that the relativistic matter, the void, the universal gravitational field maintains balance with all matter in transformations of quanta. So, the geometry of the universe is a dead issue. The energy density of free space could not be different from the average energy density of matter, the critical density.

Granted, if the electron's fine-structure constant deviated from its value only in the sixth decimal place, as it does for the muon, life would not exist as we know it. The muon is no surrogate for the electron. It cannot withstand the test of time but transforms into an electron.

The anthropic principle is teleological, all right, but the goal of cosmic evolution is not our existence. We are not here to perceive the universe due to the natural constants' being what they are, but rather the reason for our existence and the reason for the values of the natural constants is the same: the least-time quest for balance.

This least-time imperative materializes in the beings of any given time, including us. The reactions of the current era require contemporary particles; others are left in the margins. We cannot find exotic particles except under special conditions; elsewhere, they are unviable. At the beginning of time, the conditions were different, and in the future, they will be different again.[75a]

At the dawn of the modern era, we abandoned the anthropocentric view as we came to understand our place in the cosmos. So now, we should discard the anthropic principle as we understand that everything ultimately comprises photons.

Smolin thinks that not only the constants but also the natural laws might evolve as the universe ages.[75b] Are there any signs in the cosmos that even the principle of least time has shifted through time? Is there anything solid upon which to establish our worldview?

Science will not advance if we reject a thesis as absurd just for seeming absurd to us because our tenets influence how we interpret observations. So, we cannot simply disregard critics of modern physics and opponents of evolutionary theory by referring to data. Our stance might change. It has changed before. Nevertheless, the scientific community snubs qualms, sometimes senselessly so. When this coterie of kindred spirits looks upon counterarguments as amateurish or pseudoscience, the watchdog of orthodoxy is on duty. It barks at anyone who veers away from the 'right' track.

WHY IS THE DISTANT HORIZON UNIFORM?

Same cause, same consequence.

The temperature of the pitch-black sky is virtually the same in any direction. The most distant galaxies, too, are scattered nearly evenly in all directions. Cosmologists are baffled about this *uniformity*. How could the temperature of space and density of matter on the opposing sides of the universe possibly have become equal when there is not enough time for light to traverse from end to end? This puzzle is called the horizon problem.

The renowned theoretical physicist and cosmologist Alan Guth encountered the horizon problem as a young researcher when listening to Robert Dicke lecturing in 1978. A couple of years later, Guth proposed cosmic inflation as a solution.[76] According to this scenario, the very earliest universe ballooned faster than light for a very short time. So objects that were initially close to one another, and hence in the thermodynamic balance, maintained correspondence even when flung far apart. In this way, the early differences in temperature and density would have been significantly smoothed away over the expansion, although not altogether.

Indeed, the cosmic background radiation temperature deviates from the 2.725 K average by only about ten in a million. However, is cosmic inflation a tenable explanation of the large-scale uniformity? Does it qualify as an integral part of our worldview?

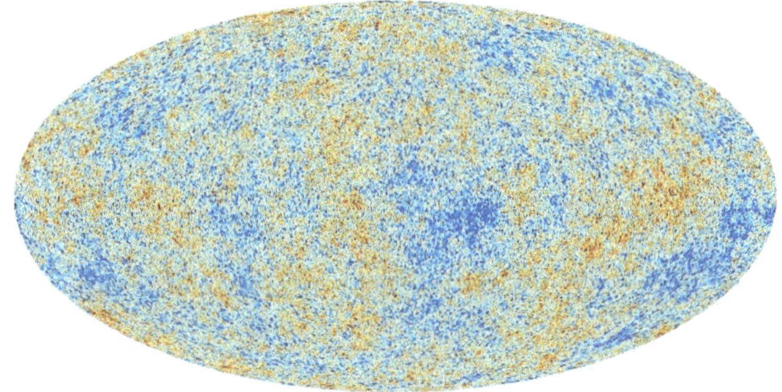

The deep sky temperature is almost the same in every direction. The variation (from blue to red) is only a few ten-millionth of a degree. The high homogeneity follows from the universal principle of least time: the larger the difference, the faster it decreases. (Image: ESA)

Inflation theory has been criticized, among others, by Paul Steinhardt. The professor at Princeton University initially fostered the inflation theory but has since backtracked.[77] Inflation does not solve the horizon problem. On the contrary, ephemeral variations, known as quantum fluctuations, are supposed to have seeded not only galaxies but also multiple universes. If matter is distributed among them, the outcome would hardly be homogenous. Besides, is inflation a theory at all? How could it be falsified when its parameters allow for not one but many different models? Why bother to tune parameters toward observations, as the fit does not even explain the phenomenon? Although this criticism is on point, it is all the more important to understand the reason for the large-scale homogeneity.

The idea that the universe was born out of nothing arose in 1969 when the renowned British cosmologist Dennis Sciama was giving a talk. The whole audience was amused when all of sudden, young American physicist Edward Tryon blurted out loud an insight: "Maybe the universe is a vacuum fluctuation?"[78] Tryon's inspiration came from Sciama's just having concluded that the universe's total energy is zero, with the proviso that its positive mass-energy is equal to its negative gravitational potential energy. Because the vacuum energy is equal in magnitude, but opposite in sign, to the gravitational energy of the whole universe, perhaps the two forms emerged from mere nothingness simultaneously. Lawrence M. Krauss, an American-Canadian theoretical physicist, elaborated on this unfathomable theme in his book *A Universe from Nothing* (2012).

Whatever the truth, quantum fluctuations and cosmic inflation are beyond our own experience. We cannot comprehend what the jargon is all about. So, back to square one. After all, it is reality that we are here to grasp.

Harvesting grapes

Even if two things manifest themselves concurrently, like the same temperature at two far-apart locations in the distant sky, it does not mean that one follows from the other. Likewise, a headache is not caused by wearing shoes during sleep. Neither are grapes ripe in one bunch because they are ripe in another. As they say, correlation is not causation.

The uniformity across the horizon follows naturally and necessarily from the universal quest for balance in the least time. Namely, Newton's second law of motion says that the bigger a force, the faster the change in motion. This means, for instance, that the higher the temperature difference, the faster the rate of cooling. Thus, there will be only minute temperature differences in due course, regardless of how massive the differences were in the beginning. When the root cause is the same, the consequences manifest alike even without a causal connection. Grapes, independent of each other, are ripe at the same time because of a common cause, namely sunshine.

Although the early differences in temperature and density have by now nearly flattened out, the process is still ongoing. A large difference in energy decreases rapidly and a small difference evens out slowly. For example, the most massive stars shine the brightest and for the shortest time, well below 100 million years; small stars can glow over 100 billion years. Thus, over time, there will be only small differences — as is observed. Ultimately, everything will become the same, regardless of what the earliest universe was like.[23,24]

The inflation theory devalues physics, the bedrock of science. By its special-purpose proclivity, the superluminal model stretches the limits of normal science, thereby exposing the graveness of the trouble with contemporary physics.

In the past, too, fiddling around with the rules preceded a revision of the worldview.[11] Still, in the 1750s, honorable professors thought electricity was a fluid to be even bottled. Finally, Benjamin Franklin grasped that the so-called Leyden jar, a device for storing electricity, is, in fact, a battery. Science works that way. No matter how fiercely solutions are sought, only a new stance sheds light. Even then, many intellectuals are blind to the long-sought explanation parading in front of their eyes, for it is too simple, unexpected.

GENESIS

In the 1940s, George Gamow, a one-time student of Alexander Friedmann, understood that the cold sky is a relic of the hot, early universe.[79] By now, the cosmic furnace has cooled down so much that the nascent short-wavelength photons have aged by lengthening to the micrometer-sized rays of feeble heat. Gamow's view on the evolution of the universe started to gain support when Arno Penzias and Robert Wilson detected microwave background radiation in 1965.[80] The physicist and the astronomer recorded one hundred times more noise at radio frequency in every direction than was expected. Soon Penzias and Wilson realized that their large antenna, the 'big ear,' had heard the 'echo' of the Big Bang. A matching result had been calculated only a little earlier by the physicists Robert Dicke, Jim Peebles, and David Wilkinson. A discovery like this one is often characterized by initial confusion and final comprehension.

Given that the universe is evolving toward a balance in the least time, the cosmic temperature map should display uniqueness rather than perfect uniformity. We know this historical signature from familiar sequences of events. As an illustration, when a cake is divided as quickly as possible into ever smaller pieces of equal size, we will not slice it randomly. First, we split the cake in half, then we halve the halves, and so on. The division at hand depends on all previous divisions so that at any given moment, the largest piece will be split next. Thus, the slices of a historical process are not in random orientations relative to each other. Neither are they identical but unique.

By the same principle, the early universe did not cool arbitrarily but always fastest at its hottest site. This history is still visible in the temperature map of the deep sky.

The latest map of the cosmic microwave background has been compiled by the research team at the University of Helsinki, Department of Physics, as part of international cooperation.[81] The temperature of space is in every direction, almost, but not exactly the same. The variation is only a few ten-millionths of a degree. Moreover, one-half of the sky is slightly warmer than the other. The halves of halves and even their halves differ from each other in the same way. This regularity, dubbed the 'axis of evil,'[82] is not a mere coincidence to be ignored but rather noted and explained.

Cosmologists are also puzzled about an unusually large cold spot in the cosmic microwave background, visible in the southern sky.[83] Although it can be argued to fit within the random variation of the normal distribution, randomness is no explanation. Instead, the anomaly could be a harbinger of understanding to come.

The axis of evil and the cold spot are characteristics of causality, not coincidences. Halves and their halves of the temperature map bespeak the chain of least-time events, the 'cutting up the cake into pieces.' The pairs of galaxies and pairs of galaxy clusters found at all scales[84] result from the same least-time rather than random decimation of matter into the void.[23,24] From this perspective, the more the fictional randomness differs from the actual uniqueness, the more the current statistical model deviates from the data.

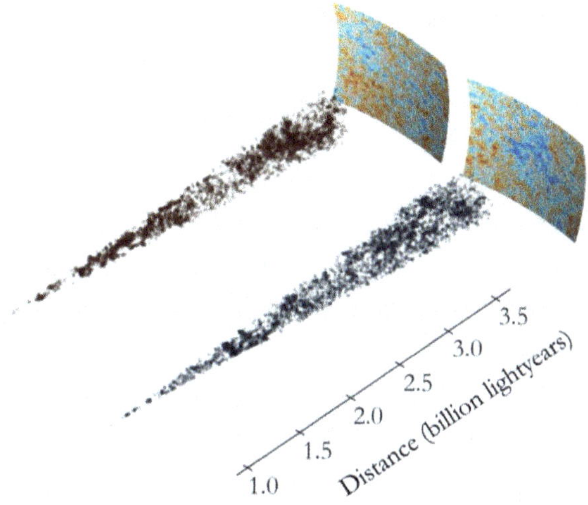

Galaxy groups are distributed in the direction of the cold spot (in the foreground, gray) in the same way as in an ordinary direction (in the background, brown).[85] The cold spot was thus cold before the galaxies emerged.

The cold spot must have formed earlier than galaxies because the number of galaxies does not differ in its direction compared to other directions.[86] The remnant from an early epoch has experienced the same evolution as the rest of the universe. As a result, the temperature map is as spotty in the spot as elsewhere. The least-time principle thus explains the large-scale uniformity and its tiny characteristic variation.

THE ROOT OF MATTER

We see with our own eyes that the Sun and other stars are transforming matter into light. But what existed before ordinary matter formed?

In the late 1960s, Zweig sketched the root of matter to his 10-year-old son: "In the beginning, there was the quark-gluon plasma, and as the universe expanded and cooled, clusters of three quarks condensed from the plasma to form protons and neutrons."[87] The boy immediately asked, "How do we know that the number of quarks in the universe was divisible by three?" When recalling this episode in 2013, Zweig admitted having no answer to the question.

The sharp-witted boy queried, in fact, why the quarks are *just* the way they are. Why is the up quark's charge exactly $^2/_3$ that of the positron and the down quark's precisely $^1/_3$ that of the electron? Based on the previous chapter, these specific fractions of the electron and positron tori can close via gluons to form stable particles that make up the stars and just about everything else.

Theaetetus, a contemporary of Plato, proved that there are only five regular polyhedrons. The simplest, the tetrahedron, turned out to be the viable geometry in the forms of the proton and neutron. In evolutionary parlance, energy flows *naturally selected* those structures that made things happen while discarding others. Likewise, we keep useful and do away with useless things. Nothing happens by itself. The forces need their carriers; reactions require their structures.

The primordial substance must have transformed into the void, just as matter does today. The history of this primal period is thus printed in the cosmic microwave background radiation.[23,24] Later, the formation of matter allowed stars to ignite. So, energy differences continued to decrease. Things went on.

A MATTER OF CHOICE

The big question of present-day physics is why and how matter gained supremacy over antimatter.[88] Physicists theorize that the Big Bang produced equal amounts of both, and soon an enormous annihilation took place. By some miracle, there would have been an excess speck of matter from which the whole universe came about.

Physicists see the slight surplus as a sign of an imbalance between matter and antimatter, wondering what might have caused it. Neither the Standard Model of particle physics nor general relativity explains it. No energy difference between matter and antimatter has been detected, either.[89] The masses and magnetic moments of protons and antiprotons are indistinguishable.[90] Still, hopes persist of finding the postulated minute inequivalence between matter and antimatter.[91]

Given that not only findings but also paradigm shifts have previously resolved great riddles, we ought to ask why we think as we do.

Reduction is a standard method in science. However, that approach does not explain the excess of matter over antimatter because it detaches the particle from its environment and history. In those unnatural circumstances, handedness, the chirality standard of substance, plays no role, for there is no hand to shake. The standard of the substance, being either matter or antimatter, is necessary for things to happen, not for stasis. In contrast, when characterizing matter and antimatter experimentally in a 'test tube,' expressly, nothing should happen to spoil the measurement. That is why the victory of matter over antimatter is indecipherable by the reductionist approach.

Now that we know the structures of elementary particles, we realize that the hypothetical massive early annihilation never took place. Even today, the particles contain both the right and left-handed arcs of tori and equal amounts altogether. Feynman's and Wheeler's discussion touches on this matter: "But, Professor," said Feynman, "there aren't as many positrons as electrons," to which Wheeler replied, "Well, maybe they are hidden in the protons or something."[92] This led Feynman to think of the positron as if it were an electron going backward in time. Indeed, the quantum coil stays as it is when its handedness is swapped, coordinates are mirrored, and the direction of time is reversed. In other words, this CPT (Charge, Parity, Time) operation preserves the symmetry of the torus.[45]

Instead, if the direction of time were not reversed, CP symmetry is broken to a small degree in certain weak nuclear reactions,[93] but not in strong ones. This bothers physicists. In quantum chromodynamics, nothing should prevent symmetry from breaking.[94] From the particle structures, we understand that the symmetry cannot be broken because the reversal of time does not change anything in the strong force (Appendix B). Only the reflection of either handedness, i.e., charge C, or coordinates, i.e., parity P, changes the structures of quarks. Therefore, the so-called θ parameter of quantum chromodynamics[95] is not free but must be zero.

According to thermodynamic theory, one of two equivalent options will inevitably win when the choice, one way or the other, paves the way to a balance.[96] That is how the least-time law explains the matter-antimatter unevenness. The victory of matter over antimatter does not allude to an imbalance. On the contrary, handedness serves evolution toward a balance. Had antimatter become the standard, we would call it matter, and we would call the current matter antimatter.

There is very little antimatter for the same reason as there are very few mirror-image forms of natural compounds. Thanks to biochemical standards of handedness, an organism can raise its energy content toward its environment much faster than a mere mixture of metabolites could possibly do.[96] The cellular synthesis proceeds swiftly using standardized molecules like industrial assembly with standardized components. Standards make things happen.

It is trivial to understand the railway track norm. It is useful. When trains run, demand in one region meets supply in another. Energy differences even out. Likewise, life's biochemical standard emerged to attain thermodynamic balance with the environment as quickly as possible. The universal standard of matter removed the imbalance between dense substance and sparse space. Although there is no energy difference between the particle and antiparticle as such, the total difference in the present surroundings, made of matter, is enormous; therefore, antimatter annihilates quickly.

At first, it may seem odd to relate the universal standard to a biochemical one, let alone a human-created norm. But data are alike. There is no distinction between the microscopic and the cosmic or between the animate and the inanimate. Like filaments of galaxies, railroads branch roughly according to the power law.[97] The pattern is the same because the cause is the same. Heedless of our naming things and thereby detaching them from each other, the world is one.

WHAT IS DARK ENERGY?

The expansion of the universe is accelerating – so they say.

The explosion of a star is a brilliant event. A supernova shines for about ten days brighter than the whole galaxy. 'Standard candles' are supernovae that flash equally brightly. Their flares have been recorded systematically since the early 1990s. Before the turn of the millennium, it transpired that the data did not fit into the Big Bang model of cosmology, where the density is critical and the cosmological constant is zero. Henceforth it was conjectured that the universe is expanding at an accelerating rate. However, its cause, dark energy, is a mystery.[98]

While cosmologists grope around in the dark, the European Space Agency plans to launch 2022 the Euclid Space Telescope in orbit around the Sun.[99] The aim is to collect data from farther than 10 billion light years away. It is hoped that the redshifts of nearly one hundred million galaxies will disclose whether there truly is dark energy or whether general relativity does not hold at very long distances.

While waiting for this big data, we can perform a small analysis using the supernova data collected so far. Rather than tailoring the standard model to fit the data, let us examine what happens to photons spreading from an exploding star into the expanding universe.

We see the light of a receding supernova getting dimmer and redder. Much the same way the siren sound of an ambulance fades in intensity and lowers in pitch. The detected amount of light is inversely proportional to the square of the (optical) distance to the exploded star. As light is in transit from the past to the present, its color shifts not only due to the speed of recession but also due to the weakening gravity of the expanding universe. This account is not the 'tired light' conjecture[100] but the gravitational effect on the color of light, also well-known from general relativity.[101]

When the intensity of light depends on *two* factors, the distance to the receding star and the expansion of space, the data do not follow *one* straight line on the log-log scale. Based on the cosmological principle's metric, the Standard Model's expectation departs more and more from the data with increasing distance and time. By contrast, thermodynamic theory leaves no room for dark energy by agreeing with the data.[4] There is no need for dark energy as there is no free variable to be tuned or tweaked. The age of the universe, about 14 billion years, is taken as given (Appendix E).

The simple explanation of the supernovae data follows the plain understanding that the universe is not expanding since a one-time explosion, the Big Bang, but due to all ongoing events. Time flows and energy differences diminish because stars, black holes, and other mechanisms transform matter into the void. The universe as a whole evolves toward balance in the least time, for its systems do so.

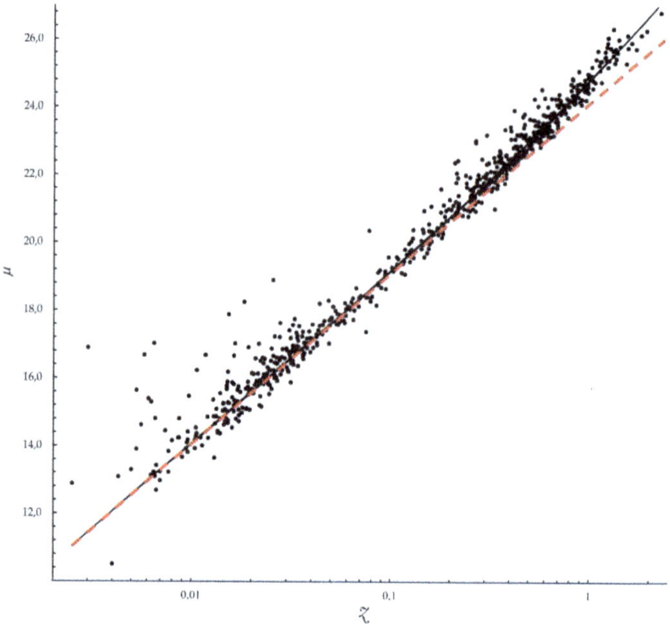

The curve calculated by the least-time principle (solid line) closely follows the brightness of Type Ia supernova, the distance modulus μ, versus redshift z (points).[102] In the ΛCDM model, the dark energy parameter accounts for the data's curving away from the straight line (dashed line).

The rate of expansion, revealed by the redshifts of the most distant galaxies, accumulates from all processes in the universe. Then the rate, known as the Hubble parameter, $H = 1/t$, could not be other than the inverse of the universe's age, t. Distant galaxies are receding because quanta move from matter to the void. As the matter is running out, the expansion is slowing down; H goes down as time t goes on. Indeed, our neighborhood of the cosmos, being the oldest, has the lowest average density of substance.[103]

Time is a thorn in the side of modern physics. The passing of all time, i.e., cosmic expansion, is particularly problematic to comprehend. For example, physicists have a hard time understanding why the hydrogen gas of the early universe seems colder in the light of the first stars than expected.[104]

While general relativity models events as changes in space-time geometry, thermodynamic theory explains events as flows of quanta. The difference between the model and reality becomes more and more evident across time. That is why standard cosmology gives different values for the Hubble parameter from observations of the oldest history of the universe and those of more recent activity.[105]

How could general relativity be wrong? After all, it explains so much. The model indeed concurs with many data, but by what means. The more a mathematical model is modified after the fact, the better it fits the data, and the less it explains what is going on.[106]

It is not necessary to account for the expansion with dark energy. The fudge factor embodies nothing. So there is no need to search for such stuff either. By the least-time principle, we have real ingredients for a realistic view. Even with them, there is a lot to explain. For example, what was going on in the past when the nascent universe was brighter than all the galaxies altogether.[107]

A COMPETENT FACULTY

The temperature variations of the cosmic microwave background (CMB) and the spatial distribution of the galaxies have also been taken as evidence of dark energy.[108] Is the reasoning convincing enough when the homogeneity of CMB and its anomalies are arcane and gravity is fundamentally unknown? Why was the standard cosmology extended with dark energy and other parameters,[109] rather than admitting its demise? The history of science gives some insight into this.

In 1905 Einstein presented in *Annalen der Physik* the postulates of special relativity: the laws of physics are the same, and the speed of light in the vacuum is the same in every coordinate system that moves steadily.[110] Einstein did not explain these claims, merely stated them, for the postulates of a theory are inexplicable by the theory itself.

The postulates of special relativity led to the mathematical transformations from a stationary to a moving coordinate system. Lorentz had found them ten years before Einstein gave the transformations the new meaning of relativity. However, Lorentz did not approve of the meaning, considering Einstein's idea of no universal frame of reference unreasonable.[111] Instead, Lorentz figured that each body moves relative to all other bodies, the universe. Every object in the universe is unique. If this were not the case, the objects would be indistinguishable, as Leibniz' law states.[72]

Indeed, the Earth moves along with the Milky Way relative to the whole of the universe. The firmament temperature is slightly higher in front than it is behind. Likewise, the air is a little denser in front of the windscreen than behind the rear window when driving forward. This pressure difference is why dirt ends up on your rear window.

Einstein's special relativity and Maxwell's theory of electromagnetism are models of the pure void. The quanta of the ideal vacuum propagate in straight lines because there are no bodies that would curve their paths. The geometry of free space is flat, Euclidean. But to cope with matter in the universe, Einstein generalized relativity. To work out the mathematical model known as space-time, he postulated that it is impossible to distinguish between the gravitational field and steady acceleration. However, Einstein did not explain why gravity and acceleration have the same effect. He simply said that that is how it is. After nearly ten years of work, Einstein presented the first version of general relativity in 1915.

As Riemann had formulated non-Euclidean geometry in the mid-1800s, his own take on gravity is instructive. He was struck that Newton argued for some substance mediating gravity instead of approving action at a distance,[112] as generally thought. Riemann emphasized this contradiction to his colleague Richard Dedekind, the German mathematician, by repeating these words from a letter Newton penned: "That gravity should be innate, inherent and essential to matter, so that one body may act upon another at a distance, through a vacuum, without the mediation of... from one to another, is to me so great an absurdity that I believe no man, who has in philosophical matters a competent faculty of thinking, can ever fall into it."[62] Einstein did.[xi]

BEYOND THE SHADOW OF A DOUBT

What could the old scholars possibly contribute to modern empirical science? After all, has general relativity not been confirmed beyond any doubt by quantitative data? Arguing in this manner, we would be committing the fallacy of taking observations as though they were independent of interpretation. Since neither Newton nor Riemann

---

[xi] Einstein, in fact, reasoned like Newton that the vacuum must be some substance in contrast to his theory which is sheer geometry.[30]

approved of action at a distance, neither should we regard general relativity as the theory of the universe but a mathematical model of gravitation. The calculations match the data, at least, by curve fitting but do not explain why an apple is falling and a distant galaxy is fleeing.

Physicists acknowledge this. Gravitation is under the magnifying glass. But is the examination thorough enough when insisting on conflating general relativity with quantum theory? Why should the unification of models be the priority? Instead, a comprehensive view opens up from a correct viewpoint. Einstein mastered this axiomatic method. He saw time as relative and built his model. We saw time as the quantum period, and the thermodynamics of time unfolded.

Einstein crystallized his insight into postulates that led to mathematical models. Boltzmann did the same. He, in turn, postulated that everything is evolving toward an increasingly more probable world. This points to the thermodynamics of time,[113] as causality is inherent in it. By contrast, Einstein's postulates are acausal. Time is only a parameter without substance in special and general relativity.

Mass defines the curvature of space-time but not unambiguously. A term can be added to the Einstein field equations without violating the conservation of energy. The term's coefficient, the cosmological constant,[6] relates to the expansion of the universe, but general relativity does not define its numerical value. Thus, the observations do not substantiate dark energy; instead, the model asks for dark energy to make the data add up.[114]

We should admit that the standard model of cosmology has been appended to rather than challenged. While it is human to believe in what everyone else believes, humanity grows by recognizing facts and standing against social pressure.

Time is not in the postulate of general relativity, yet the universe is evolving. As given by Hubble's law, the expansion gives rise to cosmic acceleration.[115] The effect is minute, below a hundred billionth of the free-fall acceleration on Earth,[116] therefore, visible only at astronomical distances.

General relativity fails with time quite visibly. Consider an everyday experience that an object appears smaller the further away it lies. Thermodynamic theory tells us the same about the angular diameter distance.[4] However, in general relativity, the same holds only as long as the body is no further away than halfway to the beginning of the universe. But[117] beyond, according to Einstein's theory, the object would look ever larger.[118] Weird, right? Shouldn't the very earliest universe appear in every direction as a tiny spot where we see nothing at all? The nascent cosmos is receding at the speed of light, as the expansion of everything is fundamentally only the lengthening of photons.

"We cannot solve our problems with the same thinking we used when we created them," is often ascribed to Einstein, but the authorship is uncertain. All the same, thinking different is science at its best.

Off course

Kepler thought that his worst mistake was to subscribe to the belief, propped up by all philosophers, that the orbit of a planet is a perfect circle.[119] An instituted scientific conviction may seem so sure that we hardly think how it ever came to power. The same applies to other walks of life. When I was a little kid, Urho Kekkonen, who ended up ruling for over 25 years, was not just the president of Finland but *The* President.

General relativity established itself as *the* theory of the cosmos so long ago that we hardly consider anymore whether physics could have developed otherwise. For a scientist, such a thought is preposterous. Even when having groped in the dark for a while, measurements would eventually have led us to the light. We could not have been off course over a hundred years, could we?

At the beginning of the last century, observations of distant galaxies questioned the worldview of the stationary cosmos. What does the redshift mean? When Hubble presented his empirical law, its interpretation by Friedmann[8] and Lemaître[10] was already there. The color of a faraway nebula is shifted to red because the galaxy is receding. For its part, the cause of the expansion was not much of an issue, naming it the Big Bang blast-off.

It was not until the end of the last century, when the observations extended so far back in time toward the beginning of the universe, that the cosmological model failed to agree with the data. It was nevertheless rescued by adding the dark energy parameter. There were no serious reasons to doubt the expansion itself. Only the acceleration was contrary to the expected deceleration.

Explaining the redshift otherwise than by the expansion[100] had been abandoned decades ago.[120] For example, Fritz Zwicky suggested in 1929 that light, on its way from the far-off but supposedly static universe, had lost energy and shifted to red. But shortly afterward, the ingenious astronomer himself concluded that the light could not lose its energy, at least not by collisions. Then distant objects would appear blurred, whereas, in fact, image sharpness hinges only on the optics.[121] There was thus no incentive to look for the cause of cosmic evolution.

In the early 2010s, Dr. Tuomo Suntola cued me in on redshift measurements. The multidisciplinary physicist awarded the Millennium Technology Prize is internationally recognized for his invention of atomic layer deposition. In his book *The Dynamic Universe* (2009), Suntola adopts a critical stance toward quantum mechanics and general relativity.[122] For instance, he has interpreted supernova redshifts without dark energy.

When gravity changes, the color of light shifts, as the American physicists Robert Pound and Glen Rebka proved in 1959. In their experiment, a photon climbed a good 20 meters (yards) from the basement of Harvard University's Jefferson Lab up to the attic.[123] The light shifted slightly toward red because Earth's gravity in the attic is slightly weaker than in the basement. Had it not shifted, the difference in energy between the photon and its environment would have changed without a cause.

The gravity of galaxies, like that of the Earth, causes a redshift. It was measured in 2011 by Radek Wojtak and his group at the Niels Bohr Institute. The light coming from the center of a galaxy group is, on average, slightly redder than the light emitted from the rims.[124] For the same reason, light shifts to red on its way from the early dense cosmos to the present sparseness. Hence, the shift is indicative of the velocity of the receding galaxy and the decreasing gravity of the expanding universe. So you see, the theory of time explains the supernova data without resorting to dark energy.[4] The universe is not accelerating but decelerating as it expands.

Michelson and Morley carried out their experiment in 1887, concluding there is no ether.[125] What would Maxwell, who died eight years before, have said about the interpretation? Having established the connection between the ether and light, would he have pointed out that the void itself is light? Neither speculations nor the authorities are of help. We had better understand the matter by ourselves.

It is time to face the facts. Cosmologists are looking for something nonexistent in space, whereas they should be looking for a theory about the substance of space. It is high time to unknot the tangled thoughts and burst the bubble of modern physics.

WHAT IS DARK MATTER?

There is no need for the unknown when the known suffices.

In the 1960s, the American astronomers Vera Rubin and Kent Ford noted that the orbital velocities of stars and gas clouds *increased* from the galaxy center toward the fringes.[126] In contrast, the orbital velocities of planets *decrease* from the Sun outward. For instance, a year on Mars is almost twice the length of Earth's year. Thus, it was posited on the law of gravity that the surplus orbital motion of galaxies must be due to additional yet unseen matter. Dark matter must mostly reside far from the center. A supermassive black hole in the galactic core cannot explain the high orbital velocities at galactic rims.

Zwicky was already offering the dark matter hypothesis in 1933 to account for the fact that galaxies move relative to one another at great speeds.[127] Our neighboring galaxy, Andromeda, is approaching us ca. 110 km/s. The high-speed attraction between galaxies calls massively for this mysterious form of matter. Moreover, the Andromeda Nebula, the beautiful spiral visible to the naked eye, spins so swiftly, just like the Milky Way, that it must contain much more dark than luminous matter.

All observations of dark matter so far have been only indirect. What could dark matter possibly be since it is so hard to come to grips with?

It was initially suspected that dark matter might be neutrinos, but they are too few and too light to explain the extra gravity. So expressly, weakly interacting massive particles (WIMP) were postulated.[128] They, too, have been trawled for, but nothing has been caught.[129] The axion has also been proposed as a dark matter particle. If it were found, the problem of unbroken CP symmetry tormenting quantum chromodynamics would also be resolved. But there is no trace of the axion.[130] Thus, other candidates are being searched for[131] and other ideas besides particles cultivated. For instance, Erik Verlinde swerves away from conventional thinking by considering gravity as an emergent phenomenon. In this manner, the Dutch theorist accounts for the observations without resorting to dark matter.[132]

Dark matter is difficult to detect, especially when we do not know what it is. Whatever it might be, its impact should vary with the seasons, reasoned physicists Andrzej Drukier, Katherine Freese, and David Spergel in 1986.[133] Data have now been collected for over a decade. Indeed, readings peak in June when the Earth is orbiting parallel to the Sun's motion around the center of the Milky Way. Similarly, readings bottom out in December when the Earth is orbiting antiparallel to the Sun's motion.[134] The Earth seems to be moving through some substance. But the physicists are whistling in the dark as they have no clue what that stuff *is*.

While the dark matter theorists cling to the hope of a big breakthrough in catching the culprit particle, we can make a small point. The galaxy's orbital velocities do not escalate endlessly but seem to settle on a constant value far away from the galactic center. Out there, the orbital velocity depends on the galaxy's mass and a small acceleration, which also shows itself in the velocity dispersion of galaxies.[135] Miraculously, the tiny acceleration equals the speed of light divided by the age of the universe (approximately 10^{-10} m/s^2). Is this universal acceleration just a cosmic coincidence, or does the universe have something to do with the motion of galaxies?

This is a critical question, as coincidence is no explanation. We face something incomprehensible. What is the causal connection between the tiny universal acceleration experienced by the galaxies and the expansion of the universe? It will not go away even if we find more matter in the future, as we probably will with developing detection methods. However, cosmologists believe that not much is missing now that astronomers have got glimpses of hot plasma filaments between galaxies.[136]

LIKE A NATURAL LAW

Dark matter is only needed when the gravity of ordinary matter drops below the universal acceleration. Nonetheless, ordinary matter ordains the distribution of dark matter in a law-like manner.[135] How is this possible? Would not the visible and dark matter have separated from each other, say, in collisions of galaxies and at the birth of stars?

David Merritt regards this as serious but overlooked evidence against the dark matter hypothesis.[106] The professor of physics and astronomy at Rochester University of Technology quotes philosophers of science but relies on astronomical data in reasoning. He deems dark matter and dark energy to be auxiliary hypotheses recruited for the sole purpose of dealing with the awkward situation when the observations falsified the cosmological model of the era. Rather than abandoning the paradigm, it was appended with *ad hoc* factors.[137b,138] This looks like sleight of hand. The history of science is replete with instances where falsifying evidence was discounted, disregarded, or used to modify a predominant paradigm.

The American astronomer Stacy McGaugh is amazed that dark matter is inseparable from detectable substances. The union holds even in the dimmest dwarf galaxies, which seem to comprise almost exclusively dark matter.[139] But then, there is hardly any evidence of dark matter in globular star clusters comparable in size to dwarf galaxies.[140] Moreover, it is unclear why simulations of galaxy formation deviate from the observations[141] when run without the universal acceleration[142] and why dark matter distributes far and wide compared with the ordinary matter within galaxy groups[143] and dwarf galaxies.[144] The standard cosmology is brought into question, as it should be if we are to uphold the credibility of science.

The speeds of stars and gas clouds, from dwarf to massive elliptic galaxies, follow the power law familiar from big data of diverse kinds. There is a lot of matter near the galactic core, and since the acceleration is high, there is no need for dark matter. Farther and farther away, more and more matter is modeled as dark. This is to say, the more we look at the data through standard cosmology, the more it seems that dark matter exists 'for us' and becomes more 'in itself.'[145]

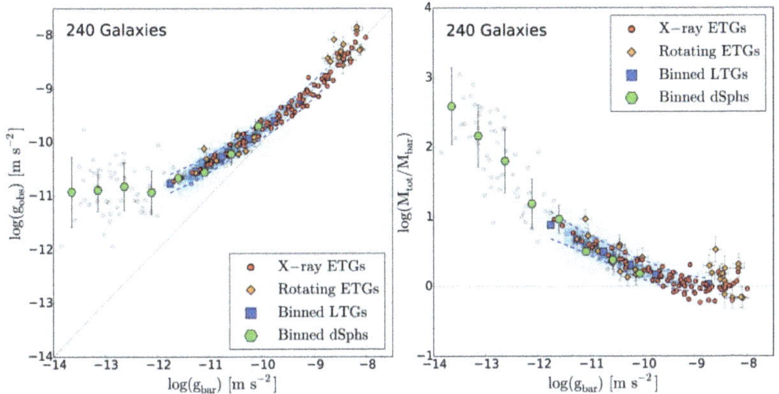

The further away stars and gas clouds orbit the galaxy center, the less they are subject to the galaxy's gravity (left). However, the observed acceleration (g_{obs}) does not decrease in step with the acceleration computed from the visible matter in the galaxy (g_{bar}), as shown by the data's deviation from the straight line. This happens, regardless of the galaxy class, when the acceleration drops below the universal value of 10^{-10} m/s^2. It looks as if the galaxy were holding more and more matter in total (M_{tot}) per visible matter (M_{bar}) when the acceleration calculated from the visible matter (g_{bar}) becomes smaller and smaller (right).[135] The deviation can, however, be explained without dark matter by the gravity of the expanding universe.

Dark matter resembles the epicycles that were added one after another so that the geocentric model could agree with ever more precise observations. Francis Bacon admonished men of learning against blind faith in conventions: "And if any neglected or unknown instance occurs, the axiom is saved by some frivolous distinction when it would be more consistent with truth to amend it."[146]

THE UNIVERSAL GRAVITY

Let us liken a galaxy in the universe to a cyclone in the atmosphere to realize that the galaxy rotates not in emptiness but in the expanding universe. A remote hydrogen gas cloud orbiting beyond the shining outer rim of the galaxy's stars experiences gravity not only due to the galaxy but also due to the expansion of the universe.

This explanation, prima facie, for the 'missing' mass may seem too simple. How could the galaxy possibly experience any force due to the rest of the universe, given that the matter on the largest scale is spread evenly about the galaxy? The thing is that although matter is distributed evenly in every direction (isotropic), the distribution is not flat (homogeneous). Namely, matter is sparse in the nearby universe, whereas it is dense in the distant early cosmos. There is thus a shallow density gradient across the expanding space from the nearby present to the far-off past. This universal gravitational field acts everywhere in a law-like manner.[147] As stated in Hubble's law,[135] it causes flows of the void to which bodies couple; hence far away galaxies move away from us and nearby galaxies move toward us.

The universal energy gradient from the dense past to the sparse present is a feeble force apparent only on the galactic scale.[33,37,106,135] For instance, the solar system's orbital velocity about the center of the Milky Way is remarkably high (240 km/s) to stay in balance with both the universal and local density gradients. Thus, there is no reason to impute the 'extra' rotation and translation to the gravity of dark matter (Appendix F). All matter is, in fact, visible, only its gravitational field, the paired-photon vacuum is dark, transparent.

While dark matter can be employed to model many phenomena, it does not explain why stars, gas clouds, and dwarf galaxies orbit in the plane of a spiral galaxy.[148] The astrophysicist Marcel Pawlowski from the University of California, Irvine, reminds us that "Science progresses through challenges," and continues, "… this gigantic ring [of dwarf galaxies] forms a serious challenge to the standard paradigm."[149] Indeed, it does. In contrast, the least-time principle explains that small galaxies circulate in the plane of a central giant for the same reason as the planets orbit in the solar system's invariable plane, the rings of Saturn are in the same plane, and the spinning Earth is flattened. The reason is the void's striving for balance.

The recent discovery of an ultra-diffuse dwarf galaxy holding hardly any dark matter was also a bit of a surprise to cosmologists.[150] Its spread-out globular clusters move relative to each other much more slowly than the clusters in many other small galaxies of ordinary density. The correct explanation lies in the fact that the clusters of the diffuse dwarf galaxy are not at a standstill. They are moving coherently at about 300 km/s relative to a nearby sizable elliptical galaxy. In the same way, Magellan's clouds orbit the Milky Way in the gravitational field of the expanding universe.[151] The dark matter hypothesis is also refuted by visible matter accurately defining the limit velocity for a star to escape from a galaxy.[152]

The void, the gravitational field of the expanding universe, disperses lumps of matter into a smooth cosmic structure and grooms the evolution of galaxies into those familiar forms of nearly lognormal distributions, logarithmic spirals, and power laws.[153] The *Grand Regularity* also displays itself in the universal orbiting time of the outer rim. It is the same for all galaxies, from the smallest dwarfs to the largest spiral nebulae.[154]

BEHIND ITS TIME

A top stays upright when spinning fast, and after having slowed down, it tips over. Why? The textbook says that the spinning top remains upright because its angular momentum is conserved. But no one knows why it is conserved, just as no one knows why linear momentum is conserved. There are equations without explanation.

Wolfgang Pauli and Niels Bohr (right) marveled at a spinning top in 1954 at the Lund Institute of Physics opening. (Wikimedia Commons)

Now that we know the essence of the void, we know that the raison d'être for the top's spinning and falling down is the same: the quest for balance. The top spins steadily because its free-energy minimum position in the universal gravitational field is optimal, balanced.

For the same reason, dwarf galaxies orbit in the plane of a spiral galaxy instead of in random inclinations. If the top is poked, it quickly reorients to restore its balance with the void. If a spiral galaxy is punched by another galaxy, the deformed disk of stars will realign to regain balance with the void.

Eventually, the top slows down in spinning relative to all bodies in the universe, and the local gravity of Earth reigns over the object, tipping it over.

A swing is another familiar object whose behavior is dictated by universal and local gravity. In 1851 Léon Foucault showed a crowd of people that a pendulum suspended from the ceiling of the Pantheon in Paris maintains its direction of swing relative to the universe. So it is the planet that revolves beneath the pendulum.

Conservation laws, the cornerstones of physics, define the condition of balance but do not explain how the state of balance is attained; that is, the dynamics. While modern physics invokes virtual forces to reconstruct data by calculations and simulations, in reality, it is the ether that forces the balance.

In big circles

"There is no falsification before the emergence of a better theory,"[137c] Lakatos asserted. Only when we come to regard space-time merely as a model of the physical vacuum will we be ready to abandon general relativity as a worldview.

In the 1850s, it was thought that an unknown planet caused Mercury's elliptic orbit to precess some 43 arc seconds in a century[155] on top of the gyration due to other planets. It was a reasonable assumption because deviations in Uranus' orbit had earlier hinted at the existence of Neptune.[156] The hypothesized celestial body was baptized Vulcan and searched for. Then Einstein came, saw, and explained that the extra precession is not due to the fictitious planet but curved space-time.

General relativity describes the geometry of space, not its substance. By contrast, thermodynamic theory attributes the precession to the gravitational field embodied by paired quanta.[157] By measurement, the orbit is genuinely longer in the Sun's gravitational field than in free space because the density of quanta is higher near the Sun than in free space. Likewise, the path of light through glass is longer than through air. The speed of light depends on the surroundings, including the strength of gravity, as Einstein and Dicke reasoned too.[110,158]

General relativity also accounts for the motion of the spinning Earth's gravitational field, measured with the Gravity Probe B satellite mission 2004–2005.[159] The least-time principle also reproduces these results.[40] This concurrence is not surprising. In the Earth's modest gravity, the satellite was essentially in balance on its orbit. In the absence of time's arrow, general relativity is an excellent model of gravitation. Thermodynamic theory and general relativity are expected to differ only upon the happening of something.

My interest in galaxy rotation stems not from a desire to challenge the dark matter hypothesis but from a will to test the theory of time in big circles. So I asked myself whether the motions of galaxies could be comprehended like other motions. They indeed can, taking the gravity of the expanding universe into account. Yet, no single observation validates the universality of the least-time principle. Even one could falsify it. So a theory should be tested until its limits of validity are found.

Reviewing Peers

The results of scientific research are subjected to peer review, but peer review itself is rarely rigorously evaluated. Properly weighing a study is a challenge; gauging one questioning a dogma is a challenge of another order. It is hard to avoid classical logical fallacies.[xii] Many comments stray from the evaluation of the manuscript itself to the defense

---

[xii] *Argumentum ad antiquitatem:* A claim is true because it is thought to be true. *Argumentum ad hominem:* A claim assails the author instead of the thesis. *Argumentum ad lapidem:* A thesis is discarded without arguments. *Argumentum ad metum:* A claim is intensified by fear. *Argumentum ad nauseam:* A claim is repeated to make it true. *Argumentum ad numerum:* A claim is true, as many people think it is true. *Argumentum ad passiones:* A claim appeals to the emotions. *Argumentum ad verecundiam:* A claim makes an appeal to the authorities. *Argumentum ergo decedo:* Critique is not tolerated. *Audiatur et altera pars:* A claim is not supported. *Ignoratio elenchi:* Missing the point. *Non sequitur:* A conclusion is not supported by its premises.

of the current viewpoint. Some of them deviate from the substance to the assessment of the author or even his imagined motives.

I sent a manuscript explaining the rotation and velocity dispersion of galaxies by the gravity of the expanding universe to the prestigious *Frontiers in Physics*. The publisher of the series says that peer review needs to be centered on objective criteria of the soundness and validity of the work presented, and has to be rigorous, fair, constructive, accountable, and transparent for everyone involved.[160]

Three reviewers commented on the suitability of my manuscript. 1: "No. The hypothesis of the article is unacceptable and does not prove anything" (*argumentum ad lapidem*). 2: "Yes." 3: "No. Paper themes are well known. There is nothing innovative" (*audiatur et altera pars*). The first reviewer wrote, among other things, the following: "Based on the title, the paper seems standard physics, but it is not! The author proposes new (very improbable) gravitational physics. Standard Gravity Law could be abandoned only after many evidences are given and a plausible theory has been expressed. In the paper under analysis, this does not occur! The hypothesis proposed is crap. If accepted it would cause a vulnus to the journal, which will become a place for cracpots!" (*argumentum ad metum et passiones*). The second reviewer requested clarifications on a few items. The third reviewer said that dark matter already explains the observations (dogmatism).

I asked the first reviewer to be specific in their comments. I also pointed out references where the bending of light, the anomalous precession of Mercury, and the passage of time are explained using the same least-time principle. After the revision, the second reviewer recommended publication. I replied to the third reviewer that many researchers question the dark matter hypothesis because there is no direct evidence of such a substance. Without further ado, the editor rejected the paper (*argumentum ad verecundiam*).

I felt that the decision was unjustified and contacted the *Frontiers* administration. After a while, the editor replied. He considered my result on the galaxy rotation to be incorrect because the advancement of Mercury's perihelion, as calculated using the least-time principle, 43.09 arc seconds per century (arc/cty), published in the *Monthly Notices of the Royal Astronomical Society*, differs from the estimate, as calculated using state-of-the-art methods, 42.98 arc/cty less than 0.3% (*ignoratio elenchi*).[157] I felt that the comparison was unreasonable. It is impossible to determine the precession so precisely without resorting to the sophisticated model with many more variables[53] than the least-action calculation. Therefore, the fit is marginally better, but the explanatory power is significantly worse than that of my simple calculation. At that point, the Editor-in-Chief decreed that this 'error' justified the rejection of my manuscript (*non sequitur*).

Peer reviewers are, of course, not in a peer-to-peer position *vis a vis* with the author but in an influential position of power.

BENT RAYS, BENT REASONING

A ray of light bends when going through a gravitational field, similar to passing through a glass lens. However, the bending due to the visible matter of a galaxy is much larger than is expected by general relativity. Zwicky had already spotted this discrepancy in the 1930s. But from such observations in the 1980s, it was concluded that dark matter makes up about four-fifths of all mass in the universe.

According to the least-time principle, light bends almost five times more than predicted by general relativity. Thus, dark matter is not needed to explain the bending of light (Appendix E).[4]

How could general relativity err? Has the theory not explained the bending of light coming tangential to the Sun from a distant star?

Eddington was the first to make that demanding measurement during the total eclipse of 1919. Today, the effect of gravity on the photon path is most precisely measured by how much a radio signal is delayed when passing by the Sun to, for example, Venus and back. The photon's travel time is longer when passing through the Sun's gravity than in free space, just as it is in glass compared to air. The least-time principle gives the correct value of 195 microseconds for light passing by the Sun.[4]

As we know, general relativity also gives this result. But in its case, the delay is calculated by a different equation than the one for bending.[101] Why does general relativity need *two* different equations for calculating the passage of one and the same photon?

The philosopher of science, Paul Feyerabend, advised us to question the measurement itself as well: "Taking experimental results and observations for granted and putting the burden of proof on the theory means taking the observational ideology for granted without having ever examined it."[161]

The bending of light is determined by the difference between the line of sight to a star during an eclipse and the night-sky line half a year later. Although obvious, it is essential to note that the ray ends up at a different point on the ground when it has passed through the gravitational lens than when coming directly from the night sky. So to trace the same ray, the telescope should be displaced by the distance corresponding to the change in direction between the two conditions.

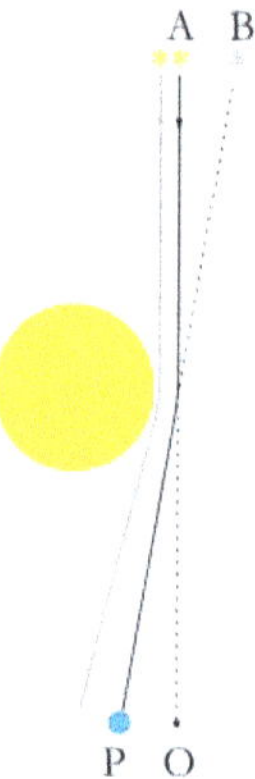

When passing by the Sun, a ray of light (black line) from a star (*) bends away from the direction of the night sky (AO) to the direction (BP). In addition to the star's apparent displacement (AB), the change in the observer position (OP) must be taken into account in the amount of bending. In other words, the ray that goes through the gravitational lens and arrives at a given position on the Earth is not identical but parallel to the ray that comes from the same star at night. The closer the ray passes tangentially by the Sun, the more strongly it bends (gray line). Therefore, a star cluster is not as focused during an eclipse as at night.

It is easy to grasp the relation between the change in direction and the telescope's displacement. For example, looking at your finger with one eye and then switching to the other seems to shift it relative to the background. If this phenomenon, called parallax, is to be averted, the position of the eye, like the location of the telescope, has to be changed to match the angle from which the object is seen. Unfortunately, Eddington did not consider this. So the bending appeared smaller than it was.

Einstein lucidly understood the relationship between experiment and theory. No number of experiments will ever prove a theory right, but even one can prove it wrong.[162] Could this one made by Eddington already have been a fatal one for general relativity?

Furthermore, images of two galaxy groups, known as the Bullet Cluster, have been taken as evidence of dark matter.[163] The photos suggest that one has penetrated the other like a bullet. The galaxies are sparse enough that hardly any of their stars collided, only slowed down a bit. However, the impact is visible as a shock wave in the gigantic clouds of gas surrounding the galaxies. The images of the objects are also slightly distorted due to light's bending when passing by the galaxies. These distortions were initially taken as proof of dark matter moving along the galaxies, as most of the visible mass was purported to be in the gas clouds left behind by the collision.

True, the gas clouds have lensing power, but the amount of distortion may be difficult to gauge, especially after a collision.[164] When more detailed images were obtained, scientists backpedaled from their earlier conclusions, acknowledging that the visible and dark matter did not get separated from one another at the collision after all.[165]

According to the least-time principle, light bends by gravity nearly five times more than general relativity (Appendix E). Therefore, the distortions in the images of the Bullet Cluster do not signal dark matter, either.

Out of Context

In 1981, the Israeli physicist Mordehai Milgrom suggested adding a small universal acceleration to Newton's law of gravity.[166] Then, galaxy rotation could be modeled without dark matter. The professor at the Weizmann Institute postulated that tiny acceleration acts everywhere but is visible only far away from the galaxy center. Although this one-variable MOND model tallies the data amazingly well, many researchers deny its validity. The worst weakness is that MOND, unlike general relativity, is not a model of the whole cosmos. Besides, the total energy is not conserved when the auxiliary factor is put in.[167]

However, does not acceleration, the very characteristic of gravitation, expressly mean a change in energy, such as expansion? Then again, unless we know what causes and carriers the small, universal acceleration, it is not a better explanation than dark matter. In any case, why does Milgrom's model of universal acceleration work so well?[168]

That MOND is a well-known model became clear to me when I participated in the 175th-anniversary colloquium of the Finnish Academy of Science and Letters in 2013.[169] Professor Lars Bergström from Stockholm lectured superbly about the mystery of dark matter. I asked him, "Isn't it the expansion of the universe that causes the tiny acceleration? Would that force not explain the rotation and motion of galaxies?" These issues were new to this member of the Nobel Committee, but the tiny acceleration related in his mind to the old MOND model. So instead of answering my question, he rehashed the above-stated weaknesses of MOND. The professor found it difficult to see that galaxies are so big that they experience noticeably the gravitational field spanning from the early to the present universe. Hubble's law, manifesting this force, was well-known to the attendees, of course. Still, in the context of galaxy rotation and velocity dispersion, its effect appeared to them to be out of context. This is understandable since standard cosmology is mute regarding the reason why distant galaxies are receding and nearby ones are approaching.

An intellectual crime

The search for dark matter is big business. A layperson might get the impression that the scientific community would be embarrassed by the revelation explaining the observations without dark matter. The expanding universe itself causes a small gravitational force everywhere. Perhaps a few truly are embarrassed. However, science is not about protecting one's reputation or that of an institution, nor about avoiding the pain of admitting the fruitlessness of years of work, but about finding the truth. In due course, appreciation follows from providing proposals that correspond to reality.

An observation can falsify, but none can validate a theory because a measurement requires a presumption of what is being measured.[114,170,171] Numerical values alone do not mean anything; only after interpretation do they gain meaning.

Contemporary physics is inconsistent because, on the one hand, the standard interpretation of astronomical data leads to presumptions of dark matter and dark energy. On the other, the Standard Model of particle physics does not crave such concepts. Such doublethink calls for newspeak. Wave-particle dualism, entanglement, action at a distance, space-time, dark matter, dark energy, and inflation are used not to comprehend observations but to uphold the framework of modern physics. Lakatos' comment hits the mark: "Blind commitment to a theory is not an intellectual virtue: it is an intellectual crime."[137a] To judge sharply, we need to juxtapose the approved image with an alternative outlook. A view without a contrary one is as fuzzy as a photo without contrast.

Our view of reality would be less deluded without the fictitious stuff. Thus explanations without dark matter are getting into the limelight. I received an invitation to give a talk on dark matter from the perspective that it does not exist. This seminar of the Ursa Astronomical Association was held in Lapland's Utsjoki to celebrate a nearly total eclipse on March 20, 2015. Juhani Harjunharja, long-time local chair of the association, planned an exchange of ideas between a representative of the standard cosmology and me. I liked the idea that a debate is in place when stances differ. However, my invitation was canceled. Harjunharja must have gotten into an awkward situation where he was forced to choose between an orthodox scientist and a heretic.

Such incidents worry me. The silencing of a dissenter is unacceptable, for freedom of expression ensures the development of society and its institutions alike. It is intolerable to shut down a scientist who thinks differently. We do not know what dark matter is or whether there is any, and we shall not as long as we are certain we do.

WHAT IS A BLACK HOLE?

It is a star among stars.

If a star was so massive that even its own light cannot escape the pull of gravity, we would not be able to see it, reasoned John Michell in 1783.[172] Such a star could nonetheless be noticed from the motions of its companions. The pioneer of astronomy, gravity, optics, and geology assumed that there were many dark stars in the universe but was so much ahead of his time that he died in oblivion.

Nowadays, such a dark star is called a black hole. Journalist Ann Ewing typed this catchword in the headline of *Science News Letters* in 1964 when reporting on a scientific meeting held in Cleveland.[173] A few years later, the name gained popularity in John Wheeler's talks about a star that had imploded according to Einstein's theory.[174]

We do not think of a black hole as a star, as Michell did, because general relativity is not about substance but space-time. We do not talk about the surface of a dark star but the event horizon of a black hole. We do not zoom in on a dark star's nuclear reactions but on a black hole singularity. However, is it not incoherent to posit that a black hole is a whole different thing than everything else?

Moreover, modern physics asserts that information cannot vanish,[175] while it is an everyday experience that information disappears when its representation is wiped out. For example, a symphony was lost when the only manuscript went up in flames before the first public performance.[xiii] So the information loss in the black hole, appearing to annihilate everything, does not seem like a paradox but a fact. But physicists reject this piece of common sense because quantum mechanics asserts that information is conserved. But how could information be conserved if entropy increases invariably, for one thing, and for another if both information and entropy are defined by disorder? As they say: an ounce of common sense is worth a pound of theory.

According to general relativity, a star could collapse by its own weight into an incredibly dense object because its curvature would go on increasing without limit. Subrahmanyan Chandrasekhar's first estimate was that nothing would prevent a star about one and a half times more massive than the Sun from collapsing. For these pioneering conjectures, the young Indian physicist, working in Cambridge in the early 1930s, was publicly humiliated. Eddington himself deemed the collapsing star a crackbrained idea. After this infamous repudiation, Chandrasekhar distanced himself, assuming a new position at the University of Chicago.[176] When black holes were discovered decades later, he was awarded the Nobel Prize.

[xiii] Symphony No. 8 was Jean Sibelius' final major work, but it never premiered, as he burned the score and associated material. (source: Wikipedia)

Today we know about these invisible objects by more than mere calculation. For example, the spinning rate of the black hole in the center of the Milky Way is known. It was measured by researchers from the Finnish Centre for Astronomy with ESO and their many collaborators.[177]

When a star becomes more and more compact, its matter transforms into neutrons. If the mass of a neutron star remains less than three solar masses, light can still escape from the surface, whereas it cannot come from yet more massive stars.[178] When the density in the black hole core increases by two orders of magnitude above that of the atomic nucleus, gravity outstrips the strong force, and the neutrons break down (Appendix G). Under these circumstances, general relativity no longer holds true but points to a nonsensical singularity. Thus, we need quantum theory to explain that matter cannot condense endlessly. But the quandary is that quantum mechanics leads to the information paradox.

We can, however, escape the mathematical singularity and dodge the information paradox by acknowledging that the quantum's measure, Planck's constant, does not change under any circumstances.

THE GRIP OF GRAVITY

When the atomic structures of compounds were uncovered about two hundred years ago, chemical reactions were understood. Likewise, today the structures of particles and the void (Chapter 3) allow us to fathom what happens in nuclear reactions.[179]

When the mass of a black hole exceeds the mass of a few tens of Suns, gravity overpowers the strong force.[180] As the paired photons of the gravitational field become as short as gluons, sthe gluons might as well disconnect from the quarks and connect to the surrounding photon fluid. As the neutrons break apart, the quarks of opposite handedness juxtapose and annihilate: matter transforms into the void.

Data from black holes is understandably scarce to justify one tenet or another. Nevertheless, the little there is calls for an interpretation. Where did the bubbles of gamma and X-rays come from that the Fermi Space Telescope detected in 2010 and eROSITA in 2020? These incandescent bulbs of high-energy rays extend tens of thousands of light years from the center of our spiral galaxy. Did they form over the eons from the intermittent jets that the spinning super-massive black hole at the galaxy's core emits? High-energy rays[181] like-wise squirt far out into space from the centers of other spiral galax-ies.[182] So, a giant cavity filled with short-wavelength photons emerges from eons of combustion.[183]

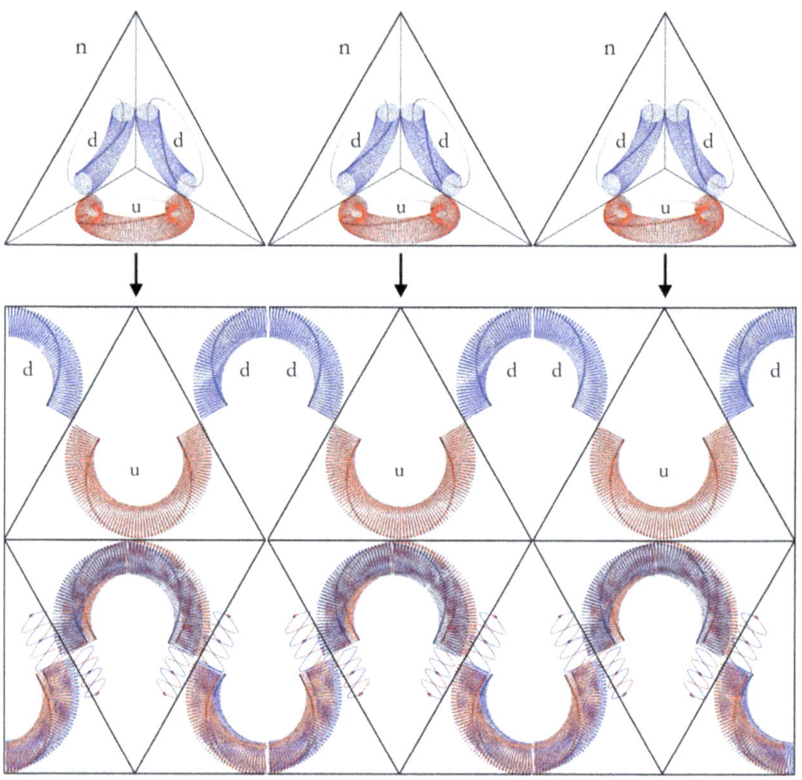

When gravity outstrips the strong force in a black hole, the tightly packed pyramid structures (top) of neutrons (n) open and flatten into the layers of quarks (center). The up quarks (u, red) and the down quarks (d, blue) of the adjacent layers align on top of each other (bottom). These arcs of the tori annihilate or discharge into waves of paired quanta that jet out from above and below the planes where the curvature, i.e., gravity, is least.

Since the spectrum of the Fermi bubbles is very high in energy,[184] the radiation could stem from hadrons.[185] The photons could originate from the flattened, disintegrating neutrons because, in the bubbles, the number of photons decreases inversely proportional to the energy squared, characteristic of a planar source. By contrast, the number of photons around spherical sources, such as ordinary stars, decreases in inverse proportion to the cube of the energy.

The black hole devours substance and produces high-energy radiation, mostly in the form of the paired-photon vacuum.[186] In turn, light may materialize, as theorized already in the 1930s.[187] As concretized more recently, vortices of the void, at their densest, could furl into closed strings of matter. Moreover, the double refraction tells the vacuum is turbulent due to strong magnetic fields near neutron stars.[188]

When matter breaks down in the core of a spinning black hole, the oblate top deforms into a doughnut shape. The gravity gives in most along the star's axis of spinning, along which the single, visible photons and paired, invisible quanta jet out into space.

REALITY CHECK

In his book *A Brief History of Time* (1988), the legendary theoretical physicist Stephen Hawking portrayed how a black hole evaporates away: near the event horizon, virtual particles transmute into real particles and antiparticles. When one of them escapes, the black hole loses a little bit of its energy and, in this way, wanes away altogether over eons. This conclusion is consistent with quantum mechanics and general relativity, but is it in line with reality?

Black holes seem to come in two sizes. On the one hand, small ones originated from collapsed stars with less than tens of solar masses. On the other hand, supermassive black holes in galaxy centers total more than one million solar masses, but the average density may be lower than that of water.[189] Astronomers wonder why there are only a few intermediate-sized black holes.[190]

With the detailed picture of the black hole structure, we realize that the remnant of a collapsed star cannot swell to immense dimensions. When matter spirals in, annihilation restarts whenever gravity exceeds the strong force again due to the accumulating mass. By contrast, gargantuan black holes are thought to have originated directly from massive gas clouds at an early stage of the evolving universe,[191] perhaps even at such an early stage that the substance of that time and its reactions – or non-reactivity – are unknown to us.

Even though the giant munches matter moment by moment, it burps out radiation only from time to time because the reactions that grind elementary particles will not ignite until gravity exceeds the strong force. And even thereafter, the reactions must massively transform matter into photons before the giant gravity weakens sufficiently for the quanta to jet out from the galaxy core.

Conversely, when light exits and matter refills the core, gravity regains. As a result, the jet shifts red, slows down, and shuts up. This spectral evolution of highly collimated jets is difficult to explain by accretion disk processes.

The whole universe is, in a sense, a black hole. Light cannot escape from the enormous mass, even though the universal density is very low, much lower than the supermassive black hole density.

G R A V I T Y   W A V E S

Neither eyes nor cameras can see the dark shine of paired quanta that a black hole ejects. Still, the emitted density can be detected as a gravitational wave. Such a density surge, a gravitational wave, was registered for the first time in September 2015. It started off when two black holes merged after spiraling about each other.[192] It happened safely at about one and a half billion light years away. By the time it struck the high-resolution detectors at the LIGO observatory, the huge tidal wave had dwindled to tiny wavelets. The sensitive instrument registers even daily variations in density.[193]

When the gravitational wave arrived, the laser light, reflecting back and forth in the detector's four-kilometer vacuum tube, momentarily passed through the wavy void. The principle of this measurement is a familiar one. By placing a glass plate in the path of light, the time of flight will increase. The glass is, of course, much denser than a gravitational wave. So the effect on optical length is much stronger than that due to such a wave. In contrast to this common-sense explanation, general relativity maintains the tube itself contracted a bit when hit by the gravitational wave.

It is difficult for us to deduce from the data alone which description is true, but we must consider the whole. Why should we explain the gravitational wave differently from other waves? Why would it shorten the detector's tube rather than change the vacuum density therein? Why would a wave of the void be, in essence, different from a slab of glass?[66b] Questioning the interpretation of general relativity in this manner makes us realize that the mathematical model is no explanation as such. The more narrowly we focus, the less we see the bigger picture, and the less we understand the thing in question, the way we understand other things.

LITTLE GREEN MEN

On August 17, 2017, the detectors registered gravitational waves again. Two seconds later, the Fermi Space Telescope picked up a powerful gamma-ray burst. The optical telescopes were quickly pointed in the direction of the signal. Fifteen hours later, the highly radiant matter was seen at a distance of about 130 million light years, where two neutron stars had collided.[194]

The present understanding of neutron stars traces back to a meeting of the American Physical Society held at Stanford University in the winter of 1933. Walter Baade and Fritz Zwicky presented that a neutron star could emerge from a supernova explosion. The idea was taken with a pinch of salt, even after Robert Oppenheimer and George Volkov calculated the properties of such a star in 1939.[180] Only when Jocelyn Bell Burnell discovered in 1967 a variable radio frequency transmission was it reasoned that the signal could come from a rapidly spinning neutron star. For a moment, the graduate at the University of Cambridge and her supervisor Anthony Hewish speculated that a distant civilization was sending the evenly paced signal. The object was hence coined LGM-1 (Little Green Men-1). After that, the idea of a dense star was gradually accepted.

The bright beams from these beacons in space can swipe across the sky up to hundreds of times per second. It was thought that spinning produces radiation out of the pulsar's magnetic field. But discovering a slowly spinning neutron star lacking accretion material suggests that the jet is neither rotation nor accretion powered.[195] Perhaps the source of power is, after all, in the star's interior, as in other stars. Werner Becker, a professor at the Max Planck Institute, admits the continued lack of certainty: "The theory of how pulsars emit their radiation is still in its infancy, even after nearly forty years of work."[196]

The explanation of a phenomenon should be founded on facts that are known better than the phenomenon itself. Now that we know the structures of particles, we can explain the signals from dense bodies originating from nuclear reactions rather than generated when matter spirals into an unknown abyss within the star. This new view might well be needed. The assumptions underlying contemporary models do not correspond to the observations in all aspects.[197] The neutron structure could hint at how quanta are released when matter becomes denser and denser (Appendix B). A mere shortage of fuel could explain why some pulsars shut down for long periods of time.[154] The spinning of a pulsar mirrors the dwindling of the Earth's rotation in a power-law manner (Appendix E). This is yet another expression of the *Grand Regularity*.

BREAK WITH CONVENTION

It may seem amazing that our exploration has taken us all the way into the depths of black holes and the beginning of the cosmos. But this is how we thoroughly test out a thesis. We are trying to find at least one phenomenon negating the idea that quanta make up everything.

Our goal is not to disprove earlier accounts but to make use of the structures of particles and the void in understanding observations. For example, treating the core of a star like an atomic nucleus may seem simplistic. Yet, it is the best approach until we know better. At an earlier time, when the details became clearer, Bohr's model of the atom displaced Thomson's raisin bread model. The future will reveal the relevance of contemporary science – or its irrelevance – although the history of science has already unveiled quite a bit.

It takes a bold character to venture into new areas, but that kind of courage is expected of a researcher. I remember from years ago how succinctly Ilkka Kilpeläinen, a professor of materials chemistry at the University of Helsinki, summed up what it takes: "Only unwarranted self-confidence is good enough." That's the way it is. When we fail to make headway in a conventional way, we must break with convention and find another way. This new way aims to consistently comprehend observations arising from events, flows of quanta, irrespective of what those events might be.

After all, we have not yet made it far from the theory's starting point: everything is quanta. We have only studied elementary particles and the ether and their reactions, nothing more complicated. Nonetheless, if the evolution of the universe is nothing more than the transformation of matter into space, how could any complexity within the universe be incomprehensible when governed by the universal principle of least time?

IS THE WORLDVIEW COMPLETE?

Modern physics, categorically overlooking ontological questions, is now entangled in them. What is dark matter? What is dark energy? These questions were rendered pointless by our answering the primary questions: What is matter? What is energy?

Now after acquainting with matter and the void and the evolution of the universe, let us review the main challenges of physics.[198]

1. *Combine general relativity and quantum theory into a single theory that can claim to be the complete theory of Nature.*
   The task is no longer meaningful. General relativity and quantum theory, as mathematical models, make no sense of causality, the central character of reality. By contrast, thermodynamic theory explains measurements consistently and within our own experiences.

2. *Resolve the problems in the foundations of quantum mechanics, either by making sense of the theory as it stands or by inventing a new theory that does make sense.*
   The wave function and the quantum field alike are mathematical models, not of the particle itself but of the vacuum quanta undulating about the particle.

Thermodynamic theory tells the particle apart from the vacuum yet describes both as comprising quanta.

3. *Determine whether or not the various particles and forces can be unified in a theory that explains them all as manifestations of a single, fundamental entity.*

Yes, the unification is at hand. The particles and the vacuum embody quanta. In its most familiar form, the fundamental element is the light quantum. The strong force takes the form of a short-wavelength photon, the gluon, the weak force has the form of the weak bosons, the photons carry electromagnetism, and the paired photons carry gravity.

4. *Explain how the values of the free constants in the Standard Model of particle physics are chosen in Nature.*

The constants of the Standard Model include particle masses, so-called CMK matrix elements, coupling constants of gauge fields, and properties of the vacuum.

- The mass of a particle denotes how strongly the void's paired quanta couple with a particle's structure.
- The CMK matrix is a model for the transformations between quarks and the MNS matrix between neutrinos. The quantum of the first family is the simplest. The second family's quantum is more curved and hence heavier. The quantum of the third family is even more curved, and hence the masses of the top quark and the tau neutrino are high. The matrix elements merely parametrize the balance between the three conformations of a particle in a given condition.
- The value of the Fermi coupling constant relates to the reaction where the muon torus opens up to the W^- boson and closes up anew as the electron torus.
- CP violation by the weak force means that the W^- boson will not stay as it is unless the direction of its quantized flow, i.e., time, is reversed along with swapping the charge (handedness) and inverting the

toroidal structure through a point (parity). By contrast, the mirroring of a gluon and a photon does not affect the strong force and electromagnetism.

- The vacuum's properties follow from its quantum structure, which is in a dynamic balance with matter.
- The coupling constants of gauge fields relate to the magnitudes of the fundamental forces, which in turn reflect the structures of particles and the vacuum.

The values are thus not chosen but result from the least-time consumption of free energy.

5. *Explain dark matter and dark energy. Or, if they don't exist, determine how and why gravity is modified on large scales.*
   There is no dark matter. Instead, the expansion gives rise to a tiny acceleration across the universe, from the past to the present, apparent on a galactic scale. No dark energy is needed either, but matter transforming into the void explains the expansion. Thus, the density of the universe could not be other than critical, and the far-off cosmos other than homogeneous.

A THEORY ABOUT THEORIES

Dark matter, dark energy, entangled particles, the multiverse, and action at a distance seem like flights of imagination compared with the reality we experience. These illusions have bedeviled us for a long time now. After talking for hours with Bohr about the Copenhagen interpretation, Heisenberg came to ask himself, "Can nature possibly be so absurd as it seemed to us in these atomic experiments?"[199]

Hardly. The universal patterns of data speak for a single reality. The unified theory should, therefore, describe and explain everything in the same way, the way we experience reality.

As we begin to ponder comprehensively, testing the atomistic axiom about quanta, many a problem in physics becomes pointless or not a problem at all. Isn't it a relief that dark energy and dark matter are not needed to explain astronomical observations when nothing remotely like them has ever been discovered?

The greatest difficulty may well be that thermodynamic theory does not concur with current theoretical conceptions, even though its interpretations tally with the data. The atomistic tenet does not deliver, for instance, spectacular symmetries or elaborate equations. Instead, it grants concrete comprehension. Henri Poincaré saw that many truths assumed of scientific theories are, in fact, mere practices.[200] Can we question the current paradigm? Do we have the guts to acknowledge that contemporary cosmology with dark matter and dark energy is a darker phase in the history of science, just as ancient cosmology with its epicycles and deferents once was?

Today's mastery of modeling does not establish the solid foundation of a comprehensive worldview but elevates us further away from such, even into midair.[201] Our own experience is the best guarantee of reality.[202] We see the quantum of light with our own eyes, sense it with our own skin and feel it with our own bodies. Is there any truer account of reality than that?

A new viewpoint has often been more rewarding than a new result. Now the thermodynamics of time provides the basis for examining causality beyond the conventional scope of physics. In the subsequent chapters, we are not so much after new findings but a coherent philosophy. Let us examine matters from the proposed perspective and relate unfamiliar phenomena to familiar facts.

SOMETHING RATHER THAN NOTHING

When we say that quanta embody everything, we assert a worldview that includes everything. However, it is incomplete. According to the incompleteness theorems of Kurt Gödel, the foremost logician, it is possible to pose questions about the universe that have no answers within the same universe. Ludwig Wittgenstein famously concluded the same in *Tractatus* (1927): "That whereof we cannot speak, thereof we must remain silent." Although it makes no sense to talk about what cannot be addressed, those matters are inexorably raised.

We may ask why there is something rather than nothing, but we know nothing about nothing. We cannot know anything beyond the universe. The photons are all there is. We may ask why the universe exists, but we have no means to answer. The dawn of the cosmos was an event, but causality does not reach beyond the first event. Asking what preceded the beginning, we wonder where the quanta originated. But our experience only tells us that they *exist*. This contradiction exposes our inability to comprehend below the quantum and beyond the universe. We cannot prove the truth or falsity of statements beyond our deductive system. The universe is the realm of our reality; the multiverse is an absurdity. Thermodynamics has no room for fanciful many-world interpretations.

We may also ask why the universe comprises precisely the number of quanta it does (about 10^{120}),[33,34] but we cannot explain this. Nonetheless, we may assume that a universe smaller than ours would not have turned on. Other events, too, such as the boiling of water, the birth of a star, or the collapse of a neutron in the core of a black hole, require crossing a boundary. We may also suspect that the universe could not have been born bigger than it was, as then it would have begun to expand ahead of its time. Georges Lemaître thought that the nascent universe was a primeval atom that evolved and is still evolving by fragmenting into pieces.[202] Now, knowing the structures and reactions of the particles and the void, we understand such events in detail.

Our suppositions are as vague about the fate of the cosmos as they are about its beginning. Nevertheless, we may surmise that, ultimately, all quanta in looped forms of matter will have straightened out into the paired quanta of the void. When there are no longer any qualitative differences, what then shall be the meaning of energy and time? In the universal balance, everything will have already happened.

Explanations must be of some substance and relate to our own experiences to qualify as explanations. Even if this sense-making implies that we cannot dissect any deeper into reality than the quantum, the atomistic axiom provides a comprehensible worldview.

KEY POINTS

- The universe is expanding flat because quanta of dense matter transform into quanta of sparse space.
- Imbalance sets the void in motion.
- Bodies couple and move with the void displayed as gravity.

5. What Does Mathematics Mean?

Mathematical notation is precise,
but precision as such means nothing.

The *Grand Regularity* of Nature is readily discernible in various spirals, such as a snail's shell, the scales of a cone, the tentacles of an octopus, and whorls of flowing water. It is also easy to spot similarities in branching patterns: a tree trunk branches out into copious twigs just as a river forks into countless tributaries. However, we won't notice the similarity between shapes in most cases until we collect data and analyze it. Then we see, for example, that mainlines of a district heating network divide into numerous radiator terminals in the same way as veins branch out into many capillaries.[1]

In the same manner, we can spot the regularity emerging through time, for example, from the similarity between the rise of carbon dioxide levels in the atmosphere[2] and the spread of new words in a language.[3]

The similarity between spirals is evident.[4] The regularity, also in its other forms, is apparent when data have been sorted and plotted.

Data of various kinds disclose that spirals, distributions, and growth curves have approximately the same mathematical form (Chapter 1). Why is the *Grand Regularity* reflected in certain mathematical forms? Why is Nature mathematical in the first place?

THE MATHEMATICAL UNIVERSE

Galileo described the meaning of mathematics: "Philosophy is written in this grand book, which stands continually open before our eyes (I say the 'universe') but cannot be understood without first learning to comprehend the language and know the characters as it is written. It is written in mathematical language, and its characters are triangles, circles, and other geometric figures, without which it is impossible to humanly understand a word; without these, one is wandering in a dark labyrinth."[5]

In the early 1960s, Eugene Wigner expressed amazement at how effective mathematics is in science.[6] The Nobel Laureate in physics did not see a cogent reason why a mathematical representation of one phenomenon often leads to a stunningly accurate account of a whole set of phenomena. Richard Hamming, in turn, maintained that we only choose or create mathematics for the purpose at hand.[7] The pioneer of information technology argued that we interpret the world in mathematical terms in order to see what we are looking for.

The Swedish-American cosmologist, Max Tegmark, regards the universe as thoroughly mathematical. Everything is a mathematical structure: "The enormous complexity we observe is an illusion in the sense that the underlying reality is quite simple to describe, and what requires close to a googol bits to specify is just our particular address in the multiverse."[8b] Hermann Weyl, German-born mathematician, theoretical physicist, and philosopher, was convinced of the same simplicity in the early twentieth century: "I am bold enough to believe that the totality of the physical phenomena can be derived from a single universal world law of the highest mathematical simplicity."[9] While numerous processes are complicated, the general principle of Nature is nevertheless trivial.

For the ancient Greeks, geometry meant measuring terrain. The whole universe, too, has a shape. Even though we might never find out why a substance is geometrical, string-like photons, we may learn what the essence of mathematics is. Let us take a concrete look at mathematics, for such a standpoint may help us see something that might have escaped our notice.

Husserl, a one-time student of the mathematician Karl Weierstrass, extolled Galileo's method, which parses an experience or an experiment into an equation. After this mathematization, a theory cannot acquire new ingredients from reality[10] because mathematical equivalence ensures that the theory is closed. So the panorama of reality through the lens of the theory cannot alter even if we later find in the equations heretofore unnoticed implications about reality. For example, Everett discovered that quantum mechanics implies parallel universes, and Weyl found that general relativity entails wormholes.

There is no way to mend a mathematical theory when it fails to tally with the data. Additions to equations only degrade the theory into a malleable model. The way out of the discrepancy is to mathematize one's own experience into a new law. That is what we did; we mathematized the experience of time's passing into the equation of motion of quanta (Chapter 2).

In the revision of a worldview, new wider vistas arise from outside the formulas deemed to be the essence of the discipline. For example, time as operational quantity counts the number of repetitions over a standard cycle. Consequently, it may be difficult for many a physicist to think of the flow of time the way Galileo did, as the flow of water, or more profoundly, as the flow of quanta.

LANGUAGE LESSONS

At first, the language of mathematics is like a foreign language. If you don't understand, you might get scammed. That happened to me in Tokyo on a conference trip in 1998. My colleague Piero Pollesello, multitalented Global Brand Manager of Orion Pharma, also gifted in languages, impressed us illiterates by reading the names of railway and subway stations and other signs. During our long flight from Helsinki, he had learned Japanese syllabic writing, hiragana and katana.

As we ascended by the escalator to a crowded station hall, a guard shouted something incomprehensible just as I passed by him. The announcement echoed throughout the hall. I then turned to Piero and asked with a bit of trepidation, "What did he bellow?" Piero replied, "Did you not get it? He just pointed out that your zipper is open!" After checking the situation, Piero's reputation for extraordinary language skills plummeted. Otherwise, his valuation climbed even higher.

Even if language can be abstruse, as mathematics often is, there is no reason to believe that a clause invariably holds a nugget of wisdom. Then again, a simple equation can carry a great deal of truth. For example, it is no coincidence that the geometric square, such as in Einstein's formula $E = mc^2$, and those well-known squares of the Pythagorean theorem $a^2 + b^2 = c^2$, is found in many equations of physics, e.g., Kepler's laws. The square is the Euclidean norm, in a sense, the optimum. For instance, mass, $m = E/c^2$, compares a particle's curvature to the void's flatness, the ultimate reference. That is how we read mathematical notation.

Whatever we equate had better denote a true balance; otherwise, we would violate the equals sign '=,' foremost among math symbols. The solid basis for the conservation laws in physics is the constant number of quanta rather than energy.[11]

When delving a second time into a profound text, one tends to discover more than at the first reading. For example, the mathematics of modern physics has been perused over and again to uncover even the most extraordinary insights, such as superposition, entanglement, and nonlocality.

In science, the more a tenet has been subject to and survived doubt, the more we can trust it. When a theory complies with the observations, the domain of verification will be expanded to other kinds of data in order to find the theory's limits of validity. Eventually, suspicion grows if the calculations deviate from the data, and the theory can be overturned.

When physicists tell you from the mathematics of quantum mechanics that a particle goes simultaneously along two or more distinct paths, one struggles to swallow the story. Similarly, when physicists spell out from the mathematics of general relativity that the passage of time is an illusion or that an object shrinks as it moves fast or is hit by a gravitational wave, one's intuition raises a red flag. Without math, we would not have these strange images, let alone delusions.

The thermodynamic theory of time, too, can be analyzed down to its root maxims to verify that it describes S-shaped growth curves, skewed distributions, spirals, and chaotic motions consistently and coherently with the data (Appendix A).

WHAT EQUATES WITH CHANGE?

When we know how to read mathematical notation, we can relate an equation to reality in the same way we relate a sentence to its meaning. Specifically, the flow equation can be formulated but not solved for a river that cuts its course through the terrain. As the erosion progresses, the flow changes, which affects the erosion, and so on. Since everything depends on everything else, the future cannot be predicted.[12] Even when the change in the surroundings is marginal and can be discounted to a good approximation, such a solvable mathematical model is never quite truthful.

The famous Navier-Stokes equation describes a variety of flows from a steady stream to a raging torrent.[13] We can infer from the formula, as there is no general solution, that the course of events can be capricious, chaotic, turbulent.[14,15] The mathematical characteristics of imbalance are clear: a change in geometry is associated with a change in mass, dissipated as heat. This applies equally to a river with banks, a nuclear reaction with reactants, and the all-embracing vacuum.

Plane geometry corresponds to balance because there is nothing more even than a plane. The Pythagorean theorem expresses this. For instance, the arc lengths of quarks in the electron, neutron, and proton are squared in the proportions of $3^2 + 4^2 = 5^2$ (Chapter 3). All matter may thus transform into the plane of the void as the universe attains perfect balance.

In contrast to the Pythagorean theorem, there is no integer solution for the equation of non-Euclidean geometry, $a^n + b^n = c^n$, when $n > 2$, according to Pierre de Fermat. In 1637 the French mathematician noted down in Diophantus' magnum opus of arithmetic that his marvelous proof is, unfortunately, a bit too long to fit into the margin of the page.[16] That might be so, as the first valid proof, assembled by Andrew Wiles and published in 1994, is 400 pages long.[17]

We can make sense of Fermat's last theorem, too. By plotting the equation with increasing values of n, we see that the shortest curve ($n = 2$) is the ellipse. It is the optimum. For example, a planet in its elliptical orbit is in balance. At the optimum, the number of constituents is fixed, and hence the system is divisible evenly into its parts. The essence of Wiles' proof is that the ellipse is divisible by an integer, whereas superellipses ($n > 2$) are not.

The ellipse is periodic despite its continuity because any point on the perimeter is both a starting point and the corresponding endpoint. Thus, unless a curve comprising parts divides evenly by an integer, it remains open and spirals outward like a nautiloid. This is how ancient Greeks inferred the link between mathematics and physical reality, concluding that substance is periodic, atomic. Is that sufficient reason[18] why mathematics makes so much sense of Nature?

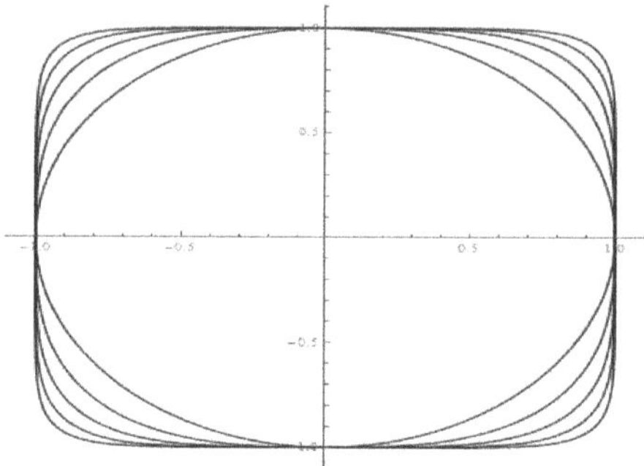

The ellipse, the innermost curve, is the shortest among the curves. Unlike the ellipse, the higher-order curves, the superellipses, are not periodic.

GEOMETRY AND SYMMETRY

The connection between a change in geometry and a break in symmetry is tangible when the structures are known. For example, when the electron torus opens and one of its loops, a neutrino, breaks loose, the W^- boson emerges.[19] In general, whenever a gap is introduced into a closed curve of quanta, symmetry breaks and curvature alters, i.e., mass changes. Specifically, the renowned Yang-Mills mass-gap[20] problem[21,22] links mass with symmetry. The photon wave and the neutrino loop cannot transform from one to the other smoothly, continuously without a gap, only by opening and closing.

The wireframe models of particles also reveal the kind of rotations, reflections, and reversals that leave the structures as they are. These operations that change nothing define a symmetry group. The symmetry group of electromagnetism is U(1), that of the weak force is SU(2), and that of the strong force is U(3). The three form the symmetry group of the Standard Model U(1)×SU(2)×SU(3).[23]

On the one hand, the Standard Model of particle physics says that a massless particle cannot be torn asunder. The photon is eternal. On the other, modern physics concocts virtual photons that exist for only a fleeting moment. The fact that the virtual-particle model is used solely for calculation does not cure the inconsistency. If the calculation does not correspond to reality, what is the point?

The nautilus shell closely traces a logarithmic spiral.[24] This growth pattern is among the most common forms of natural processes.

The conservation of momentum troubled Newton and his contemporaries because the substance of the gravitational field was unknown to them. Now that we have this in hand, we understand that the same substance, released when a body is moving away, gets reabsorbed when the body is coming back. In other words, physical quantities are conserved when a body in a gravitational field, or a charge in an electromagnetic field, does not lose in changes of momentum per cycle anything more than it gains.[11] Conversely, asymmetry entails imbalance driving irreversible evolution.[25]

Beauty is about optimality. Our preference for certain relationships, such as the golden ratio, is no coincidence. A sequence of golden ratios spirals out logarithmically, yielding the familiar form.[16] We regard symmetry as beautiful, but perhaps perfect repetition is not the most beautiful form after all. The atoms in crystals are stacked in flawless arrays, but what is free from variation is devoid of development. The balance of forces is harmonious, but we shun stagnation. Everything keeps changing.

WHAT IS MATH TELLING US?

It is often the case that a matter is truly understood when the words expressing it have been found. Likewise, a theory is not clear until its mathematical form is at hand. Darwin's tenet, the narrative about the survival of the fittest, remains somewhat abtruse[26] until expressed as the equation of evolution.[27]

However, a mere equation can also be ambiguous in its meaning. For example, Planck discovered the quantum from an equation, but it was Einstein who suggested that Planck's law is about the quanta of light. Likewise, Dirac found the positron from another equation, but it was Weyl who suggested that the second solution of the Dirac equation is the electron's antiparticle, rather than the proton, as Dirac himself thought initially.[28] In mathematics, there is room for interpretation, albeit none for error.

It is a different thing to interpret mathematics as reality than it is to rationalize reality in mathematical terms. For instance, when deriving thermodynamics from one's own experience, it becomes apparent that Planck's constant does not mean the quantization of energy but rather the invariant measure of the quantum. Moreover, the two solutions of the Dirac equation do not mean point-like particles but tori of quanta, the electron and the positron.

In any case, mathematics keeps a theory within its postulates. An alternate view, therefore, requires different foundations. Rather than considering a falling apple, Einstein inferred that he would not sense gravity if he were to fall freely. Only after the mathematical form for this insight was found in Riemann's non-Euclidean geometry could

Einstein test his thesis. The idea of the free fall seems to ring true, doesn't it? However, no one is free from the gravitational field of the expanding universe. Without the formalization of general relativity, it would have been difficult to notice that the observations, particularly on a galactic scale, do not match Einstein's intuition precisely.

Mathematics makes our doubts more definite than mere words. If Schrödinger had not put his model of a system into a mathematical form, we would have no specific object to doubt. Now we have. We can find a solution to the Schrödinger equation[29] in which time does not exist at all. The famous formula models endless circulation. But the world is constantly changing. The wrinkles on my face are all too real. I've lost quanta irreversibly. Likewise, unless the flow of time was put in the form of an equation, we could not test whether Parmenides' idea of the indivisible element is correct.

Many years ago, when recording molecular spectra, the decaying signals showed me that the atomic nuclei were spinning out of sync. So, a damping term added to the Schrödinger equation's endless rotation mimics the increasing decoherence.[30] That model matches the data of many other processes as well but explains none of them. It fails to reveal the quanta the decaying system is losing. It would have been better to describe the evolution toward balance with the universal equation of change rather than reworking the equation of balance in hopes of making it a model of change.[31] Often, a small shift in viewpoint is insufficient to clear things up, but you have to take the opposite stand to grasp reality.

OUT OF THIS WORLD

Quantum mechanics is a model of periodic motion, such as that of the quanta that form the electron. The circulation can be in any phase so long as we are free to define the point of observation. This randomness is neither strange nor unique to the microcosm, let alone evidence of entanglement or the multiverse. What the second hand is pointing at when you glance at your watch is similarly arbitrary.

When mathematics is not properly cognized, the equations can even be misunderstood; particles are imagined as virtual or envisaged as excitations of the vacuum's fields. Without mathematics, we would hardly think about such oddities. Then again, the exactness of mathematics forces us to judge modern physics either as real all the way up from its foundations or as unreal all the way down to its roots.

Baggott says that theoretical physics proceeds impeccably from its postulates to its conclusions, but mathematical consistency is no guarantee that the results are from this world.[32] Even though the world changes, the steady state is held to be the norm of physics. That is why stationarity is known in precisely defined terms, such as equilibrium, conserved, commutative, computable, linear, Euclidean, and deterministic. In contrast, the full range of processes is referred to by vaguely understood terms, such as nonequilibrium, non-conserved, non-commutative, non-computable, nonlinear, non-Euclidean, and nondeterministic. Indeed, the equation of evolution seems to be out of the ordinary for seasoned theoretical physicists, like Professors Enqvist and Kajantie at the University of Helsinki, who spotted this difference right away from the formula in my presentations.

OUT OF TOUCH

Plane geometry, familiar from school math, is a special case, for we live on the globe's curved surface. Gauss not only studied non-Euclidean geometry but considered space itself as curved. It is said that in 1820 he climbed the three mountains Hohenhagen, Brocken, and Inselberg in Central Germany and measured the adjoining angles of the three peaks.[33] Their sum was 180 degrees: light travels in straight lines. The Earth's gravity is too weak to make a visible difference relative to free space.

But the idea of curved space stuck. In 1854, Gauss' student Bernhard Riemann formulated a curved geometry,[34] which Einstein adopted as the mathematics of general relativity.[35] The Einstein field equations delineate how space-time curves about a celestial body. But, the equations are derived from the Einstein-Hilbert action using Lagrange's stationary rather than Maupertuis' evolutionary principle of least action.

The German mathematician David Hilbert, unmatched in many areas of mathematics, worked on general relativity and had a hand in quantum mechanics, too. Modern physics is thoroughly mathematical.

It is mesmerizing, but sophistication does not make its import real. Big words do not make a tale true, either. Have physicists lost their touch with reality on their way to the world of mathematics?[36]

Initially, Einstein made his theory agree with a steady-state universe. So even when space-time stretches, its symmetry stays the same because the metric of general relativity is such that the dimensions of space and time are opposed to each other.[37] In reality, symmetry stays put only in balance. Symmetry is thus not an inexplicable initial condition but rather a natural outcome of events.

What is solvable?

In school, we are taught that probabilities always add up to one. This unit norm is also the criterion for solving the renowned Riemann zeta function.[14] This condition of balance is the natural way to prove the Riemann hypothesis, for the zeta function occurs in many models of stationary systems, such as nuclei, crystals, and synchronized systems called condensates.[38]

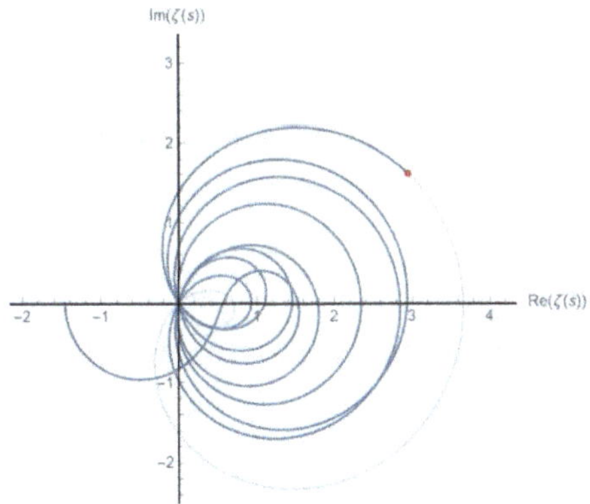

The Riemann zeta function loops through zero when the real part of the complex function is ½.[39] This condition of the unit norm is equivalent to a physical balance.

Euler figured out how to write the zeta function as a product of its factors. His formula associates a prime with each state of balance. Hence these zeros are unique, non-trivial. Finding a new prime is as surprising as the emergence of a new compound or a new species. The analogy may appear unwarranted, but as Marcus du Sautoy, a professor of mathematics at Oxford University and a well-known figure on BBC TV, says, "Primes are the atoms of arithmetic."[40]

If you wish to contribute to the search for the biggest prime, you may donate your computer's capacity to an international project.[41] So far, the largest prime, found in 2018, has over 24 million digits. It is a lot but still short when compared with the number of quanta in the universe, about 10^{120}.[42]

Since the 1950s, the largest known prime has stepped up almost in a power-law manner. This trend is consistent with the thermodynamics of time. We can, therefore, presume that it is not possible to predict the primes, not even from the condition for zeros of the Riemann zeta function.[14] There is no algorithm for finding primes because the search for factors changes the search itself: a newly found prime will be a factor of yet greater numbers. This conclusion is also propped up by the fact that the first digit of primes, like that of numbers in sequences, is 1 in about 30% of cases and 9 in less than 5% of cases.

According to thermodynamics, numbers do not exist without physical form. Likewise, new words and words altogether do not exist until spoken. From this perspective, reality is defining mathematics rather than mathematics determining reality.

I am ashamed to admit that I deemed researchers who piqued interest in numbers somewhat bizarre. Not anymore. We may live in a far more ordered universe than we ever thought.[43] It is possible that numbers, especially primes and natural constants, such as Planck's constant and the fine-structure constant, are loaded with truths about reality. It is for us to figure out how to extract them.

Eternity and Continuity

John Wallis designated infinity with the number eight tipped over (∞).[44] The English mathematician needed the concept of infinity when pondering the age-old problem of forming a square with an area equal to that of a circle. In 1655 Wallis squared the circle with an integer expression extending to infinity and approximating Pi (π).

As is well-known, π is not a ratio of two integers. The discovery of such an absurd, irrational number shocked the Pythagoreans. It implied that substance is not divisible into elements but continuous. Continuity, an endless division, Wallis logically denoted with $1/\infty$.

Even though Newton invented differential calculus based on continuity, he did not consider a curve as continuous but rather points in motion.[45] Differential equations, such as the Navier-Stokes equation and Newton's second law of motion, are convenient to analyze, but substance is, after all, divisible into the indivisible quanta. Water feels continuous but comprises water molecules. A stream of light seems continuous but consists of photons.

In the well-known story, Achilles is supposedly never going to overtake a turtle, as the turtle will always draw a little forward in the interval when Achilles sprints to the point where the turtle had just been. Although a distance could be divided infinitely in this way, the total length remains finite. Shortening the step size of a calculation without any limit below the quantized reality leads to conceptual problems of continuity and infinity, as encountered by the ancients.

Of course, Achilles will catch up with the turtle. Likewise, matter will stop being squeezed endlessly into the abyss of a black hole. But in quantum electrodynamics, distances are deemed to diminish without any bound, and hence interactions increase without limit.[46] One ought not to divide by zero. This is acknowledged and dealt with by mathematical renormalization, but many scientists think there must be a more realistic way. Dirac was frank: "This [renormalization] is just not sensible mathematics. Sensible mathematics involves neglecting a quantity when it turns out to be small – not neglecting it just because it is infinitely great and you do not want it!"[47]

As long as we realize that the vacuum quanta do not shrink unrestrictedly around a finite particle, quantum field theory works fine as a model of the void. Neither do the particles in a black hole contract endlessly but unwind into the quanta of the void. The quantum itself is the constraint that prevents calculations from running off to infinity. Renormalization is pointless, for the quantum is equal to Planck's constant under all circumstances.

Physicists are flummoxed why the vacuum energy density is, in reality, 10^{120} times lower than the value by the standard theory.[48] Such an egregious error is made because the standard theory allows energy to increase to infinity over an infinitely short distance. This cannot be true. Since modern physics is short of substance, the calculation is terminated at a theoretical length, Planck length, whereas the quantized particle and the vacuum impose the actual limits. The Planck units are derived from the properties of free space alright, but without understanding the vacuum structure. When difficulties stem from the foundations of a theory, they cannot be overcome within that theory.

Despite the seductive models, let us face up to the truth as Tegmark does: "We have no direct observational evidence for either infinitely big or infinitely small."[8c] The universe is not infinitely vast. Even after its expansion, there are always the same 10^{120} quanta as there were at its birth.[42] The quantum, the least of things that exist, is not infinitely small, either. Numbers are finite, as the Dutch mathematician and philosopher L. E. J. Brouwer reasoned. They become ever more exact as more digits reveal themselves in producing values with greater and greater precision.

THE REAL PROBLEMS

From a historical perspective, the fundamental problems of science are not mathematical. It is the foundations of science that are problematic.[49]

At the end of the 19th century, some foremost physicists thought particles were ether vortices. The issue with this view was that there could be too many kinds of vortices. Present-day physicists imagine that the particles are strings, but the issue is that there could be too many kinds of them.[50] The theory of supersymmetry, in turn, does not seem to match the measurements,[51] and the inflation theory[52] is in trouble because it lacks ways of falsification. As a rule, the more obscure the basis of a tenet, the wider its spectrum of conceivable worlds. By contrast, ancient atomism, in the form of the quantum as the fundamental element, constrains our imagination by spelling out the facts that could not be otherwise – the truths of nature.

Mathematics governs modern physics for a good reason – the equations lay out unambiguous tests. Then again, mathematics imprisons physics within its problems when it is insisted that new thinking must be mathematically equivalent to the old. Most notably, evolution does not fit into the equation of equilibrium.

Mathematics verbalizes the world, but the wording can also be irrational. That fact shocked the ancient Greeks, to whom numbers were geometric entities, elements. So, what is there to count on if not mathematics? Contemporary physicists, too, are inclined to contemplate what sort of reality mathematics represents and believe incomprehensible results for a while.

Paraphrasing John Hunter, the Scottish surgeon and initiator of applying the research-based method to medicine: knowledge is important only to the extent that it leads to the principles.[53] Some principal questions about mathematics are:

- *What is mathematics for?*
  Mathematics emerged from the bookkeeping of items. It advanced to formalize experience and experiments into the laws of nature.

- *What is the relationship between mathematics and reality?*
  Even if a concept is mathematically consistent, there is no guarantee that it is consistent with reality.

- *What does mathematics miss from reality?*
  We can write equations in the same way we write sentences, but a tale, just like a formula, however precise, remains only an account of reality.

- *Why do we theorize?*
  A theory is a guide to reality, helping us to live.

Silly consensus

Tegmark examines the nature of existence, including human nature, in his book *Our Mathematical Universe: My Quest for the Ultimate Nature of Reality* (2014). He is disappointed by physicists' sheep-like conforming to doctrines. Praise of dissidents and rebels tends to be the mere doling out of empty phrases. The narrow-minded milieu poisons the profession. Tegmark says that those interested in the biggest questions will soon find themselves working at McDonald's. To make his point,

Tegmark cites an email he received from a senior fellow who reprimanded him for crackpot publications and warned him against jeopardizing his career.[8a]

If a merited professor's insightful research is seen as crazy only because it differs from the mainstream, the scientific community has lost its ability to revise its positions. Then it is no longer seeking the truth in an upright manner but bowing to prejudices. Let us recall that reality is problem-free; our perceptions of reality are problematic.

A theory's truthfulness should be judged not only against observations and measurements but also by studying the message itself, especially in its mathematical form. That is what Tegmark does and how he justifies his work: "To me, the key point is that if the theories are scientific, then it's legitimate science to work out and discuss all of their sanctions even if they involve unobservable entities."[8c]

It is precisely the mathematical analysis of modern physics that has led us to extraordinary conclusions.[54] In quantum mechanics, the order of cause and effect can be swapped in an experiment that seems to display action at a distance.[55] While such calculations agree with the data of a stationary state, the conclusions are nevertheless strange. The world we know is evolving.

HOW TO MODEL REALITY?

One tends to walk the length and breadth of an unfamiliar supermarket. A traveling salesman faces a similar problem.[56] This is one of the Millennium Prize Problems that the Clay Mathematics Institute listed in 2000.[57]

Curiously enough, there is no algorithm to work out the most inexpensive travel plan for the salesman from one city to the next and finally back to their hometown. Nobody stands a chance of solving the problem because the calculation changes as it is conducted.[58] When the salesman arrives at a city, the number of cities still to be visited decreases by one. This affects the optimization of the remaining route and so on. In the worst case, the cost of each conceivable travel plan must be evaluated all the way to the endpoint. Hence, there is no effective rule for the calculation.

Problems are wicked not only because of complexity but also because everything hinges on everything else.[59] Even problems involving only three bodies are unsolvable, as the motion of one body, say the Earth, affects the forces that act on the other two, say, the Moon and the Sun, and vice versa. Although the trajectories of bodies are unpredictable, not just anything can happen. Only those paths with free energy can be taken.

The hard way

In the mid-18th century, Thomas Bayes reasoned that the present follows from the past in the same way as the latest judgment follows from previous perceptions.[60] The English reverend's idea of conditional probability was radical compared with the concept of static probability that Descartes and Fermat formulated for gambling. Bayes understood that knowledge accumulates from subjective inference. Objective reality is not only unknown to us, but it does not exist. In other words, it is not only hard in practice to find out all the necessary minutiae, but perfect knowledge is, even in principle, impossible because the act of knowing changes the object to be known.[31]

Today, artificial intelligence uses Bayesian probabilities to model alternative courses of events but without comprehending causal connections.[61] Neural networks produce output by correlating input with input from a training set, whereas we relate what we see with what we have experienced. For example, learning algorithms routinely tag objects based on textures, while we humans spot objects by their shapes.[62] We have learned to link causes with consequences the hard way. Likewise, only when a robot blames itself for bad decisions will the responsibility of free will affect its decision-making.

It is also fashionable to simulate courses of events using so-called Markovian chains as if events were coincidences. In reality, events lead to further events. Consequently, the acausal simulation does not find its way to balance. Instead, it must be steered into a predetermined final state. So, even a perfect match with the data does not make sense of the data.[63]

The mathematician Aleksandr Lyapunov from Saint Petersburg, a good friend and colleague of Andrei Markov, understood the least-time nature of balance: the further a system veers from balance, the faster it eventually moves back.[64] We should thus expect that the forces of nature at a global scale shall strike back all the more fiercely the further away we perturb the planetary balance by our way of life. At worst, we might have already passed the point of no return, whereafter unprecedented forces are brought forth.

A TRUE CHALLENGE

The Millennium Prize Problems are famous, but fame does not automatically signify the genuineness of a challenge. Perhaps it is just difficult to free ourselves from our persistent but misleading patterns of thinking. So how do we learn to ponder differently? The answer may well require departing from prevailing preconceptions about thinking.

A layperson finds many a mathematical problem unimaginably abstract. In turn, a mathematician may find it unthinkable that a mathematical problem could have a real-world counterpart. And yet, it may be precisely the concrete correspondence that helps us puzzle out, for example, why the traveling salesman problem cannot be solved and realize that many other issues cannot be settled for the same reason.

The wordings of Hodge's conjecture, one of the Millennium Prize Problems, may seem at first glance incomprehensible. The essence of the conjecture is nonetheless plain: a system comprised of subsystems adapts to the surrounding system, and concurrent changes in the subsystems lead to a change in the surrounding system.[20] For instance, a community of people adjusts to the natural habitat so that changes in habits go together with changes in the habitat. Thus, an ecological niche is both produced and occupied. Likewise, an electron and the surrounding void make up an inseparable pair, a so-called dual, where one is affected by the other.[65] In contrast to physical realism, Dirac reasoned from Maxwell's equations that the dual of the electric charge would be the magnetic monopole rather than the void. Similarly, physicists consider today dark photons as the dual of dark matter,[66] since, for them, the vacuum is vague and its dual, matter, murky.

Richard Hamming summed up the art of practicing science in ten rules.[67] Among other things, the American mathematician, whose work had a major impact on telecommunications, advised leaving the door open for ideas. Hamming based his rules on how he had seen prominent scientists working. In the Manhattan project, Hamming was assigned to double-check calculations done by physicists. On one occasion, he asked what the computation concerned – and got for an answer: "It's a probability that a test bomb will ignite the whole atmosphere."[68] The numbers checked out all right, but the physical grounds were uncertain. A friend comforted the horrified young mathematician: "Never mind, Hamming, no one will ever blame you."

Hamming did his life's work at the Bell Laboratories with Claude Shannon and John Tukey. Hamming recalled the atmosphere as inspiring: "We did unconventional things in unconventional ways and still got valuable results. Thus management had to tolerate us and let us alone a lot of time."[69a]

Today, after company mergers, Nokia owns Bell Labs. There is always a shortage of big ideas. Hamming advises us to focus on key questions and reminds us that understanding often comes unexpectedly. The mathematician condensed the meaning of mathematics: "The purpose of computing is insight, not numbers."[69b]

REALITY CHECK

We think of science as a study of Nature, but in fact, it merely produces interpretations of reality, often based on mathematics. Such inferences are not solely founded on the observations at hand but mainly on our previous conceptions – or misconceptions. Sometimes it is thus more profitable to study our reasoning per se than to go on with studies posited on our reasoning.

Often, new thinking is dismissed by saying that "extraordinary claims require extraordinary evidence," but the same mantra should then apply to instituted thinking. Naturally, all new seem speculative, but seemingly established quantum entanglement, wave function collapse, action at a distance, space-time, dark energy, dark matter, cosmic wormholes, cosmic inflation, or multiverse are sheer speculations. All we have is mathematics interpreted as if it was real.

Mathematics is a formalism akin to a natural language. It describes events with astonishing accuracy, quantum by quantum. The precision is necessary to account for all causes and effects. When we understand reality as it is, we also recognize those mathematical models that are unreal. The reality check of ground-laying science is about realizing that modern physics is only a mathematical model of data rather than an authentic account of reality.

KEY POINTS

- Mathematical consistency is no guarantee of verity.
- Continuity and infinity are convenient concepts but unreal.
- Equations of equilibrium can be solved, whereas those of change cannot.

PART II
THE SIGNIFICANCE OF THE WORLDVIEW

The basic aim of academic inquiry
is to help us develop
wiser ways of living, wiser institutions,
customs and social relations, a wiser world.

Nicholas Maxwell

6. How did life originate?

We search for the origin of life.
But do we have any evidence of life?

We classify Nature into living and lifeless by habit, but data do not show such a division. Distributions of all kinds are comparably skewed. Big fish are far fewer than small ones; big boulders far fewer than small stones. From the data alone, we cannot say whether the distribution represents fish or stones. Numbers without units, curves without labels, and graphs without legends do not distinguish between living and non-living. We make up that distinction ourselves.[1]

Distributions are skewed, whether for populations of particles, animals, plants, or stars. Growth curves are S-shaped, whether for molecules, bacteria, companies, or cities. Data of all kinds closely follow power laws giving us no clue whether it is about molecules, species, or consumables.[2] The universal patterns suggest that animate and inanimate are one and the same (Chapter 1).

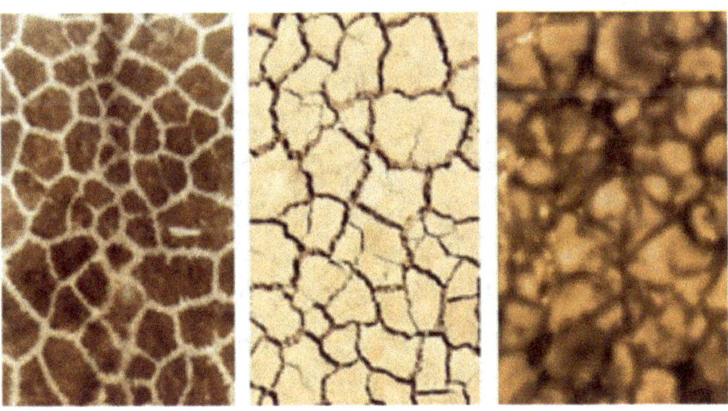

The similar patterns of giraffe fur, dry soil, and the surface of the Sun suggest that they display the same principle. (Pinterest, Stock Photo 16010939, Hinode JAXA/NASA/PPARC)

When we make a distinction despite there being none, it may well be that we are pondering mysteries that do not exist. The problem of life's origin comes to nothing by recognizing there is no division between the animate and inanimate. All processes comply with the same natural law. Everywhere the quanta move so that imbalances level off in the least time (Chapter 2). For instance, both speciation and erosion progress in the same way. The universal principle is one; only the mechanisms are many.[3]

A THIN LINE

Even though it is practical to classify things into living and non-living, there is no principal difference. The animate and inanimate display the same statistical characteristics. For example, trees and firms grow alike. At first, it will take a couple of years before a seedling begins to grow quickly. After that, the growth runs its course, and finally, the tree ages. Similarly, a company will prosper after the initial difficulties have been overcome. In the long run, the growth will level off to a relatively staid subsistence when the market is saturated. Erosion proceeds in the same manner.[4] When the S-shaped curves for each of these growth patterns are plotted on a log-log scale, they mostly follow straight lines, a hallmark of the *Grand Regularity*.

Already Galileo noted that the weights of various mammals present the same mathematical relation to the thickness of their bones. The flight speeds of insects, birds, and airplanes are likewise proportional to their weights.[5] This power law ranges from about 10 milligrams for a housefly to more than half a million kilograms for an Airbus 380 aircraft. In this fashion, D'Arcy Thompson illustrated the similarity between biological and mechanical forms in his peerless book *On Growth and Form* (1917). The Scottish biologist and mathematician was convinced that a more general principle than evolution by natural selection directs development into the ubiquitous patterns.

Another hint toward the fundamental equality of the living and the non-living is the unresolved difficulty in defining life.[6] Is a virus alive, given that it cannot reproduce without a host? Then again, what can? An animal needs an ecosystem; a local store needs a neighborhood. Reproduction and speciation are not exclusive hallmarks of life, neither adaptation nor differentiation. For example, there is an increasing diversity of household items that are fit for various and changing habits, just as there is a diversity of animals and plants fit for various and changing habitats. In every respect, an economy shares the characteristics of an ecosystem.

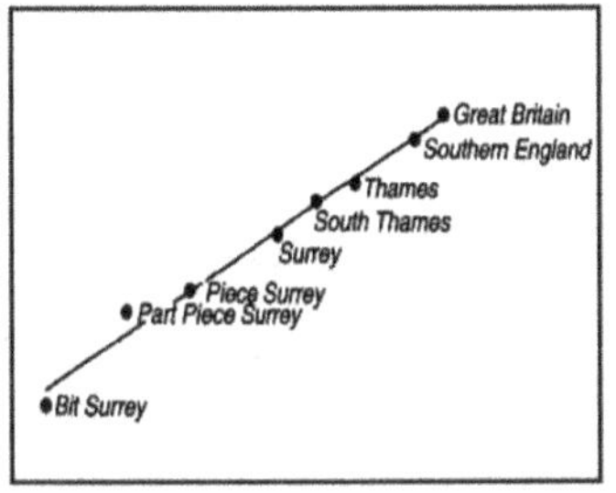

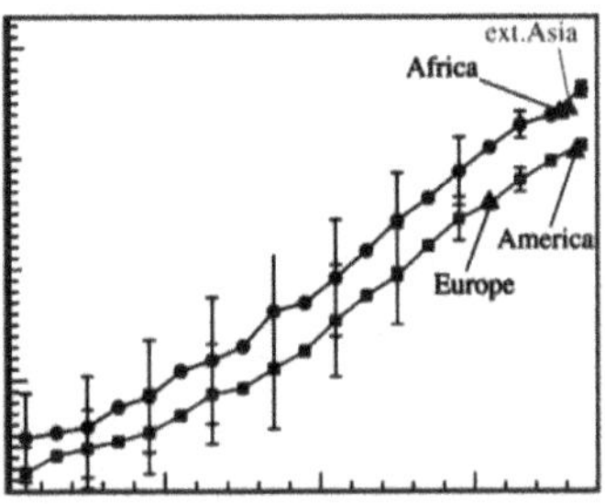

The British botanist Hewett Watson discovered in 1859 that the larger a land area, the more plant species it holds (left).[7] Likewise, the larger the area, the more languages are spoken (right).[8]

Unity has no clear-cut boundaries but engulfs everything; hence, reductionism does not make sense. Already Darwin was keenly aware of the ambiguity entailed in categorizing: "I was much struck how entirely vague and arbitrary is the distinction between species and varieties."[9] Today we know that even one quantum is enough to move an object, say a particle, from one category to another. For example, a single quantum of light may trigger a sensation by converting a retinal molecule in your eye from the *cis*-isomer to the *trans*-isomer. Moreover, it is difficult to define a cell unequivocally, for we can remove substance from a cell, in principle, quantum by quantum, without encountering any cell-specific limit. This form of Zeno's paradox exposes that the central concept of biology, the cell, is too vague to bear the weight of the solid worldview.

A thin line separates the living and non-living, say, being active and passive. Organisms are not only dynamic but also dormant for long periods, of which viruses and bacteria, as well as bears, are prominent examples. Likewise, simple substances, such as hydrogen and oxygen gas, are inactive or highly reactive depending on circumstances. We do not always even voice a difference between living and non-living.

For example, we say that an old car is at the end of its *lifespan* and that good old times *live* in our memories.

Taxonomy is not just boring sorting. We grasp reality according to the way we group it. Changes in classification entail changes in our worldview. Since the natural and synthetic are both outcomes of processes, there are no impermeable borders to protect us. For instance, could artificial intelligence evolve like our own intelligence? Perhaps it could eventually rival and even quell us. From this holistic perspective, the future is only a matter of time.

Ancient atomism spells out a seamless unity of living and non-living. Many a mythology tells the same story. Neither does thermodynamics make a difference between the two. While contemporary science has come to hold the animate and inanimate to be the same, it still upholds disciplinary boundaries.

Not only living but everything is in evolution toward a balance. A plant grows toward light because, in that way, matter seeks balance with light. A stone falls to the ground because, in that way, the vacuum tends toward equilibrium. "The aim of science," as Henri Poincaré, mathematician, physicist, and at one time the head of the French Academy of Sciences, put it, "is not things in themselves, as the dogmatists in their simplicity imagine, but the relations between things."[10]

Customarily, we think of a substance as alive when it gains quanta and dead when it gives away quanta. But, in fact, it is the environment that dictates the course of events. Light powers the growth of a seed into a plant, whereas the plant will die in the dark. Conversely, a stone will erode into sand, but the sand will melt into magma and solidify to become stone again when it is hot enough.

Even so, when contrasting the Amazon rainforest with the Atacama Desert, it may be hard for us to imagine that substance simply adapts to the prevailing conditions. While the quantitative differences are enormous, there is no qualitative difference. Matter has evolved into diversity on Earth, whereas it has stalled in scarcity on Mars.

Not only organic but also inorganic matter on Earth has evolved. Now that circumstances have become amenable, much has happened in a short time. For example, elevated manufacturing temperatures have transformed silicon, the second most common element in the

Earth's crust, into the main component of artificial intelligence. Similarly, carbon once took up its position as the main component of life. So we see that the development of technology parallels the evolution of life. Artificial intelligence, having some 100 million transistors packed onto a square millimeter,[11] is evolving along the lines of force as natural intelligence, having an estimated 10 billion protein molecules in a cell,[12] has already done over eons.

PURPOSE

What do we see when we see purpose in life? Are we impressed by the strength, adaptability, or perhaps beauty of living beings? Are we amazed by the modularity, symmetry, or some other attribute of an economical feature in the creature? Do not these and other characteristics only tell us about the organism's ability to consume free energy, that is, evolve along with the lines of force?

We tend to associate purpose with design since we design things, tools, and machines for their intended uses. However, are our objectives any different from the universal aim for balance? For example, the weights and fuel consumption of four-wheeled vehicles, from a beach buggy to a mining truck, are distributed in the same skewed manner as the weights and metabolic rates of different mammals.[13] Crane trucks and giraffes are few, whereas cars and antelopes abound. The *Grand Regularity* is self-evident. Is it too obvious to ask for its cause or seek its explanation?

We regard a product as poorly designed if it does not fulfill its purpose. Such goods and gadgets are dumped, and others are tried out until a good one is found. Does not everything develop in that way, too? A species evolves without knowing where, when, and how to attain balance in its environment; a rivulet winds its way to the sea without knowing where the sea is.

Sometimes animates are considered exceptional because they exhibit irreducible complexity; the sum is more than its parts. If something is taken away from the organism, it dies or at least loses a lot of its functions. Is not a chemical compound equally irreducible? When cleaved into parts, the parts do not have the compound's properties. In each cleaving, some quanta invariably escape into the surroundings.

There are several examples where similar functional forms have evolved from different starting points. For example, birds, bats, and insects fly but use different kinds of wings. The course of evolution is not always evident, but some forces always drive it. Chemical energy has guided and still guides the development of bats and birds as well as other creatures. The bat has its flying membrane, and the bird has its wings to get food. However primitive, the primordial flipper was better than nothing; however simple, a prototype is much better than nothing in meeting demand.

The convergent courses of events are typical not only of animates but also apply to the evolution of household goods. Glass and plastic bottles, as well as aluminum cans, serve much the same purpose, while each of them is optimal for particular needs. Similarly, the meanings of some words are converging, while others are diverging. Evolution is not a random process but takes its course along the lines of force. As the motion affects its own course, this teleology is neither goal-oriented nor designed.[14]

A FEAST FOR THE EYE

The eye was not designed for seeing but developed to allow an organism to access food and other resources in light. Conversely, the eye became unnecessary when nutrition was found only in the dark, of which some bat and cavefish species are glaring examples. In this light, we see that evolution, at all levels of Nature's hierarchy, moves along the lines of force – irrespective of what those forces are.

Darwin devoted an entire chapter in *On the Origin of Species by Means of Natural Selection* (1859) to the development of the eye. Even a slight ability to sense light was better than nothing. So, minor improvements, one after another, eventually converged into an advanced eye. In flatworms, the light-sensitive detector is, at its simplest, merely a nerve cell surrounded by pigment cells.[15]

It stands to reason that the eye, as a sensory organ, developed before the brain. What would the brain have had to process if there were not already sensations? Even today, signals propagate from the eye of an octopus directly to its tentacles so that no time is wasted in processing visual perceptions while the prey might escape.[16] Signals from the optical cells of elevator doors are likewise directed to the door control so that no time is wasted in processing while someone might get caught between the doors. We, too, act unconsciously, by reflex, to make a narrow escape.

Thermodynamics makes sense of evolution on all scales by covering the whole spectrum of changes in one direction or another. Energy differences, of whatever kind, will decrease in the least time. Many different mechanisms expedite free energy consumption, such as organisms with eyes and doors with photocells.

It is easy to see that biophysics, the discipline of my professorship, unites the living and lifeless into one. I knew this from the start, but I did not fully understand its grand implications right away. It struck me only in the late fall of 2001. The days were already chilly when a flock of long-tailed tits flew to a birch tree in our yard. I had not seen them before, although they are not exactly rare, either. As birdwatchers know, long-tailed tits are actually not tits but belong to the order of sparrows and the genus named by Aristotle, *Aegithalos*.[17] The squeaking and swirling flock came like a shower of hail all of a sudden and was quickly gone. I was left wondering why the motions of birds and rain are similar, universal. It is this fluttering flock that awakened me to the *Grand Regularity*. I soon realized that many had noticed the same pattern. Its reason has nonetheless been shrouded until now.

WHAT IS EVOLUTION?

Evolution is merely a sequence of events.

Charles Darwin's cousin Francis Galton already understood that evolution by natural selection is a statistical theory by nature, but it remained unclear until today that statistics summarizes causal rather than random processes.[18]

Evolution not only creeps forward little by little, one mutation after another but also leaps when opportunity knocks. Stephen J. Gould and Niles Eldredge referred to this intermittent course of events as punctuated equilibrium.[19] Punctualism contrasts with gradualism, the view that evolution is a gradual process. The truth lies on both sides of scientific controversy. Evolution leaps when free energy pours into the system; conversely, evolution stalls when free energy runs out. The sudden changes are analogous to the phase transitions between solid, liquid, and gas, while the piecemeal changes parallel continual processes, like increasing temperature in a phase.

Imbalance, in one form or another, is the cause of events. It drives the synthesis of chemical compounds, growth of organisms, development of ecosystems, and evolution of the biosphere in its entirety. Evolution is not geared toward any predetermined state of balance but consumes free energy in the least time, which results in *Grand Regularity*. When the balance of forces is attained, the purpose is fulfilled, and evolution ends. The Finnish academician Oiva Ketonen understood that ideological movements also terminate in mere maintenance, bureaucracy, having achieved their goals. Presumably, science will also culminate in such a state, having explained everything.

Realizing that biological evolution is no different from other courses of events, we can understand life more profoundly than in the limited terms of biology. Various processes happen in the same way but only employ different mechanisms. "Analogy frees us from the pain of thinking new things and the even greater pain that uncertainty may cause,"[20] as Maupertuis wrote in *Vénus Physique* (1745), the bestseller of his time.

A beneficial mutation, like a technological innovation, makes things happen more rapidly than before. For example, the primary form of an insect was a major innovation at the beginning of the Devonian period. This success story is repeated in the multiple and diverse insects that have followed. Similarly, the airplane was a significant invention. Its triumph is reiterated in many kinds of aircraft. A new viable tenet, too, renders new resources available.

In Denial

"Nothing in biology makes sense except in the light of evolution,"[21] claimed the Ukrainian-American evolutionary biologist Theodosius Dobzhansky in 1973. So what is evolution all about?

Leibniz called kinetic energy a living force (*vis viva*) and potential energy a dead force (*vis mortua*). Émilie du Châtelet, also a notable natural philosopher of the era, was acquainted with Leibniz's writings and translated Newton's *Principia* into French. She recognized that energy changes in every change, in a developmental phase, just as in an evolutionary step. These flows of energy, carried by quanta, from an organism to its environment or vice versa, are typical but not unique to the living. Maupertuis' principle of least action mathematizes these transformations. His comprehension is no coincidence. Du Châtelet and Maupertuis wrote ardently to each other about the foundations of physics, even after their love affair had ended.[22a] The life of the Marquise du Châtelet has been memorably retold in the opera *Émilie* (2009) by Finnish composer Kaija Saariaho.

Maupertuis is recognized as a pioneer of evolutionary theory. He maintained that survival is neither random nor prearranged but a consequence of selection by Nature. Maupertuis was among the first to examine populations instead of individuals. He understood the principle of heredity, the evolution of species, and statistical variation. Moreover, Maupertuis noticed that dogs and mice alike could become immune to scorpion poison, a piece of knowledge from which Louis Pasteur benefited a century later. Maupertuis was convinced that reproduction is a manifestation of the innate ability of substance to organize itself.

Maupertuis must have thought about the spontaneous birth of life, abiogenesis, for he asked the French naturalist Georges-Louis Comte de Buffon, working with the British scientist John Needham, to put under a microscope not only samples containing microbes but also just plain water! An excerpt from a letter to the academician and old friend Charles Marie de La Condamine discloses Maupertuis' holistic thinking: "[Based on Needham's work], Here we can say that the structure of the tiniest insect is more marvelous than that of the whole planetary system."[22b]

Maupertuis understood biology, but did biologists understand Maupertuis' reasoning? Some hundred years later, when *The Origin of the Species* appeared (1859), biology began to distance itself from natural philosophy and became a subject in its own right. Evolutionary theory's principled and persistent objection has not hit the mark, but the prevailing doctrine is in denial, too. It is curiously circular reasoning that the most viable forms are naturally selected to continue life. By contrast, Maupertuis straightforwardly argued that evolution is neither ruled by a divine force nor a random choice but takes the course of least action.

THE LOST TELEOLOGY

Biology is too narrow a subject to win controversies over worldviews by its own means. In truth, there is nothing in biological evolution, as such, worth defending because the *Grand Regularity* emerges from all kinds of processes. Popper praised this universality: "Simple statements, if knowledge is our object, are to be prized more highly than less simple ones because they tell us more; because their empirical content is greater; and because they are better testable."[23]

The puzzle of life's origin is not about life; it is about missing teleology. We have been taught that evolution results from random mutations and that the fall of a rock is a deterministic event, not that both are manifestations of the same law. Without holistic comprehension, contemporary science focuses on means rather than motives. So we ask how life came about rather than why it emerged. No theory of history or evolution can do without teleology, an explanation (*logos*) through purpose (*telos*).

It is not a chance but a cause that explains a consequence. Therefore, ideas with Aristotelian elements have reemerged. The systems biologist Denis Noble realizes that life is not reducible to genes, but natural selection occurs and operates everywhere. Genome and environment have an impact on each other.[24] This interplay between a system and umwelt was the very point of Alfred Russel Wallace, the man evolution left behind, and D'Arcy Thompson, the man ahead of evolution.[25] Moreover, James Shapiro, professor of microbiology at the University of Chicago, as an insatiable critic of orthodox evolutionary theory, reasons that evolution is a total rather than genetic response to the demands imposed by the environment.[26]

Thermodynamics expresses the same insight. The striving toward balance in the least time is a teleological explanation, although the goal cannot be set or seen in advance, for everything depends on everything else.

In his book Mind and Cosmos (2012), Thomas Nagel maintains that the explanation of evolution results from reorganizing the foundations of science. The weakness of materialistic reductionism in the laws of contemporary physics is that time has not but should have a direction, an ultimate goal, *telos*, since the whole cosmos is evolving. The final cause at the heart of Aristotle's thinking is more recently found in the analyses of the Finnish philosophers Eino Kaila and Jaakko Hintikka.

Nagel, a professor of philosophy at the University of New York, says he is regarded as a heretic for indicating the shortcomings of the prevalent mindset. In truth, there are no grounds to accuse Nagel of ignorant pseudoscience. He instead sees the faults of modern science in sharp focus. Nagel analyzes arguments and counterarguments with professionalism. He calls for a renewed reverence for common sense instead of blind willful worshiping of scientific doctrines. Only then can we find a comprehensive and coherent worldview. The fundamental questions remain open precisely because the science of our times neither acknowledges purpose as the least-time imperative nor embodies the flow of time with substance. The very need to comprehend reality is also a manifestation of natural teleology.

VITALISM

The existence of a special force of life, *vis vitalis*, was challenged when Hermann Kolbe, a founder of modern organic chemistry, succeeded in synthesizing acetic acid from inorganic starting materials in 1828. Thereby he established that the ingredients of life could form independently of plants and animals. Today, strands of DNA are industrially produced for diagnostics and research. Even infectious viruses have been assembled from their ingredients.[27]

Nonetheless, the spirit of vitalism is still alive. To all appearances, chemistry and physics fall short in explaining life, most notably its key characteristics of self-organization, complexity, and emergence. Is our view of reality thus flawed? Are we not seeing the woods for the trees? Does the old dichotomy between the animate and inanimate, no matter how decisively denied, reassert itself insofar as physics fails to explain the flow of time? All in all, balance is fleeting; evolution is lasting.

In this regard, Terrall recaps Maupertuis' thinking: passive is superficial; active is fundamental.[22c]

Random variation is a core concept of evolutionary theory. But is there truly random variation without any cause? This would require a change without any force, that is, a miracle. Near the balance point, the forces are small and motions alike but still not arbitrary.

Then again, a course along the lines of force may give an impression of a preset goal, a fixed plan, a specific purpose. However, are not the courses of Nature selected for expediency rather than an ultimate goal? Don't we, too, see many a process as more meaningful when it happens faster? The same logic applies to molecules. For example, the carbonic anhydrase enzyme makes the innately fast conversion of carbonic acid to carbon dioxide and water even faster.

Mutation in DNA may seem like a pure coincidence, but there is a reason for it, too. For instance, natural background radiation may modify the molecular structure. A mutation itself is a change in energy at the DNA level but more meaningful when it alters the organism's ability to consume free energy. Evolution naturally selects its course so that free energy is consumed in the least time. As life evolves, the conditions for life keep evolving, too. Hence the balance, or goal, cannot be known beforehand. This is teleology without a preset final goal.

The Nobel Laureate biochemist, Jacques Monod, examined the tension between randomness and purposefulness in his book *Chance and Necessity* (1970). Monod spoke in favor of purpose but assigned teleonomy exclusively to living beings. The American philosopher David Hull argued that the difference between teleology and teleonomy is only ostensible.[28] All processes, animate and inanimate alike, display a sense of direction, or purpose, by consuming free energy in the least time. A variety of forces drives diversification and specialization.[29] Only in balance, there is no purpose. The goal is reached.

In the soup

Darwin pondered over the origin of life in a letter to his friend Joseph Hooker, a botanist and explorer.[30] Publicly, however, he refrained from suggesting abiogenesis. His contemporaries had a lot to digest in the world-shattering — or rather, world-uniting — idea that all living beings have common ancestors. For many, it might have seemed all inconceivable that light and electrochemical reactions in a little warm pond would have engendered simple compounds, leading to ever more complex ones. Substance worked its way over the eons toward the full spectrum of life.

The idea of the primordial soup as the cradle of life was not forgotten. Alexandr Oparin argued in the 1930s that there is no profound difference between living and lifeless matter.[31] The Russian biochemist reasoned that the characteristics of life are products of matter in evolution. Rather than speculating, the chemistry graduate Stanley Miller decided to experiment with the autonomous genesis of life in the laboratory. His supervisor, the Nobel Laureate Harold Urey, shouldered the costs. Since discovering deuterium, a hydrogen isotope, Urey had moved on to contemplate the conditions that might have prevailed on the young Earth. The experiment carried out in 1952 was intended to mimic them.

After a week of electrical discharging into a mixture of gases, typical of an early Earth's marine atmosphere, complex compounds, such as amino acids, formed from simple organic substrates, but nothing more alive than these building blocks of proteins were found in the reaction chamber. What, then, was realistically expected?

The quest for balance drives chemical reactions forward, just like other events. The conditions of the Miller–Urey experiment were adequate only for what was produced. Many planets lack materials sufficient even for this outcome. The Finnish philosopher Ketonen reasoned that the time for a living cell to evolve from a soup of amino acids could not be shortened to the laboratory scale. From that standpoint, life emerged not soon but as soon as it was possible. Along the least-time path, no force is without some effect.

The nucleic acid moieties DNA and RNA, as well as metabolic ATP molecules, all central to life, absorb sunlight exceptionally well.[32] This makes sense in thermodynamic terms. Such compounds are enriched because they quickly raise the energy content of earthly substances toward that of sunlight. In universal terms, natural selection does not necessitate genes and their mutations but follows free energy consumption. Nucleic acids developed into powerful mechanisms, genes, to consume free energy. So, genes are not the prerequisites but the products of evolution.

The history of life is imprinted in the molecules of life as the past of an ancient culture is engraved in pieces of its pottery. The origins of letters are still discernible in a language as the origins of life in the bases of DNA. For instance, the letter A initially stood for a bull's head, the letter B symbolized a house with rooms, and H delineated a fence. By now, the meanings have changed. A is not just a symbol of the valuable bull; the base is no longer just an excellent antenna for photons.

Fragments from the past may never unravel the detailed course of abiogenesis nor the fate of long-lost people. How life began or how a language arose is a moot question. A step-by-step process has no definite beginning, no decisive moment.

The universal principle puts special questions into the general context. Although things remain the same, the perspective is more comprehensive than before. A paradigm shift rather than a piece of knowledge revitalizes science. When we ask how life was born, we seek something that never happened, and hence we are bemused. When we ask why life emerged, we look for the cause, wherefore we can answer: the energetic imbalance between matter and sunlight powered the emergence of biota.

After some 4 billion years of evolution, plants capture photons, and herbivores feed on the quanta bound into the plants. The herbivores, in turn, nourish carnivores. The last beings in the chain are the various decomposers. The photons arriving from the Sun fuel the entire food chain as well as generate winds and ocean currents. Earth's hot interior, in turn, powers tectonics and volcanoes. The same law of nature holds true also on neighboring planets. But, the natural process has advanced furthest here on Earth thanks to its raw materials and orbit in the Goldilocks Zone, the habitable zone around the Sun.

The Origins of Life GRC conference[33] was held in 2008 in a charming Southern Californian town called Ventura. I recall that Jack Szostak, the Nobel Laureate-to-be, began his talk by reminding us that hardly anyone studies the origin of life as their main job. Szostak himself was rewarded for investigations of telomeres. These DNA pieces at the ends of the chromosomes affect the cell's longevity. Even though life's origin is a great riddle, it alone is not a potent enough driver because scientific work must progress like any other work. It is fascinating to speculate but challenging to show how life started. Szostak discussed a variety of hypotheses.

The age-old question of life's origin enticed me already as an engineering student. If I only knew molecular structures precisely enough, I thought I could understand how inanimate matter molds into living beings. After two decades of investigating protein structure, I set aside the whole issue of how life came about and instead asked why it came about. This challenge latched me onto foundational physics.

THE FOLDING PROBLEM

For quite some time, it was thought that genetic information alone controls cellular functions. Since genes translate into proteins' chemical structures, such as those of enzymes, by the same logic, the chemical structures should, in turn, determine how proteins fold into their functional structures. But it does not happen in this way. Moreover, how does a protein actually find the proper form, since according to calculations, the process should take more time than the age of the universe. Consequently, protein folding is seen as one of the foremost theoretical problems of biophysics.[34]

Alas, along with many other profound scientific questions, the protein folding problem follows from a misconception. We have no grounds to assume that the protein chemical structure dictates the folding. Instead, we should be asking *why* the protein folds.

When examining protein folding with my colleagues Vivek Sharma and Ville Kaila, it became apparent that not only the chemical structure but also the surroundings have a say in shaping the molecule.[35] Proteins lose shape in hot surroundings, such as on a frying pan. Many adopt their functional form only in the cell, where other proteins catalyze the folding process. Since all this is well-known, it is remarkable that folding is nevertheless regarded as a major theoretical problem.

As genomic information is not even enough to determine how a protein folds, how could it possibly be enough to determine the characteristics of an organism, the phenotype? It seems that scientists have difficulties freeing themselves even from patently false beliefs.

An unfolded protein, an unstructured polypeptide chain, is commonly believed to have numerous different forms. Not so. The array of alternatives is only imagined because the unfolded conformations are similar in energy, next to indiscernible. Thus, there is no astronomical set of alternatives to be explored by trial and error along the folding pathway. Any one of the similar structures qualifies for the right one. So it does not take an inordinate amount of time to find the free energy minimum state.

The similarity between equivalent alternatives is apparent, for example, when bricking a chimney. The minute differences among the bricks are of no importance for making or using the vent. Even if the bricks were numbered so as to be able to count the possible ways (the microstates) of laying the blocks, the number of permutations would be irrelevant. Equivalent options do not change the course of events. Instead, history follows from true choices.[3]

Folding begins when even the slightest force moves the polypeptide in a direction such that the imbalance decreases. Then, the pace increases with more possibilities to consume more free energy. Finally, the residual imbalance decreases slowly when the protein is almost fully folded. Protein folding thus proceeds along an S-shaped curve, like other processes. The pattern is the same for an adolescent looking for a mission in life. An early, seemingly insignificant interest may usher in greater enthusiasm, which eventually becomes a dominant force in their life. Protein folding cannot be predicted precisely, nor can any other sequence of events, because the process itself keeps changing its driving forces.

On balance

Boltzmann struggled for thirty years to find the equation of evolution but paradoxically failed to discern the dynamic because he knew the end state. In 1872 he had succeeded in deriving the expression for the balance of gas molecules. However, that equation does not have any trace of the forces that brought the gas to the thermodynamic balance in its surroundings, for at the balance, the sum of forces is zero.

The root of the problem was noted by Boltzmann's friend Josef Loschmidt. The professor of physical chemistry wondered how an equation that is symmetric with respect to time could possibly describe the flow of time. Furthermore, the German mathematician Ernst Zermelo remarked that, according to Boltzmann's equation, a system that had once been in a state of imbalance would return to the same state of imbalance. Such things do not happen. The issues raised by Loschmidt and Zermelo concern likewise other equations in which energy is constant. Such equations do not explain the leveling of imbalance but only model the condition of balance.

Lecturing in 1899 at the decennial celebration of Clark University in Worchester, Boltzmann himself admitted his failure in relating the second law of thermodynamics to the principle of the least action.[36]

WHY ARE THE MOLECULES OF LIFE SINGLE-HANDED?

The quest for balance eradicates the mirror-image enantiomers.

Curiously enough, many natural compounds lack mirror images. For example, all essential amino acids are left-handed. Many natural sugar molecules have no mirror-image counterparts, either, whereas an industrial synthesis yields a mixture of right and left-handed compounds. Single-handed compounds are therefore regarded as a signature of life.

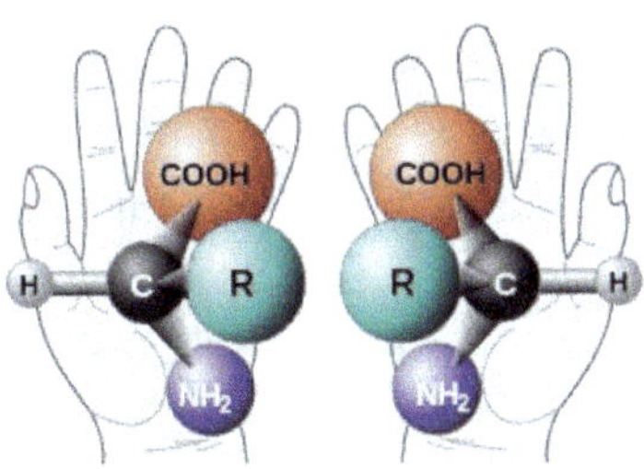

Usually, Nature exhibits only one form of any pair of mirror-image compounds. Expressly, amino acids are almost exclusively left-handed.[38]

The origin of handedness is seen as a mystery.[37] While hypotheses are at hand, none of them offers complete comprehension. So, before puzzling the problem out, let us get acquainted with how single-handedness, a uniformity of chirality, was discovered.

In 1831 Jacob Berzelius, a founder of modern chemistry, asked the German chemist Eilhard Mitscherlich to synthesize and study tartaric acid salts, anticipating that their crystals would separate into two chiral forms. But Mitscherlich did not produce results. So years later, Jean-Baptiste Biot took up the challenge. The French physicist showed that pure tartaric acid isolated from a natural source rotates polarized light. By contrast, the racemic mixture of the two tartaric acid enantiomers in equal parts did not.

These results nudged Mitscherlich to state in 1844 that the chemical compositions and crystal forms of pure tartaric acid and its racemic mixture are identical. Biot, however, intuited that there must be some difference since the optical properties differ. So he assigned his apprentice Louis Pasteur to delve into the matter. Pasteur looked at the shape of the crystal, just as the respected Mitscherlich had done, but as a novice, not laden with expert knowledge, he went beyond the mere surface appearances. Pasteur imagined two types of tartaric acid molecules in the racemate. They mirror each other, just as the right hand is a mirror image of the left hand. Fame followed.

Pasteur continued his work with single-handed compounds, but in 1856 he left the professorship at Strasbourg for the position of Dean at the University of Lille and left chemistry for biology with well-known achievements. "Chance favors the prepared mind,"[39] is Pasteur's most popular nugget of wisdom.

Pasteur regarded homochiral compounds as a characteristic of life. Thermodynamics recognizes single-handedness not only in compounds but also in shaking with the right hand and driving on the right-hand side of the road. Bolts, too, are usually right-handed, as are corkscrews. Our society is imbued with copious standards to make it function efficiently.

In general, the stronger the network of interactions in a system, the higher the degree of standardization.[40] For example, the global economy becomes more and more standardized as it becomes more and more integrated. Yet, integration is not an end in itself. By its very nature, standardization whittles down diversity, the means to make things happen. As everything affects everything else, the optimal degree of standardization cannot be known but has to be sought.

We are used to explaining things by relating the unknown to the known. The practice is empirically motivated by *Grand Regularity* and theoretically justified by thermodynamics. So we understand that the primordial standardization of biochemistry is, in principle, no different from the harmonizing of contemporary economies with industrial standards. Whether it is two molecules or two animal species, two human beings, or two products, services, or companies, only one will survive unless either develops away from consuming a common source of free energy.[41] It is like a tug of war: the winner takes it all. Thus, the homochirality of natural compounds does not suggest a single origin of life. Instead, it is an indication of unity in the biosphere.[40]

Many a problem results from isolating the system of interest from the whole of which it is a part. The question of why natural substances are single-handed arose and persisted because the object of study is detached from its environment and history. In the test tube, the energetic difference between the compound and its mirror image is negligible. Under those reduced circumstances, there is no cause for homochirality, and hence its occurrence seems a puzzle.

Industrial synthesis yields a mixture of chiral compounds unless provided with elements promoting handedness. Similarly, the austere environment of the primordial Earth was indifferent to handedness. The need for consistency grew as the network of interactions, e.g., food chains, emerged and extended. Single-handed systems gained ground by moving faster toward a balance than mixed-handed systems. The sped-up processing called for further standardization.[40,42] Since cause and effect are inseparable, we cannot work out the origin of biochemical standards, just as we cannot predict future industrial standards either. "In this great chain of causes and effects", as the naturalist Alexander von Humboldt wrote, "no single fact can be considered in isolation."

After eons of evolution, the right kinds of molecules are edible, whereas the wrong types are toxic. For example, some compounds produced by cyanobacteria act as poisons by blocking metabolism in the liver.[43] Likewise, it did not matter much in the past on which side of the road one drove with one's Ford Model T since no one was coming the other way. Today, driving on the wrong side may halt a highway. As a case in point, the Swedes switched to the right-hand traffic standards in 1968 to better integrate with Continental Europe.

The chirality problem may look like a discipline-specific question of *how* molecular standards emerged. But, it is solved by asking *why* any standard came about. Similar to molecular handedness, there is not much certainty about how the events led to right-handed traffic. And even less, we know about those events where matter instead of antimatter became the standard of substance. The chains of events extend to the beginning of the universe. Yet, the cause of standardization is clear and certain: the quest for balance in the least time.

The fork of which way to go involves the whole system. It is like a path to a critical point, along which fluctuations in fluid density increase in a power-law manner until they span the whole system on the verge of a phase transition when fluid separates into a liquid and a gas. The data are fundamentally no different. Still, drawing parallels between various systems may at first seem dubious. However, is it only our beliefs that block this logical train of thought?

Everyone knows that …

I was pleased to grasp that single-handedness follows from the least-time evolution toward balance. After all, the origin of homochirality is a well-known scientific problem. We submitted the study to several journals before getting it published. As usual, the editors of the most esteemed forums replied that there was not enough space to publish the work. From *Chirality*, a journal specializing in handedness, we got memorable comments: "Work is the strangest I have ever read. Everyone knows that entropy decreases with increasing order, such as the emergence of homochirality." This comment, unlike most others, was not expressing arrogance; it reflected genuine astonishment along with a classical error in reasoning (*argumentum ad populum*).

In physics, increasing entropy is associated with inanimate processes, decreasing with animate self-organization. On the other hand, such a division is not made in chemistry. Irrespective of whether it results in order or disorder, any reaction proceeds toward balance where free energy is at the minimum and entropy is at the maximum.

Since I was not locked into one or the other doctrine, I was able to sort out the entropy concept. As the data does not show any difference between living and non-living systems, we can deduce that when free energy decreases, entropy increases *without exception*. By consuming free energy, matter organizes into living structures,[44] such as cellular arrays, as well as into lifeless orderly structures, such as convection cells in the atmosphere and on the Sun's surface.[45]

A theory is particularly impressive when explaining in the same way supposedly opposing phenomena, such as evolution into order and disorder, standards and diversity, similarities and differences among species, as well as galaxies moving away and toward us.

IN THE NAME OF SCIENCE

In the early 2000s, James Kay and Stanley Salthe caught my attention for having inferred the least-time character of the second law of thermodynamics. Then, in their recently published book, *The Evolutionary Imperative* (2017), Charles Beck and Louis Irwin identified the universal patterns as typical, or even stereotypical, of a general principle and assembled a holistic worldview for humankind to navigate toward a sustainable way of life.

After all, it seems we will understand why the world is changing and our responsibility and role in it. Moreover, Ladislav Kováč advocates this holistic worldview in *EMBO Reports* by asking: Why does something happen? Why are there events in the universe?[46] The professor of biochemistry at Comenius University sees the second law of thermodynamics as naturalizing perennial questions of philosophy: What is this unidirectional movement heading toward? What is the place of humans in the universe? Will humanity last forever?[47]

It is perhaps easiest for those who have studied chemistry to absorb thermodynamic theory. After all, the idea of an indivisible element, the atom, is at the heart of the discipline. Moreover, the emission or absorption of quanta is learned not only by the book but also by hands-on experimenting with those events that chemists call reactions. So, there is little new in thermodynamics for a chemist apart from its universality.

Erkki Kolehmainen also recognized this tenet. The emeritus professor of organic chemistry at the University of Jyväskylä asked me to give a course at its Summer School. This traditional, international multidisciplinary event was familiar to me as I had taught spectroscopy there in 2002. My 2015 course on chemistry was presented in the flyer with the same content as the course I used to give to biochemists at the University of Helsinki each year. However, in the end, events took a different path. My course was retracted.

Apparently, in a faculty council meeting, a representative of the Department of Physics brought up a critical comment to an article about my work on the unity of reality, which the magazine of the University of Helsinki had published in 2014. However, he did not show the council members the responses by the Editor-in-Chief and me, in which we both urged the discussion of results; instead of expressions of trite and irrelevant sentiments. In the name of science, the representative's intention seems to have been far from legitimate.

Universities enjoy public esteem. We rely on academia to ensure that the truth is not dictated but searched for by questioning prevailing paradigms, too. Prejudice by scientists themselves jeopardizes this invaluable trust.

WHY IS THE GENOME CLUTTERED WITH DROSS?

Things pile up until they do more harm than good.

The entire human genome was mapped in the early 2000s. The sequencing of over three billion base pairs crowned the work that began in 1953 when Francis Crick and James Watson deciphered the structure of a piece of DNA. Back then, the beautiful double helix pointed at genetic determinism. But today, we know that genes do not determine the phenotype of an organism alone, but also the environment exerts an epigenetic, heritable impact.[48] Moreover, development is nondeterminate. It cannot be predicted, not even in principle, for everything depends on everything else.

Already in the 1940s, the British biologist Conrad Waddington reasoned that the epigenetic landscape imposes changes on the genome that, in turn, mold this landscape of development.[49] For example, changes in culture may facilitate the evolution of the genome, and its evolution may subsequently precipitate cultural changes.[50] Causes and consequences intermingle as cooperation asks for the ability to communicate, and that ability supports collaboration.

Some 70 years later, "biological research is in crisis," Sydney Brenner put it bluntly. We are drowning in data, hence thirsty for a theory that organizes all observations. The Nobel Prize-winning South African biologist went on in his *Nature* paper: "Although many believe that more is better, history tells us that least is best."[51]

In their legendary paper, Crick and Watson wrote: "It has not escaped our notice that the specific pairing we have postulated immediately suggests a possible copying mechanism for the genetic material."[52] The complementary DNA strands hinted at how the genome duplicates. Similarly, the mirror-image structures of the electron and the positron imply how matter annihilates with antimatter (Appendix B).[53] One knows right away to put one's hand in a glove and one's foot in a shoe. While complementary structures propose pairing, many forms of Nature are too complicated to disclose at first glance their purpose; in other words, what will happen.

Crick and Watson took for granted that DNA is an instruction. It remained only to crack the code, to figure out which set of bases corresponds to which amino acids.[54] This mission was completed by the end of the 1960s. After that, the cell seemed like a machine, undeniably complicated but basically controlled by genetic information. At least, many wanted to see the cell that way.

However, the modern worldview had been challenged even earlier. In the late 1950s, biochemist John Kendrew and molecular biologist Max Perutz succeeded for the first time in determining the structures of two proteins with almost atomic precision. But the outcomes were not at all as expected. Unlike DNA, the structures of myoglobin and hemoglobin did not disclose how these molecules functioned. Crick understood the gravity of the problem straight away and left for California and switched from molecular biology to brain research.

It is difficult to figure out molecular function from a single structure, even from the structures at different stages along the chain of events. When all of them have finally been worked out, what is left for a theory to predict?

Also, genetic determinism was already in trouble in the 1960s. It turned out that the genome not only consists of instructions for the cell to synthesize its peptides but also seemingly superfluous segments. These sections were quickly dubbed *junk DNA* by Susumu Ohno, an American-Japanese geneticist and evolutionary biologist.[55]

Genes make up less than two percent of the human genome.[56] What is all the rest for? It is also puzzling that even though the genome's size varies between organisms by over a thousand-fold, it is independent of organismal complexity. This C-value paradox demonstrates itself so that the genomes of some protozoans are much larger than ours and, correspondingly, the proportion of genes much smaller. However, such a comparison might well be as irrelevant as the comparison of vocabularies of vastly different languages, which are simply products of history.

The genome is not just genes and junk.[57] There are sequences to regulate gene expression and elements needed for replication, most notably, centromeres in the middle of chromosomes and telomeres at the ends. Moreover, DNA not only supplies codes for proteins but is also transcribed to RNA molecules with their own functions. Our genomes house not only our own DNA but also sequences from bacteria and viruses as well as pseudogenes, broken cast-off genes.

Many genes are in pieces flanked by non-coding segments. These introns are excluded before the gene is expressed as a polypeptide, but their removal influences expression. This alternative splicing provides, among other things, a vast spectrum of antibodies against a wide range of pathogens. Some genes express enzymes that only remove and re-integrate their genes repeatedly in the same place. In a sense, nothing happens. In that continual circulation, there is no net effect. Such a steady flow is an example of *Grand Regularity*. So is a planet orbiting year after year and an engine droning one rev after another.

Besides introns, there are other moving elements, too, such as transposons. Barbara McClintock found jumping genes in the 1940s.

This unexpected finding stymied the scientific community. As reactions were even hostile, the pioneer of cytogenetics forwent publishing her work on that subject. It was only in the late 1960s that her groundbreaking work began to be acknowledged. After retiring, McClintock told her colleague Oliver Nelson how painful it was to confront experts with stubborn prejudices.[58] With its mission to advance human knowledge, it is appalling that the scientific community tries to trip up those who dare take landmark leaps. McClintock was awarded the Nobel Prize in 1983.

CALL FOR CAUSALITY

From the thermodynamic perspective, the genome is like any other system.[59,60] It is not an inert instruction but a living system. For instance, gene products in the genome are responsible for most cellular metabolism in the same way that trees in a forest are responsible for most of the energy metabolism of an ecosystem. While the fallen trunks do not poleaxe us any more than pseudogenes, figuring out all the functions of the detritus in a forest would be as tedious as revealing all the purposes of pieces of the genome. Junk is marginally functional but functional nonetheless, as everything participates in something. Polypores, lichens, and fungi live on the trunks and branches of trees, dead and alive alike. Likewise, various independent and moving genomic elements live next to genes.

Our genomes differ from each other as one thicket differs from another. Tiny differences are numerous. Nevertheless, scientists imagine some DNA snippets cause common diseases. By contrast, they would hardly think variations in twigs would be responsible for pests invading one forest while leaving another intact. Instead, they would consider factors beyond the forest, such as the weather, to find the cause of the plague. Indeed, it has become apparent that individual differences between genomes do not mean much, if anything, in causing common diseases. It is the whole that matters.

Jonathan Pritchard questions the direct connection between genetic variation and phenotype.[61] The professor of genetics at Stanford University maintains that the prevailing perception of the relations between genes and diseases is incomplete. Subtle nuances in the genome explain only little about why one catches a disease while another does not. Pritchard and his colleagues point out that genetic details, such as those accountable for human height, distribute in the same way as many other quantitative traits that affect diseases, e.g., diabetes and autoimmune diseases or cholesterol levels.

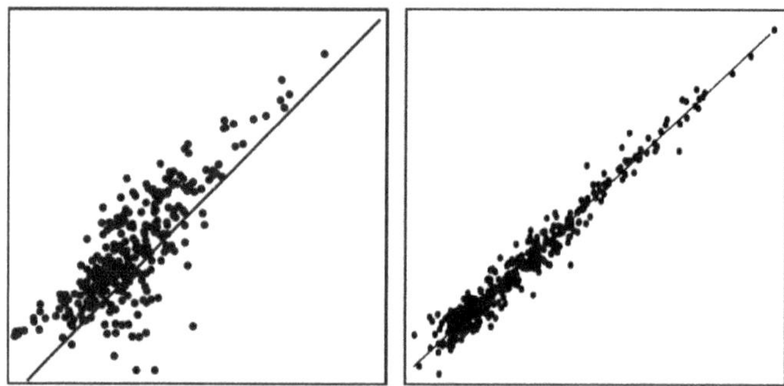

The heredity of Crohn's disease is higher the larger the disease-related single-base diversity (left).[61] The value of the city's production is higher the larger the city's population (right).[63] Causes and consequences are many, yet they all defer to the same law.

"Currently, biology exists in an era of abundance of data but no theories,"[62] says Sui Huang about the scholarly state of bioscience. The professor at the Institute for Systems Biology in Seattle upholds that correlations, trends, and models do not explain causal relationships. By contrast, a force identifies with cause and ensuing motion with a consequence, as a metabolic imbalance identifies with a disease.

THE BOOK OF LIFE

Genetic diversity underlying a trait or a disease alike manifests the least-time imperative. For example, the height of a human being would not be a central characteristic unless associated with many factors. A disease would not be common unless it implicates numerous factors. Gross National Productivity would not indicate the state of the economy unless involving many factors. In all their messiness, the relationships are nonetheless causal.

Genomic diversity is as comprehensible as biodiversity in an ecosystem or diversity in an economy. There are many kinds of forests, as there are all sorts of genomes. Species richness in tropical thickets is as uncharted as the contents of large genomes. Conversely, dry heathlands are as scant in diversity as bladderwort's genome, where only a few percent seem to be superfluous.[64] Both the abundance and scarcity of species in a forest, genome, and economy are consequences of the prevailing conditions and history.

When we ask "why," we are looking for a cause, that is, a force. When we ask "how," we are looking for a mechanism, thinking the mechanism is more important than its aim.

The overplayed role of genes[60] is apparent to Professor Keith Baverstock. After a career at MRC, he moved to WHO and investigated the Chernobyl and Fukushima disasters. He maintains that genes in a cell are like books in a library. For example, a cell needs genetic information for both actions and reactions. So, the same information can even be used for opposite purposes.

While information is unarguably necessary to forge a cell, a DNA sequence does not alone ordain the outcome, no more than a built building fully matches the model. Genes do not dictate everything but facilitate free energy consumption as everything else. Thermodynamics gives us a ground to relate what we are puzzled about to what we have already puzzled out. For example, the susceptibility of a metabolic network to genetic defects is analogous to the vulnerability of a telecommunications network to faults. A virus capturing a cell's resources is like a thief robbing an identity. The cell has the means to recognize foreign agents but not flawlessly. There are data security flaws in many other systems, too.

A piece of DNA seems like a verse of poetry. Since we managed to decode it, we expected to read the entire book of life soon. Instead, we faced a spectrum of genomic life. Over generations, the elements mount up in a genome like mementos on a mantelpiece. One may thus accumulate a rich genome, just as one may acquire a vast inheritance. For example, our genes hold hints of famines that past generations endured.[65] Perhaps now our lifestyle is engraving therein a mark of overabundance.

When we perceive our genome as our inheritance, we understand better what it is and what potential it has. We know what we have from an inventory, a mapped genome, but not how it works. So, it is not apparent what a mutation does or whether it does much at all.

While yearning for the new, we are also frightened by it. For example, we are concerned about genetic manipulation, just as we fret about the abuse of databases. We are sifting out seeds and manipulating genes to find high-yield crops, but by doing so, we are narrowing down our future options. Faster-growing variants are populating ever-larger areas, but their super-productiveness is vulnerable. If all our eggs are in one basket, a collapse could truly be catastrophic. Fostering diversity ensures robustness all around.

WHY REPRODUCE SEXUALLY?

Evolution employs all means.

Despite the comprehensiveness of evolutionary theory, Darwin felt that natural selection, as such, does not explain sexual selection.[66] An attractive appearance not only favors mating but also invites predators.[67] For example, the bird of paradise signals its fitness with a long tail, making flying arduous. However, energy is the universal measure even for seemingly incommensurable items in the pans of balance.

It seems as if sexual reproduction is associated with an additional cost. If only females give birth, what need is there for males? Although mixing parents' genetic material by recombination produces variation for natural selection, the male of many species seems redundant after fertilization. It could even be a burden on its own offspring by consuming resources. As a matter of fact, the females of some spider species eat the males after mating. Of course, females and males must meet to mate, which slows down reproduction by taking its time. Against such a backdrop, it is perplexing why sexual reproduction has survived throughout evolution.[68]

Using the theory of evolution, it is also difficult to explain why worms have both female and male gametes or why many plant species have pistils and stamens in the same individual. More remarkable than hermaphroditism is that female wrasse can transform into males as they grow old. The purpose of sexuality seems more diverse than mere reproduction. For instance, many primates like us humans exhibit sexuality for solidarity but also use it for suppression and savagery. Sexuality is perplexing in its pervasiveness and ambiguity. What do we talk about when we talk about sex?

The theory of biology or physics, or any other subject for that matter, does not improve from the proliferation of concepts in the drive toward ever more detailed mechanistic descriptions of phenomena. Laplace clarified this point: "The simplicity of nature is not to be measured by that of our conceptions. Infinitely varied in its effects, nature is simple only in its causes, and its economy consists in producing a great number of phenomena, often very complicated, by means of a small number of general laws."[69]

We can understand the complex spectrum of reproduction, from genes to behavior, through thermodynamics.[70] Recombination and genomic duplication, i.e., mitosis, as well as mating and social interaction, are all processes. It depends on the circumstances, whether speed or endurance is vital, whether variation or specialization is needed, whether wisdom or some other factor is required for things to happen. Even the slightest gesture, just a loving glance, can release tremendous forces toward fulfilling one's dreams.

Species at the top of food chains reproduce sexually. There they need a comprehensive ability to consume various resources. Humans have diversified into the most varied environments, whereas species at the bottom of food chains do not have many assets to vary. For them, sexuality would be more of a disadvantage.

All in all, sexual reproduction, just like the asexual kind, is a mechanism to let things happen. Yet, there is no universal way of easing up all tensions. For instance, some germ cells of plants are naturally sterile[71] because, in that way, sexual cellular proliferation and asexual mitochondrial propagation are in a dynamic balance, like a bow and a string. So if we only examine one aspect, we see the imbalance and stare at the tension in wonder.

AT DINNER
The environment influences an organism's mode of reproduction. A prime example is the aphid's annual shift from asexual to sexual

reproduction and back. Over the spring and summer, the asexual phase prevails. Maupertuis marveled at this hectic prolific phase, where no one has time for romantic candlelight dinners.[20] In the fall, aphids switch to sexual reproduction. Mating supplies eggs with genetic variation and the aerial display of the winged aphids yield spatial diversity as they settle down in various wintering places. Some eggs are bound to survive, and larvae will hatch again in the spring. In greenhouses and temperate climate zones, aphids reproduce only asexually. Sexual populations have vanished because energy flows faster through parthenogenesis than sexual reproduction.

The maximization of free energy consumption also displays itself in various species' ratios of females to males. This spectrum is particularly wide among insects. For many species, gender is not carved in stone but molds itself to the situation. A wrasse with mutable sex is clear proof of this situational dependence.

Reproduction is not a goal in itself. Instead, it is a means to diminish an imbalance relative to the environment. Suppose one sees a potential spouse mainly as an economic asset. In that case, one is looking for a profitable match, as Gary Becker, winner of the Nobel Memorial Prize in economic sciences, cool-headedly valued affairs.[72] Whatever the motives of marriage might be, thermodynamics maintains that they all are ultimately measurable in terms of energy. In the same holistic sense, Philip Ball compares the proportions of married and unmarried versus economic status and social pressure in society to the proportions of liquid and gas versus temperature and pressure in a fluid.[18]

There is no reason to contrast sexual selection that emphasizes reproduction with natural selection that favors viability, but all reasons to describe everything in the universal terms of energy and time. A handsome tail or an awesome sports car as such hardly charms a partner but signals that if one can afford such a luxury, a lot more must be within reach, including truly necessary things. In sexual selection, as in other choices, the reality may well turn out to be something other than what is apparent at the first impression. The partner might fly away when feathers molt or the rental contract is annulled.

IDENTITY

"All electrons have the same charge and the same mass … they are all the same electron!" Wheeler enthused over his idea to Feynman.[73] But all people are different, including identical twins. Where could identity arise if the fundamental elements of substance were all identical?

It was insightful to examine the origin of identity with my colleague Esa Kuismanen,[74] who excels at providing new perspectives by talking about seemingly different things in the same way and about the same things in different ways. As an example, electrons are not strictly speaking identical because each particle has a unique position in the universe. Likewise, each identity is built up from numerous events along the course of life, eventually all the way from the common ancestor of all organisms. Conversely, none of us is totally different from the others, either. On the contrary, we have so much in common that even a minor difference seems like a big deal.

While all species' putative descent from a particular primordial cell is a straightforward thought, it is illogical. Evolution does not have a single origin but extends from inanimate to animate without demarcation.[75] Consequently, comparisons of genes across species do not point unambiguously to any specific nascent form. The family tree does not terminate in a unique foundation but branches into roots. The flows of quanta crisscrossed along various routes just as they zigzag today by way of viruses and genetic engineering.

It is not only that our genomes are rich tapestries of interwoven ancestries but that the whole Homo genus harbors genetic fragments from other closely related but long-extinct lineages.[76] Similarly, the history of the automobile forks to a horse, cart, and steam engine. Identity, like any novelty, emerges from combinations of existing entities rather than creations out of nothing.

IS THERE EXTRATERRESTRIAL LIFE?

We need to decide what life is to know what to look for.

A big meteorite fell on September 28, 1969, near Murchison, a small village in Western Australia.[77] Over 100 kilograms of carbonaceous stones from the sky were collected. Since then, over 10,000 compounds, including 70 amino acids, have been identified, but no unambiguous telltale marker of life, say, molecular homochirality.

Despite inconclusive evidence, the meteorite brought back the old idea of panspermia.[78] Could life have spread in the cosmos by meteors, asteroids, and comets? In this way, the question of how life was born is replaced by the question of how life was sown here on Earth. The transport hypothesis also gives the impression that life is a superficial phenomenon on the face of the Earth. Thermodynamics offers a different view, maintaining that our planet is alive in its entirety.

The idea of a living planet is not at all new, either. In Greek mythology, Gaia personifies Earth. The deity was born from Chaos. The Gaia hypothesis came alive anew in the mind of James Lovelock in the 1960s.[79] At the time, Lovelock was working for NASA on a project to discover life on Mars. Based on the very different atmospheric compositions of Mars and Earth, Lovelock suggested that the Earth is a super-organism. It regulates itself as an animal regulates its vital processes. The biosphere is entwined in life-sustaining interactions with the atmosphere, hydrosphere, and geosphere.

According to thermodynamics, Mars is not entirely lifeless. Its white polar caps of carbon dioxide 'live' in tune with seasonal changes just as the geosphere, hydrosphere, biosphere, and atmosphere of our blue orb cohere with the seasons, too. A response of any kind is part of planetary processes. It is only that the interaction network on the red planet is sparse and thin compared with our home planet.

Lovelock's Gaia idea pleased environmentalists immediately but only slowly academia. Among others, the scientists W. Ford Doolittle, Richard Dawkins, and Stephen J. Gould asked, albeit later reconsidered, how natural selection acting on rival organisms could lead to a planetary balance, global homeostasis.[80,81] In the face of fluctuations, the free energy minimum state is stable in the same way as our physiological state while not being invulnerable to severe distress.

SAY IT WITH FLOWERS

Lovelock did not resort to the general thermodynamic principle but a simple computer model when defending his view of global self-regulation.[82] In this Daisyworld simulation, complied with his colleague Andrew Watson, light and dark daisies grow on the surface of a fictitious planet. The light ones reflect radiation, whereas the dark ones absorb it. The flowers are otherwise identical and grow best at the same constant temperature.

On a planet orbiting an ever-brightening star, such as the Sun, when left alone, the dark flowers initially gain ground and then cede dominance to the white ones. During this transition, the planet remains at an almost constant temperature, despite the intensifying insolation. Eventually, the planet is fully shrouded with white daisies, but even that is not enough to reflect the increasing radiation. As the temperature explodes out of control, the flowers wilt away.

Thermodynamics yields the same scenario as the one Lovelock and Watson produced. I also showed with Mahesh Karnani, a colleague of mine, that the vegetation adapts to ice ages and warm periods. After such disturbances, the system finds balance anew.[83] However, it turned out that the greater the damage, the slower the recovery. If the white flowers are cut out from the planet's surface, the course back toward balance takes the longer, the hotter the planet becomes. And the system cannot necessarily recover from devastation when it is already struggling to survive at the limits of its regulatory power.

As the Daisyworld model demonstrates, vegetation is matter's response to sunlight. Eradicating natural vegetation invites catastrophe. The destruction of the most effective regulatory mechanisms of global homeostasis at the same time as greenhouse gases are released in increasing amounts will cause insurmountable problems.[84] These conclusions are not unheard-of. But thermodynamics leaves no room for speculation by keeping track of every single quantum.

Lovelock, the environmentalist and futurist, is alarmed about the loss of biodiversity, particularly rainforests. His book *Gaia's Revenge* (2006) stresses reality. We are on our way out of the frying pan into the fire by felling forests for food and biofuel. The only issue is not the forests harvesting carbon from the atmosphere but the full effects of vegetation.[85] When heat no longer flows properly to the cold upper atmosphere, rain does not fall on large areas as before, hot deserts expand, seas warm up, and glaciers shrink.

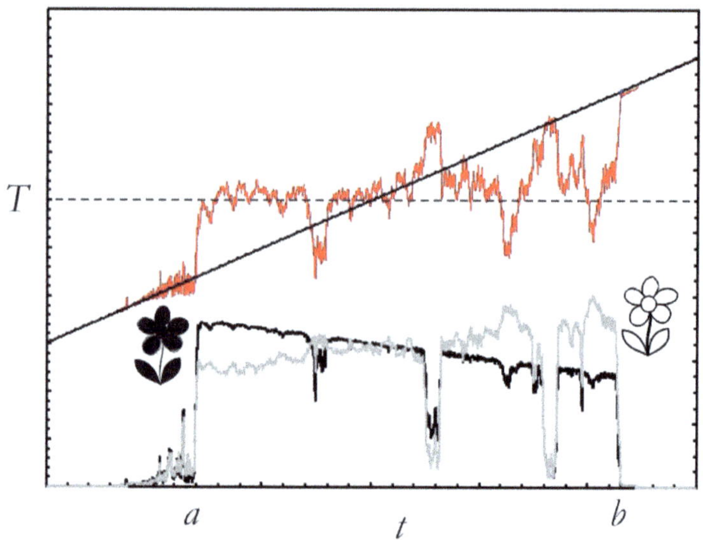

A thermodynamics simulation demonstrates that the planet is like a self-regulating organism. With time advancing from left to right, a Sun-like star shining on the planet, becomes ever brighter (rising line). At first, the temperature T of the planet (red line) goes up at the same rate. When the temperature rises above the growth conditions, at time point a, the flowers bud and cover the surface, diminishing the imbalance between the radiation and substance. There is no other maxim than the quest for balance. Initially, dark flowers (black line) dominate, as they convert starlight to warmth better than white flowers (gray line). In this manner, the temperature reaches the optimum for flower growth (dotted line). Despite the star becoming brighter and brighter, conditions remain favorable from a to b as the white flowers proliferate and the dark ones decline. The vegetation also smooths the occasional ice ages and warm periods introduced into the simulation. At last, the star shines so intensely that no reflection suffices to maintain balance. Temperature escapes from the domain of regulation at time point b. As the flowers die in the heat, the planet becomes barren.[83]

A general warning against the overuse of natural resources also reverberates in the books of Jared Diamond, a multidisciplinary American scientist.[86] Now, thermodynamics lays a solid foundation for planetary environmentalism. We can make sense of even complex flows of energy and avoid doing harm despite our best intentions. So, a devastating development may not be inevitable. We are already accessing our inextinguishable solar energy source more effectively and improving energy efficiency in many facets.

Climate change exerts an immense impact on our living conditions.[87] While many factors influence climate, the cause of the change itself is simple – an imbalance imposed by us. It is thus not necessary for us to study the biosphere scrupulously to grasp what is happening. Instead, the general principle allows us to relate other changes to climate change and learn from them. For example, we readily understand what would happen if the critical functions of the healthcare system were paralyzed. Many immediate and long-term repercussions would follow if the central hospitals failed to function. In such a situation, we would not be investigating the effects in every nicety but would rather find out what is blocking normal operations and restore them. Likewise, preserving mechanisms, specifically tropical forests, mediating the most voluminous flows of energy, are the most critical in counteracting climate change.

WHERE IS EVERYBODY?

In 2009, the Kepler Space Telescope parked in orbit around the Sun. Despite being broken, it found thousands of stars with planets among hundreds of thousands of candidate Suns. At this distance, a dozen so-called exoplanets look Earth-like. Do they harbor life?

We do not know. First, we need to decide what to look for. Since there is no qualitative difference between living and non-living, we simply must choose what we call life. That is how we go about answering this question here on our home planet, too. However, we are unable to agree on the definition because there is no unambiguous one. Would we consider an Earth-like atmosphere as evidence of an exoplanet harboring life?

Launched in 2021, the James Webb Space Telescope can identify many compounds from the light that has passed through a distant planet's atmosphere. Ozone could be a sign of life, although not an unambiguous one.[88] In any case, such an observation would suggest that the Earth is not exceptional. We are also looking for intelligent extraterrestrial life, yet, it seems, without understanding what life is.

Picking up an alien broadcast would be shocking, for we would not capture it by chance. Unfocused signals, such as TV and radio transmissions, degrade too much on their cosmic voyages. Nonetheless, other worlds are always far away in the future or in the past since nothing implies a courier faster than light. Even from orbiters of the nearest stars, messages will arrive after a generation or two has passed here on Earth. Our time is our limit. It forces us to think differently – above all, about our lives here on Earth.

Enrico Fermi was convinced of life elsewhere in the universe. He reasoned that advanced civilizations should by now have colonized the entire galaxy.[xiv] Frustrated by the fact that no one had come across an alien, he once asked out loud as he came down for lunch at the Los Alamos laboratory, "Where is everybody?" To this, Leo Szilard, a Hungarian-born physicist, replied: "They are already among us – but they call themselves Hungarians."[90]

Several Hungarian scientists settled in the United States before the Second World War. Theodore von Kármán, John von Neumann, Paul Halmos, Eugene Wigner, Edward Teller, George Pólya, and Paul Erdős were known as 'Martians.' They adapted to their new world well. Teller, though, a cocktail of scientific ingenuity with a peculiar personality, preserved his strong accent. Another planet would be another thing – for the Hungarians as well. Let the Hungarian plains, the puszta, be equally verdant on this planet, but we would not survive on them. We could not eat anything, as molecular standards would be different.

It is unlikely, improbable, energetically unfavorable that events on an exoplanet would have led to the same standards as here on Earth. Diversification begins when connections break up. Organisms evolved into distinctive indigenous species when Australia broke off from the other continents. Similarly, the Earth would share history with an exoplanet when it comes to the elements, perhaps some simple compounds, but hardly when it comes to most metabolites. The molecules and organisms over there would undoubtedly resemble but not equal to those here. The *Grand Regularity* would be on display there, too, but the details would differ. Already in 1698, Huygens brought this understanding to the fore in his opus *Cosmotheoros*.

In 1961 Frank Drake presented a formula to estimate how many civilizations there are in the Milky Way.[91] Despite discovering exoplanets has narrowed the original guess of the American astronomer

---

[xiv] In truth, Fermi's question prompted the astrophysicist Michael H. Hart to make the conclusions about the odds of extraterrestrial life.[89]

and astrophysicist, uncertainty remains sky-high. The number of civilizations could be as high as 100 million, or our fleeting existence might be quite the exception in the universe. Thermodynamics does not refine these numbers. It only underscores that systems evolve toward balance as quickly as possible. We tend to take that kind of effectiveness as a sign of intelligence. So are we intelligent?

The resources of a planet do not alone dictate the course of events since each event affects future ones. It is, therefore, not immaterial how we use our assets. Our worldview materializes in our decisions and their irreparable consequences. With the theory of time, we have a better basis for weighing our deeds.

Earth's arsenal is abundant yet limited. The scarcity of phosphorus curtails plant growth. That is why it has to be habitually spread on fields. The shortage of rare earth elements, in turn, constrains economic growth.[92] Running out of fossil fuel has worried us already for quite some time. We depend on the many resources and mechanisms that characterize the Earth. In the long run, our existence requires volcanic activity to cycle nutrients and the Earth's magnetic field to shield us from cosmic rays.

On the whole, we lack the capacity to colonize other places in the universe. Transforming a desert planet into a viable habitat is beyond our means. So, for better or worse, we are Earthlings.

We have mapped our home planet and have now focused our telescopes on exoplanets. But for us to survive, we must realize who we are. Are we atoms, molecules, cells, bacteria, fields, forests, power plants, computer networks, or even much more? Are we, in fact, inseparable from our vibrant planet?

What Does the Worldview Entail?

Through the ages, humankind has sought a holistic view of the world. Although the animism of primitive cultures appears to us as mere superstition, it provided a working worldview crafted from the ingredients then at hand. Detailed information was scant, but the balance of Nature was kept. Now the Tellurian balance has been tipped by modern humans, unrivaled invading species, harnessing ever mightier streams of energy to its needs. In our hubris, we hardly realize that we are part of the Earth. We know more details than ever but comprehend less the vital connections. The balance is difficult to regain and maintain unless we see the whole. We do not appreciate the true value of life until we understand everything in the same way.

Darwin's theory of evolution is an impressive, albeit incomplete, doctrine without mathematical form. Natural selection is an important but imprecise concept without substance. Survival displays itself in proliferation, but the evidence does not bear out the notion that reproduction alone is the criterion or expression of natural selection. Thermodynamics frees us from the idea that genetic material is a prerequisite for evolution, that natural selection can only occur at the level of populations, or that the course of evolution by mutations is random and without cause. In 1877, Charles Sanders Peirce anticipated this applicability of statistical physics to biology: "In like manner, Darwin, though unable to say what the operation of variation and natural selection in any individual case will be, demonstrates that in the long run, they will adapt animals to their circumstances."[93]

Without any sight of direction, the modern vision of evolution is blind. We talk about genetic and epigenetic inheritance, cultural evolution, and technological development as if distinct and independent processes. Instead, we ought to conclude that the selfsame law operates everywhere since the regularity of data displays itself at all levels of the natural hierarchy.[2]

The character of natural selection becomes apparent by asking *who* selects. Phrasing the question in this pointed way exposes the agent. The flows of quanta themselves choose their least-time paths along the lines of force.

Personifying Nature and seeing a purpose in it are deemed nonscientific, yet we think of ourselves as intentional agents. Instead, we would comprehend everything alike by personifying everything, just as the hunter-gatherers did. At present, as in the past, animals, plants, and mighty natural forces define our necessary conditions of existence.

Continuing our revision of the scientific worldview from previous chapters, we now understand that biological evolution is, in principle, no different from any other sequence of events. Since the living does not stand out in the data from the rest, it is logical that the theory does not make such a distinction either but rather explains everything on the same basis, namely as natural processes.

Of course, we are not only concerned with what thermodynamics teaches but also with what the theory of biology is expected to explain. There are many questions, but the following answers sum up the insights provided by the holistic worldview.

- *Explain how life originated.*
  This task, albeit interesting, is not meaningful. Without any principal difference between living and non-living, there is no singular genesis to be explained.

- *Explain at what level natural selection takes place.*
  Evolution is a mere sequence of events at all levels, a historical process where future events follow from past ones. The flows of quanta themselves select the least-time course toward a balance, free energy minimum.

- *Explain why natural compounds exist, as a rule, only in one form of handedness.*
  Eons ago, the evolution of the biosphere led to molecular standards in the quest for the least-time consumption of free energy. Today, the intensifying production of goods leads likewise to industrial standards.

- *Explain why genomes contain elements other than genes.*
  We can classify the genome into functional and dysfunctional elements. However, it is more important to understand that the genome is not a repository of information but a living system with a diversity of ingredients.

- *Explain where the rules of ecology, including skewed distributions and competitive exclusion, come from.*
  Distributions with long tails and S-shaped growth curves follow from the least-time evolution toward a balance. One of any two similar species is excluded when one is more efficient in consuming free energy than both.

- *Explain the ecological niche.*
  When a foreign species enters an ecosystem, it gets a foothold, filling an ecological niche, provided that free energy is thereby consumed more efficiently than before.

- *Explain the principle of global homeostasis.*
  The Earth evolves by its mechanisms toward thermodynamic balance with the hot sunlight and cold space. The geosphere, hydrosphere, and atmosphere were the first mechanisms of the planet. The emergence of the biosphere led to a deeper minimum of free energy, a robust planetary balance.

There are many more questions. For example, what determines the life expectancy of a species? Undoubtedly, a change in longevity due to various forces is a complicated process mechanistically. Yet, the principle is simple: life lengthens when a longer life experience contributes more than it consumes. That longevity is an adaptation has been understood for a long time; only adaptation as a mere chain of events toward a balance has remained unstated.

As we search for answers to various questions, thermodynamics urges us to relate an incomprehensible phenomenon to the phenomena we understand by experience. We can also question the questions themselves to become aware of our implicit assumptions. The most vexing problems often point to the deepest-rooted misconceptions.

KEY POINTS

- The animate and inanimate are one and the same, quanta.
- Evolution is flows of quanta, sequences of events toward thermodynamic balance.
- Natural selection means that events gravitate to the least-time paths.

7. WHAT IS CONSCIOUSNESS?

Consciousness is seen as a mystery.
But put simply, it makes things happen.

The content of consciousness is reality itself, maintained Thomas Reid, a Scottish philosopher of the Enlightenment.[1] Basically, we need to be conscious of the world to live and thrive in it. The pragmatic American philosopher and psychologist William James refined the argument: consciousness serves the purpose of knowing, and knowing is the structuring of reality in terms of its elemental substance.[2] Antonio Damasio, in turn, regards awareness essentially as an effective mechanism that makes things happen.[3] According to this Portuguese-American neuroscientist, known from his nonfiction books, consciousness complements instincts. The more conscious we are of ourselves and our environment, the more aptly we act. Thinking has selection value. Isn't that what evolution by natural selection is all about?

Thermodynamics explains consciousness in the same way as other things (Chapter 2). Consciousness becomes more comprehensible by realizing that its most prominent features, the sense of experience and subjectivity, are on display everywhere. Such a universal stance is not unprecedented. In the words of Kurt Goldstein, "All creatures have a specific nature; all represent wholes having the character of individuality. Therefore, we can obtain insight in all living forms by one methodological principle – the holistic."[4] In his masterpiece *Der Aufbau des Organismus* (1934), the neurologist and psychiatrist described the organism as actualizing its subjective potentiality as it comes to terms with its environment.

The process of thinking proves to be nothing more than a sequence of events when an event is understood as a flow of quanta. It is eye-opening to shift our focus in this manner from the brain's mechanical complexity to the elemental simplicity of thinking. We become more aware of our motives by recognizing that we orient ourselves, often by sheer reflex, along the lines of force.

While such a plain view does not inspire wild imaginings about consciousness, it does correspond with observations. We find no unique characteristics in the brain, the presumed locus of consciousness. The central nervous system displays in its operations and structures the same patterns as other systems.[5] Skewed distributions, S-curves, and power laws, as well as oscillations and chaos, manifest themselves also in the collective consciousness, for instance in the coordination of actions, in a community's norms and values, and in the herd-like behavior and civil unrest that are fomented in crises.[6] Donations, indicating empathy, also follow the power-law pattern closely, although after a disaster more generously than before.[7] Furthermore, the *Grand Regularity* exhibits itself in the distribution of phone calls, text messages, emails, and the use of apps and social media.[8]

Is consciousness more than a means to weigh various forces and act accordingly? The materialistic theory cannot but answer this question affirmatively, as the viewpoint already sets up the view. It is nevertheless of interest to examine consciousness as a mere mechanism that makes things happen. It turns out that the conclusions are not new. It is only new to derive them from materialism, though not neo-Darwinian reductionism, but holistic teleology.

There is a lot of room for theorizing in the cognitive sciences, thereby, a burning need for compelling reasoning. Especially when it is unclear what consciousness is all about, it is better to use a theory that covers more than the prime interest. A comprehensive theory explains not only more than a special theory but also more competently.

SANDPILE

About twenty years ago, Per Bak introduced a radically new cognitive theory.[9] The Danish physicist argued that the basic principle of neuronal activity is the same as that of a sandpile! Indeed, the data look alike. When sand is poured, a heap builds up. The shape of the heap stays quite the same despite its growth because slides of different sizes occur from time to time. Small ones happen every so often, big ones seldom. The magnitudes of signals in a neuronal network are also distributed in a power-law manner: small responses are frequent, large ones infrequent. Bak called this dynamic balance Self-Organized Criticality (SOC). He noticed that it was impossible to predict whether adding one grain would cause a slide and, if so, how big. Nevertheless, the chain of events is not random but regular.

In his book *How Nature Works* (1996), Bak describes earthquakes, market volatility, traffic jams, biological evolution, galactic dispersion, and neuronal activity in universal terms.[9] The statistical variation among various events is, by and large, inversely proportional to the frequency. But the origin of this so-called $1/f$ noise[10] has remained one of the great enigmas of science.[11]

At first, many experts sneered at Bak's universal vision. Later, getting acquainted with the evidence, many were struck dumb. The maverick was looking for a simple explanation for the universal patterns and did not care much about complicated models.[12] Bak, alas, did not live to see the broadening interest in his holistic ideas.

The SOC model describes events as unpredictable, occasionally even chaotic, yet conforming closely to the power laws. This *Grand Regularity*, as we now appreciate, arises from the least-time evolution toward balance. Since the data prove once again to follow the same pattern, consciousness should be understood like other things.[13]

OUT OF THIN AIR

Consciousness as a manifestation of the universal law may seem at first a far-fetched idea, just as Bak's thesis did initially,[13,14] but the functions and structures of the brain display the same patterns as other systems, only the magnitudes and mechanisms are distinct.[15] Cortical electrical activity is similar to seismic activity in the Earth's mantle.[16] Nodes of activity in neural and information networks are distributed in the same skewed manner.[17] A nervous system, like a stock market, exhibits fluctuations, shocks, damping time series, and sometimes even chaotic behavior. Neuronal oscillations resemble the vibrations of chemical bonds, cell cycles, daily and annual rhythms.[18]

The stable, homeostatic state of the brain, like the expression of genes, is strictly but not fully controlled.[19] Episodic seizures of the brain resemble avalanches of a semiconductor circuit. Weather, cellular automata, and lasers display the same instability.[20] Surprises lurk in the long tails of skewed distributions.[21] The harmony between dynamic stability and erratic change is subtle. Without regulation, our thoughts would be unstructured, without freedom, stuck in a rut. In general, order expedites flows of quanta, whereas disorder may open new, eventually even faster, pathways.

Let us ascribe consciousness to individuals and communities, as awareness at that level, too, exhibits ubiquitous patterns. Reader statistics of journals, archives, books, newspapers, and social media display skewed distributions, cumulative sigmoid curves, and power laws.[22] Only a few are in the spotlight; many are in the shadow. The very meaning of consciousness becomes apparent as we relate it to other matters.

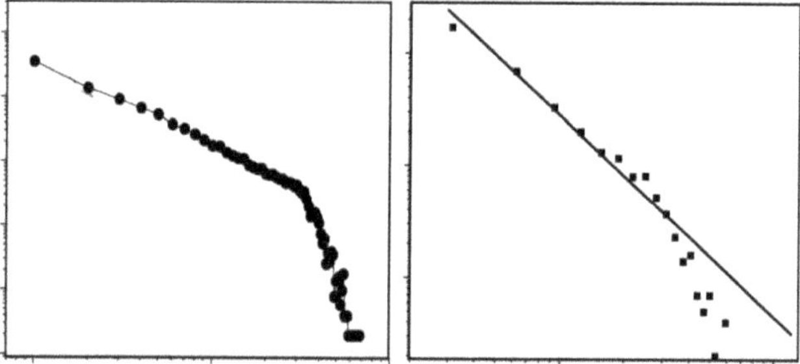

Massive electrical discharges of a neuronal network are rare, whereas small ones happen every so often (left).[23] Long-lasting electrical discharges of a molecular network are occasional, whereas small ones are frequent (right).[24] The data depart from one straight line to follow another one when the flow of quanta is diverted from one mechanism to another.

In whatever ways we may define consciousness, the data is qualitatively similar. Consciousness subsists on many levels without sharp boundaries between them. Variation is found only in quantity. This is evident to anyone who has observed animals. For example, the neuronal system of a nematode comprises 302 cells, whereas the human brain houses roughly 100 billion neurons. The basic principle is nevertheless the same.

The world, even the whole universe, could be regarded as conscious, as the same patterns present themselves in the food chains of ecosystems, in business networks,[25,26] and in cosmic structures.[27] As seen from this perspective, consciousness is a characteristic of substance. The philosophers Gottfried Leibniz and Baruch Spinoza arrived at this selfsame and logical conclusion called panpsychism.[28] Nothing is inexplicable in consciousness since nothing is unique in it. In other words, we begin to comprehend consciousness by recognizing its characteristics in other things, too.

The *Grand Regularity* suggests that consciousness evolved as all other mechanisms did. It, too, is an adaptation to circumstances, a response to forces. For instance, a baby becomes aware of one of the foundations of reality when it knocks its head against the floor. An adult has learned to take the floor for granted – until an earthquake. Consciousness accumulates from life experiences. It is not a blank slate at birth. The ingredients have also been inherited from the parents, who knows, all the way from the primordial ancestors.

Things happen faster by following conventions, for what has been experienced already has a form, a template for functioning, while what is new is yet to be kneaded. In this regard, cognitive heritage is no different from material inheritance. Like our metabolism, consciousness is conservative by nature. The present is constructed in the least time from the materials of the past. We cannot thus digest just any matter, and we cannot stomach just any theory. We are not free of our prior thoughts, even if we want to think differently. Comprehending is demanding, but changing one's views is even more demanding. It takes energy and takes time, literally, quanta.

Daniel Kahneman elaborates on the ease and effort of thinking in his book *Thinking Fast and Slow* (2012). The Nobel Laureate in economics sees our cognitive faculty as our most useful tool in its rich profusion. The essential cognitive question is how to find the most meaningful way to behave in each circumstance. In these modern times, we face situations no one has ever encountered before, and hence habits do not guarantee the proper response to the same degree as in the past when the habitat was scanter and simpler than today.

As long as consciousness has a physical embodiment, whether the central nervous system with its functions, a social system with its activities, or an information network with its processes, thermodynamic theory leaves no room for unknowns by accounting for reality with the precision of a single quantum. On the other hand, if consciousness, regardless of its physical embodiment, were all different from other things, what could that difference be, given the similarity of the data? Finally, if consciousness were without any substance, its consequences would emerge out of thin air. That would be unexplainable.

THE SELF

Ever since René Descartes and John Locke, philosophers have cerebrated how to define consciousness. However, a definition demarcates and disconnects the subject from its surroundings. That causes problems. We might focus on the mind in explaining behavior, while the truth may be that behavior reflects external forces that influence the mind.

Like any other change, a change of mind follows a force, no matter how weak or wobbly that force might be. We may not even be aware of every insignificant influence, and at times, not even the most compelling argument can change our minds. Then again, an observant individual or a sensitive community shifts its conduct swiftly yet most readily in a habitual manner.

We are aware of fatigue, hunger, thirst, etc., and act accordingly. Similarly, a corporation monitors production, evaluates markets, and adjusts its operations to meet the demand; the nation amasses statistics on its economic capacity, surveys business attitudes, and improves on its functions to meet the challenges. So, the level of awareness of an individual, community, or society rises through experience.

Consciousness builds up with integration and peaks atop a hierarchy. Up there are found the most potent agents; the regime in the case of a society and the brain in the case of an individual. Effective communication strengthens unity. As an infant grows up, nerve cells connect more strongly to one another and circuits integrate with one another. As a society develops, agents link more tightly to each other. Thus, the tax authority knows your income. Shops know your shopping habits. Do you know who knows about your health?

We are aware of intensifying surveillance. The tightening control follows from the least-time principle, the quest for more efficient use of resources, i.e., free energy. Eventually, a deviation is seen as delinquency. In this Orwellian dystopia, the most dreadful prospect is that there is no chance of change.

Sometimes the course of events is so rapid that the system cannot pass information to the top of the hierarchy before a local response is necessary. For instance, one becomes aware of a reflexive rescue only after the incident. This might suggest that the self does not exist as such. Instead, like culture, it is a dynamic state of beings within beings. Many organs of society, too, respond to incidents autonomously and report about them later. For instance, the fire department rushes to the scene of the fire and only afterward broadcasts about the situation. Sharing information creates awareness. The more conscious we are, the more comprehensively we act.

CONSCIOUS ABOUT CONSCIOUSNESS

"What you see is all there is," writes Kahneman, arguing that available information is treated as though there were nothing else to be known.[29b] In the view of this Israeli-American psychologist, it is easier for us to build a credible story from a few ingredients than from extensive evidence. The developmental biologist Lewis Wolpert says the same: "The primary aim of human judgment is not accuracy but the avoidance of paralyzing uncertainty."[30] We are not geared toward truth but action – which may, however, lead to truth.

With its two hemispheres, our brain has evolved to create choices and choose.[31] But too often, conflicting and opposing information is ignored and overlooked to make it easy to decide, even though truth, not ease, is the goal of decision-making. Even when we try, it is difficult for us to bring all the pertinent factors to our highest level of awareness. As such, it is good, from time to time, to stop for a while, step back, and critically evaluate why we act the way we do.

Similarly, the true state of a nation barely penetrates the social consciousness before finding its interpreter. The voice gains popularity by expressing the suppressed feelings of many. A system that is broadly and intensely conscious of the full scope of a situation navigates with skill through the riptides of opposing forces.

Consciousness is an advantage for an individual as well as for a community and society. Yet, it also often has an overly tight stranglehold on us. Absent-mindedness is not only a disadvantage but sometimes even an asset that leads to discovery beyond exploration. Skillful management does not interfere with activities but relies on independent initiatives. The government, too, tries to boost innovations by easing up on regulations.

The thermodynamic theory does not take a stand for or against regulation. It only maintains that the conditions for the abundance of ideas are like the conditions for the richness of species. Vegetation blossoms in a cove protected by rocks, while only a few plants take root on a bank exposed to the open sea; freedom of thought needs protection from the tyranny of doctrine.

Moreover, according to the overarching tenet, we may rightly describe signaling in the neuronal network in the same way as traffic on the road network. A bottleneck, such as a weak synapse between neurons, is as concrete as a broken bridge. To fix it, we need to identify the cause of the neurotransmitter deficiency, as we need to understand the bridge failure. We cannot grasp consciousness without suitable concepts, just as we cannot repair things without the proper tools. The power of a unified theory is that many things can be comprehended in the same way.

Mind over matter

Although consciousness has no unambiguous definition, we can quantify it, as the measurements themselves define the object. As a result, the mind is measured as substance. For instance, we may characterize awareness by registering neuronal responses. We may label the gamma waves of the brain's electrical activity[32] as a necessary, although not the unambiguous sign of consciousness, and record them. We may supplement our list of characteristics with other signals and gauge them. We may diagnose consciousness as impaired if any of the presumed hallmarks are not detected or if an unusual attribute is discovered. Regardless of how long our list of measurables is, the data display the same universal patterns, suggesting that, profoundly mind is matter.

Since we may only quantify what we register, we cannot capture consciousness through measurements, only its manifestations. Neither can we detect the electron itself, only its field; that is, its dual. We can deduce the electron structure from its field, whereas reconstructing consciousness by analysis of behavior would be an overwhelming task. We cannot even figure out what is left unexplained from grossly incomplete observations.

It may seem unreasonable, although not extraordinary,[33] for me to address both consciousness and elementary particles in the same book. It may be difficult for each of us to be fully aware of our influences and consequent motives, but it is clear to me that I was molded through the years to cross and bridge disciplines. At the Low Temperature Laboratory of the Helsinki University of Technology, matter and mind were studied side by side. Some scientists moved from one research group to the other. I, too, came to the Lab to study the mind, not condensed matter, or, to be precise, it was my brain that came under study. MD Juha Huttunen, a specialist in clinical neurophysiology, registered responses from my motor cortex with an ultra-sensitive magnetometer. The instrument records the brain-generated magnetic fields outside the skull. They are about one billion times weaker than the Earth's magnetic field – well, not only mine.

How does experience follow from sensation?

The characteristics of consciousness are not unique but universal.

Consciousness feels personal, but in science, subjectivity is considered deceptive. Is it really so? Consciousness also feels experiential, but phenomenology is judged non-scientific. Is it actually? Consciousness appears to be goal-oriented, but in science, intentionality is held to be illusory. Is that truly the case?

Is consciousness a mystery only because we think it is exceptional? Could it even be that we are unable to grasp anything that we cannot relate to something we know already? A valid theory is thus expressly that which exposes an unknown as known. Conversely, if attributes previously perceived as unique prove universal, many supposedly well-known facts also deserve to be re-examined.

THE DELUSION OF OBJECTIVITY

Science swears by objectivity, but each one of us has a first-person perspective on the world. Could there be any other perspective? An electron experiences an environment different from all other electrons, as there are no two identical loci in the universe. No measurement is objective either because interactions are between subjects. A quantum does not move from one subject to many, only to one other subject. So it is not possible to examine any one thing without changing that thing as well as the observer by at least one quantum.

In fact, what we really mean by objectivity is that we should take a multifaceted view. A many-sided analysis is preferred over a one-sided assessment because it brings up many forces. Using general concepts of physics, such as force and quantum, we understand complicated processes as manifestations of the simple principle.

Consciousness is unmistakably unique, personal. But there is not any one thing that is entirely identical to another. If we argue that two atoms are identical, we have been able to distinguish them somehow from each other. Otherwise, we would speak only about one.[34] The more complex the subject, the more it adapts to other subjects. Already a protein molecule assumes different forms as part of different molecular moieties. Each of us, too, tends to take different roles in different situations. An identity accrues from personal experiences; the story of a nation accumulates from collective experiences. A subjective mindset, an endemic fauna, or our solar system, is a unique result of history. We need comparisons all the more, the less we know.

Nonetheless, David Chalmers sees the subjective experience as an unfathomable phenomenon.[35] The Australian professor of philosophy of mind asks: how does light striking the retina produce a sensation of intense red? How an *unspecific* signal gives rise to a *specific* experience is known as the hard problem of consciousness. The issue here is how the brain's physical processes generate impressive subjective experience, qualia from a simple signal.

The light that strikes the eye triggers various processes in the same way as a catalog from Ikea, Macy's, or Marks & Spencer that lands on the doormat puts people in motion in different directions. In other words, what matters, alters. We orient ourselves along the lines of force, i.e., relevance, in the evolving free energy landscape. The meaning of intense red not only follows from the photon wavelength recorded by the retina but also from the individual's life experience and even more distant heritage. Similarly, a flag awakens the consciousness of a nation. The colorful piece of cloth symbolizes experiences shared by the people. History explains matters.

While consciousness is not objectively accurate, it is subjectively consequential. We remember relevant experiences, whereas we tend to forget, even actively, irrelevant ones.[36] We can even recall something that did not happen if it is important enough, as revealed by examining many witness testimonies. We may not even be aware of what incident left the trace in our minds and what stimulus brought it back to consciousness. The recollection process itself is meaningful in strengthening memory. The next time, the signal propagates more quickly, and so we form opinions at once. This is cognitive ease.[29a] Likewise, after water has strenuously carved its way through the terrain, the brook can flow with gusto. We become more aware of what thinking is when we liken it to other processes.

THE ILLUSION OF REPEATABILITY

Along with objectivity, repeatability is a scientific ideal. The truth is that it, too, will remain an ideal, for we can reconstruct a situation almost, but never exactly. Even if we were to excite a single atom repeatedly, with only one photon each time, those quanta of light would be taken from the surrounding sources. The circumstances would thereby change. No process is independent of the background, ultimately of the evolving universe in which it is embedded. It is a grave misunderstanding that one thing could be changed while keeping everything else as is. Although this *ceteris paribus* assumption, a background-dependent model, or an effective theory, does not apply anywhere, it is used everywhere. Repeating tests with a conscious person is not only challenging but quite impossible. The subject gets (at best) bored and thus changes.

Scientific results are expected to be statistically significant, but the scientific inquiry itself may prove to be insignificant in its impact. A true meaning relates to what has happened rather than to an arbitrarily chosen statistical significance of a random variable. For example, even though one event paves the way for another, would one actually walk a little more slowly after reading words related to old age, as claimed by a famous study?[37] Statistical significance is especially difficult to validate when the individual characteristics accentuated in consciousness are distributed far from the average. Natural variation does not fit into the normal distribution.

A PART OF THE WHOLE

A new idea that combines earlier concepts does not differ in principle from a molecule assembled from atoms and photons. In the compound, the atoms are not as they used to be when free.[38] Likewise, concepts do not remain as they used to be after being related. For example, when time is identified with a photon period, neither the notion of photon nor time remains the way it used to be.

This is nothing new. In the book *Essai de Cosmologie* (1750), Maupertuis stated that when particles come together, they lose individual consciousness and receive awareness as part of the whole.[39] The consciousness of a compound is greater than the mere sum of the consciousness of its constituents. A person is aware of themself and others but not of every particle in their body. In a cosmological explanation, things are put in relation to one another, whereas in a special theory, the thing itself is special – incomprehensible. The more comprehensive a theory we utilize, the more phenomena are understood as alike, and the less there is to be explained.

A SENSE OF MEANING

Without memory, history, perspective, we would be without a sense of meaning. As portrayed in the movie *Still Alice* (2014), starring Julianne Moore, such a fate is tragic. Meanings do not materialize instantaneously, but paraphrasing Jean Piaget, the renowned Swiss psychologist, they emerge through experiences. We do not choose a way of life but adapt to circumstances by exploring various ways of thinking. Similarly, an ecosystem does not come into existence all at once but evolves through time by exploring various ways of living. Memories, traditions, and genes alike are valuable mechanisms that speed up processes. For the same reason, robots are not only programmed to do their chores but also to learn lessons from what has happened. We should also take heed of greater history, not only of our own.

The science fiction film *Blade Runner* (1982), directed by Ridley Scott, impressed me by proposing that there is no fundamental difference between us and others, between the organic and artificial. The movie is set in an urban dystopia, a future Los Angeles. Bounty hunter Rick Deckard (Harrison Ford) is tasked with eliminating four androids, newly arrived on Earth after escaping slavery. With a lifespan of only four years, these human-like robots should not have enough time to develop feelings – and yet they have done so. Moreover, an android serving as a personal assistant has been implanted with a life story, for a being without a history is one without perspective. With that in mind, our mission should not only be about refining robots to ever closer resemble ourselves but, far more importantly, to refine ourselves to be more humane.

We can convey subjective meanings to the extent that we share collective experiences. In the same manner, both tissue transplantation and data transfer call for compatibility between the source and the receiver. Peer support and peer review draw from that which is common – and dismiss the uncommon. Yet, it is good to remember that none of us can fully know what it is like to be another person. Otherwise, one would be the other. At best, we can belong to one another.

"What Is it Like to Be a Bat?" asks Nagel in the title of his famous essay.[40] We humans have at least so much in common with the bat that we dare to ponder whether it experiences its own existence. We hardly consider an electron conscious. Nonetheless, Gell-Mann, who discovered quarks, pointed out that by assuming particles to be intentional, as if conscious, it is easy to understand the physics of particles.[41]

In effect, Leibniz had already said the same thing: when you must explain a machine, it is best to say what it is supposed to do and show how each part serves this purpose. Motives., i.e., forces, explain.

Intentionality is not only a characteristic of consciousness, but all systems gravitate toward balance. Nagel anticipates that explanation of consciousness and, more generally, that evolution follows from a universal tenet. The philosopher argues that we must think differently about what physics is if we want to explain life in terms of physics. Today's equations of physics do not account for the arrow of time, nor the course of history, relying only on blind randomness. But the world does not look like that. According to Nagel, Nature is goal-oriented. Natural teleology means for him "that some laws of nature would apply directly the relationship between the present and the future, rather than specifying instantaneous functions that hold at all times."[42] The causal is teleological.

THE STRUGGLE OF FORCES

We had better be aware of the principles that govern our thinking if we are to govern our thinking. According to the thermodynamic theory, we prefer favoring opinions and rejecting opposing points because our thinking aims at balance in the least time. This conclusion parallels Kahneman's argument: "Questioning your intuitions is unpleasant when you face the stress of a big decision. More doubt is the last thing you want when you are in trouble."[29c]

Wealth is the evolved abundance of species in the tropics and companies in Silicon Valley and the accumulated richness of conceptions. Through the ages, mental structures originating from hunting and gathering evolved to master agriculture and matured into the present-day urban culture. Diversity sprouted from diversity when social connections opened new lines of thought. In this way, we became aware of the possibilities whose realization paved the way for more opportunities. As consciousness expanded and integrated, we gained access to new resources as individuals and as a society. This increased efficiency in free energy consumption has evolved in a power-law manner.[43]

It takes time and energy for an individual to mature in thinking, just as for a society to establish its functions. Synapses are created in the same way as road junctions are constructed. Flows of quanta, such as nerve impulses and intercity buses, are selected for ever-faster connections. Neurons are covered with myelin for the same reason as highways are paved with tarmac. Along these well-trodden paths, you will sometimes even pass by the intended destination or get diverted in the wrong direction because everything happens so fast and with ease.

When choosing a line of thinking over another, the struggle of forces in the mind is often more vicious than a draft in a house, yet the law is the same. Even a small pressure difference causes a waft. Conversely, stale air has not moved for a while, and a mind set in its ways is stuck.

It is symptomatic that contemporary science is blind to the *Grand Regularity* and grapples narrow-mindedly with the broad questions of time, space, life, and consciousness. Many problems arise from the reductionistic paradigm and vanish in the holistic conception.

SECOND NATURE

Consciousness changes as the structures of the mind are renewed. Society, too, changes by regenerating its structures. However, the individual's investments in the past, say, in education, career, relationships, are so profitable that it seems a fool's errand to try something new. Just as a way of life, a line of thought has delivered such substantial returns that reforms seem no longer worthwhile. The efforts of unlearning are greater the more thoroughly the learnings have been entrenched.

Habit is second nature. We deal effectively with the conventional and ineffectively with the unconventional. So, it is harder for an expert to see their case in a new light than for an uninitiated person to see the point simply in light. For example, it may be difficult for a neuroscientist to imagine that the responses of a neuronal network would, in principle, be the same as the avalanches of a growing sandpile, even though the data look the same. Hermann von Helmholtz testified to this adversity in the Faraday memorial lecture: "It is often less difficult for a man of original thought to discover the new truth than to discover why other people do not understand and do not follow him."[44] Presumably, the German physician and physicist, who was among the first to present the law of energy conservation, spoke not only about Faraday's experiences but also his own.

When a new idea seems bizarre, the associated signals cannot easily connect to the established network. Specifically, the insight that consciousness is a manifestation of the universal law may appear to be a mere stray remark. When there are no connections, no signals are relayed, and the message is lost. So the new is often opposed without knowing it.

When there is nothing new in your purview, ongoing activities tend to be intensified, but as a result of streamlining, your abilities to engage with the new decline further. Specialization cuts opportunities to spot new ideas, and thereby consciousness becomes more and more impoverished. Similarly, when a steady-state stagnation is about to be attained, the abundance of species in an ecosystem and the number of agents in an economy decrease.

New prospects do not open up from compartmentalized thinking but from free thinking. A discipline renews itself from its foundations. Likewise, new branches of the phylogenic tree do not sprout from a specialized species but from the stem. The primary structures, such as stem cells, can differentiate into a variety of forms. Similarly, young people have the capacity to pursue a variety of jobs. A new venture does not originate from enhancing an established business but from a new idea. Comparisons are not mere metaphors but speak about the same principle.

THE TRAGEDY OF THE COMMONS

In the news, we see how floodwaters wipe out roads and tear down telephone lines. Even though we do not directly see how the connections in the brain break apart, they, too, fail. When the mind breaks down, we are not capable of performing demanding tasks. There can be difficulties even with daily chores. The same goes for communities and societies. When communication fails, ideas are not transmitted, and critical factors are missed. When social cohesion has gone, the community has lost its most vital characteristic, its identity.

Such a course of events is also known as the tragedy of the commons. Everyone who sees only their own needs tries to take advantage of the depleting resources. Yet, everyone faces the drawbacks of dwindling free reserves. The outcome is a tragedy for everyone. In his article from 1968, the ecologist Garrett Hardin lists the atmosphere, oceans, and rivers as overconsumed resources.[45]

We would avoid troubles best with a broad sense of solidarity. Still, it is difficult to establish cohesion, holistic thinking, and other efficient energy transfer mechanisms when the environs' demands vary widely.[46] A neuronal network also disintegrates when forces pull in different directions, making behavior unpredictable. Disparate objectives tear individuals, companies, and communities apart, just as a region exposed alternately to severe drought and heavy rainfall declines.

Considering consciousness as a means, among others, to make things happen, we can relate changes of mind to shifts in the zeitgeist. For example, sharp economic fluctuations resemble extreme mood fluctuations. Splitting a community or society into mutually bickering groups parallels a complex mental disorder. These and other correspondences between an individual, a community, or a nation do not solve the problems but may help us fathom what the problems are all about.

Is thinking computing?

We count on thinking, but thinking is non-countable.

Neuronal signaling and the brain's architecture bear all the hallmarks of *Grand Regularity*. The size of events and inter-event intervals, for instance, are distributed in a power-law manner, just as traffic disperses in urban areas. The intensely wrinkled and folded cortex implies that our peak cognitive activities cover large areas, just as the species at the top of the food chain claim the largest territories. An eagle has the means to benefit from its vast territory and also needs its big habitat to maintain its way of living. Similarly, our cognitive functions endow us with the means to prosper, and we need those means to support our effective thinking. The eagle is not that often at its nest, more likely soaring somewhere over its territory. A noteworthy thought is neither associated with one thing nor found at any specific locus of the brain but is widely distributed in the cortex.

The universality of patterns suggests that we should describe our cognitive machinery and functions in the same way we describe other systems.

A TRAIN OF THOUGHT

Impulses dash from the sensory organs along nerve fibers to the central nervous system, and upon arrival at the primary cortex, they somewhat unpredictably spread to other cortical areas, for these trains of thought, as they progress, consume their driving forces, the motives behind the thought.[47] The fact that we do not know precisely how thinking proceeds does not ultimately follow from the complexity of the brain but from the fact that thinking, like any other natural process, is intrinsically intractable, as everything depends on everything else.[48] The brain itself could not possibly know in advance how to think in one situation or another.[13] This experience must be familiar to everyone.

Like other quanta flows, the brain's electrical impulses seek various ways to attain balance as quickly as possible. A signal chooses its course as naturally as a river finds its route. Past events have an impact on ongoing ones, which in turn affect future events. Sometimes a jarring experience diverts the course strongly, just as a flood's peak discharge may open a new path for a stream. The capricious character of thinking is thus neither about randomness nor the ambiguity claimed by the uncertainty principle of quantum mechanics.[33] It is about the same unpredictability and chaos that is present everywhere. So we know exactly why we cannot know things exactly.

Even though our thinking is fundamentally unpredictable, we can focus our attention, even astonishingly well. In a famous study, subjects were asked to watch a video of a basketball match and keep their eyes on the ball at all times. This intense focus left them utterly unaware that a person in a gorilla costume was also strolling on the court.[49] We choose what we want to see. We select what we value. Fundamentally, we value meanings in terms of free energy, those forms of energy and time, i.e., quanta, that we can make use.

A noteworthy hypothesis regarding awareness is that signals compete for the brain's capacity.[50] Competition, as a Darwinian term, expresses the contest between alternative cognitive processes, just as other flows of quanta distribute themselves along possible paths. When we focus on a matter with rapt attention, some signals pass through while others are blocked. Similarly, nations' heads are guaranteed roads free of traffic, while other people are held up. Capital, too, gets targeted on some initiatives while others are left without funding. Signals, vehicles, and money move along the lines of force.

Let us relate our thinking to other processes to grasp how we think. An association makes us aware of the underlying similarity, enabling us to reach meaningful conclusions from surprising connections[51] and find something valuable we were not even searching for. Many a discovery is pure serendipity, as is the spread of species by daredevils and strays. Ideas are like mutations that need to be tested in reality to know their verity.

Our genetic information is far from sufficient to structure our neuronal network; adaptation to various circumstances requires plasticity. It is the experiences that shape our lines of thought. Already in a fetus, after the first signal has elicited synapses, subsequent ones strengthen them. As the impulses run faster and faster along the same tracks, learning continues until competence approaches perfection, where motivation is almost exhausted because there is only an insignificant difference between achievement and ambition. The learning curve is thus S-shaped, like other growth curves.

Motivation is a driving force, an imbalance between outcome and intention. With that in mind, talent is more about seeing room for improvement than demonstrating acquired skills. The same aspiration for perfection, yet without a predetermined goal, is displayed in the evolutionary arms race between species[52] as well as in the artificial intelligence algorithm where two neural networks compete with each other.[53] We can comprehend all motives in the same way by expressing any difference as an energy difference. Paraphrasing Leibniz, unless the difference is discerned, there is none.[34]

The thermodynamic theory of time cannot but interpret thinking as a chain of events, and hence conclusions cannot be something other than what they are. Equally, we need to ask what the premises of other theories of cognition are and how well the resulting findings correspond with observations.

MAKING SENSE OF ARTIFICIAL INTELLIGENCE

On the one hand, we can program predefined automata to execute a given formula, but such agents cannot cope with new situations on their own. On the other hand, we might devise autonomous agents. They change the world, just as we do – surprisingly and threateningly. In this regard, artificial intelligence is a manifestation of natural law, just like natural intelligence.

Instead of getting frightened about artificial intelligence, we should expect it to contribute to numerous processes by complementing our capabilities rather than narrowing our opportunities. Earlier, biological inventions became integrated into the biosphere, and later technological innovations into the world economy without demarcation. The contrast between the artificial and natural is becoming arbitrary as machines, one after another, pass the Turing test. A synthetic agent is distinguishable from a human. That does not point toward an inhumane dystopia but should make us ponder the essence of humanity: cooperation and solidarity.

Nonetheless, it is worth contemplating our lives eventually beneath the yoke of artificial intelligence, especially as we flounder to comprehend what intelligence is all about. "People worry that computers will get too smart and take over the world, but the real problem is that they're too stupid and they've already taken over the world,"[54] jibed Professor Pedro Domingos, an expert on machine learning, in his book *The Master Algorithm* (2015). We are hardly any better than algorithms, either, if we only correlate things with one another without any clue about causality.

Since correlation and curve fitting explain nothing, contemporary science cannot answer the "why" questions, says Judea Pearl, a computer scientist and philosopher, in *The Book of Why: The New Science of Cause and Effect* (2018). If the machine does not have an accurate picture of reality, it cannot behave most sensibly; resources are wasted in establishing correlations rather than invested in sorting out causation. Likewise, if our worldview does not correspond to reality, we dissipate free energy in intensifying meaningless actions instead of focusing on caring for living conditions.

WHAT IS INFORMATION?

Information is free energy for its receiver.

As we live in an information society, the definition of information, too, is only a few clicks or taps away. It comes from Claude Shannon's publications of 1948.[55] At the time Shannon was working at Bell Labs, he was theoretically interested in the practical problem of putting through as many calls as possible along the few lines available at the time.

Shannon did not care about what people were saying on the phone, only about making sure that gossip spread with fidelity. That is why Shannon defined information apart from meaning so that a message that cannot be written any shorter holds maximal information. Isn't it strange that this definition, this downright absurdity, became the cornerstone of information theory? Isn't the meaning of a message its essence?[56]

Textbook knowledge is not meant to be copied but considered, even reconsidered. So, what is the significance of, for example, a sugar molecule that a bacterium happens to take up? That message has at least a metabolic value, energy. Sugar can also signal that even more free energy might be available in the neighborhood. Information is a vital commodity because it holds free energy for its recipient, the means of making a living.[57] That is to say, information is what you don't already know and what may thus open new opportunities for you.

The information contained in a chemical compound, a kind of combination of atomic characters, is free energy. There is, through thermodynamics, no principal difference between characters and significance.[57] Since a mere piece of paper is also fuel, even a cockroach gets something out of this book. In other words, semantics and syntax are profoundly the same. The Canadian philosopher Marshall McLuhan reduced it to the crux: "The medium is the message."[58] So, a blank tweet is not an altogether meaningless statement, either.

The American philosopher Charles Sanders Peirce understood that signs, referring to an object, are interpreted in the context of the understanding – or misunderstanding – of a receiver. The elementary unit of information actualizes as "a difference which makes a difference," according to the English anthropologist and semiotician Gregory Bateson. As noise makes no difference, one has to say something wise – or really stupid – to stand out from the ordinary.

Information is distributed in the same way as other entities, approximately in a power-law manner.[59] For example, among the documents on the Web, only a few are needed frequently, whereas most are retrieved rarely. Rolf Landauer's revelation, "information is physical,"[60] therefore, seems very true. The German-American physicist understood that carrying out calculations is work. When toiling hard, the computer will also get warm. That's why its fan turns on. The quantum computer might keep on crunching numbers without loss of total energy, but free energy must decrease for the calculation to proceed from input to output.

Curiously, a computation might halt at a dynamic balance before yielding any result.[61] Alan Turing, a British mathematician and logician, proved in 1936 that this halting problem is irresolvable. Since everything depends on everything else, there is no way of knowing before execution whether the program will run forever or finish. Similarly, it is difficult to predict whether an interstellar object will voyage forever or eventually be bound into orbit around a star as an asteroid.

THE NAME OF THE GAME

Information makes things happen with its associated free energy. Do we thus measure the meanings of all messages in energetic terms? This idea may seem like a massive simplification of the intricate subject of communication. But, at least, it is not new.

John von Neumann was looking for a natural law that explains human behavior. His studies in the late 1920s laid the physical foundation for game theory as a model of behavior.[62] In that worldview, predicated on all-inclusive thermodynamics, a human being is, in principle, no different from any other being. As Philip Ball asks, "Might the enormous diversity in behavior that social science seeks to study turn out to be based on simple foundations?"[63]

Naturalism, originating with Maupertuis, was adopted by biologists and economists via von Neumann's and Oskar Morgenstern's book *Theory of Games and Economic Behavior* (1944). Also, the Nobel Prize winner Paul A. Samuelson, a late MIT professor, compared transactions in an economy to chemical reactions.[64] Furthermore, the Nobelist John Nash saw an economic balance as a chemical equilibrium.[65] In short, behavior is about balancing.

While these conclusions are mostly consistent with thermodynamic theory, it should be noted that in game theory, information is associated only with bound energy.[66] It lacks the energy difference or imbalance between the system and its environment. Von Neumann inherited this incomplete comprehension from Boltzmann, who had understood only the condition of balance, not evolution toward balance. In truth, communication is motion like any other process toward ever more probable states in the quest for balance.

Perhaps it is surprising that the least-time law was no longer properly understood at the time of Boltzmann, just over a hundred years after Maupertuis. As a measure of information and thermodynamic state, entropy had become mixed up with the incoherent concept of disorder. Leibniz's legacy that things happen because they are probable,[67] had been lost. It is time to recover it from plain observations: a bacterium swims in the direction where the sugar content grows fastest; business looks for the direction in which profit increases fastest. For every increase in entropy, there is a cause, a force.

The Beautiful Mind, which premiered in 2001, tells the story of Nash suffering from schizophrenia. I remember that it was quite a while into the movie before I realized that the main character was acting under delusions! Likewise, it may take time and energy before we recognize that our view of reality departs from reality. Otherwise, we would already be revising our positions. An erroneous worldview includes unrealistic incentives, as revealed by this realistic movie about Nash. We see problems that do not exist while remaining ignorant of those that do. A different view points differently, whence we behave differently. Nash summed up motives for behavior, whether imaginary or real, through the concept of utility.

My colleague Jani Anttila and I showed that game theory's imperative of maximizing utility means attaining balance in the least time.[66] There are gains of many types, such as rewards, and losses of many types, such as penalties. Relating incentives of any kind to forces, we realize that our behavior, like other motions, follows the lines of force. We direct ourselves toward what we value. This is the kernel of the thermodynamic calculus of social order.

UNDERSTANDING MISUNDERSTANDINGS

The physical nature of the information was clear to Shannon. Information is like any other substance because machines apply work to it.[68] And the results of data processing are no different from the outcomes of other events. The distributions of sentences, words, and letters are as skewed as those of genes and galaxies, approximately lognormal.

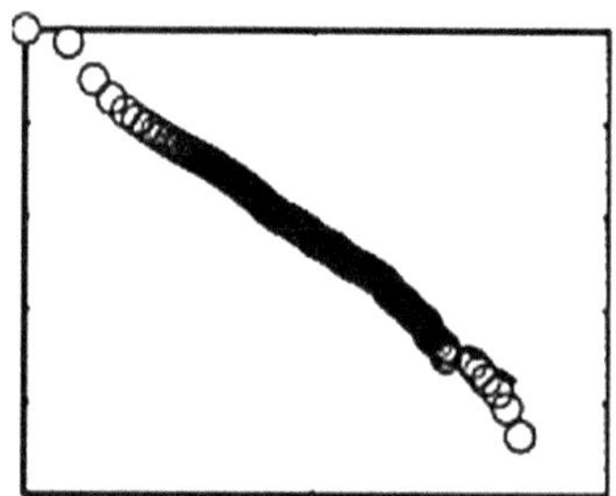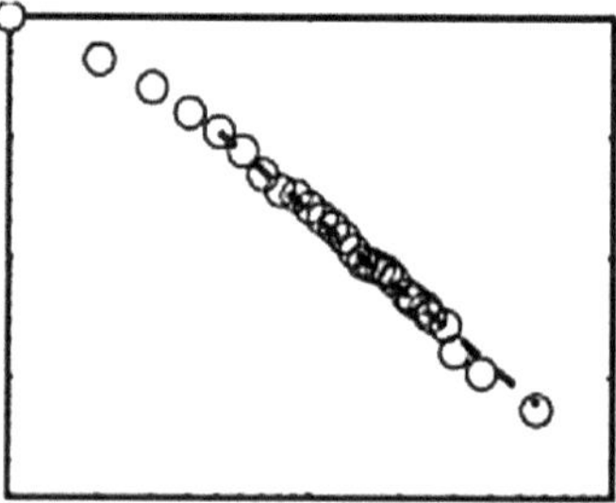

The Internet (left) and a protein interaction network (right) have many nodes that connect to only a few other nodes and only a few that link to many.[59] The number of connections complies with the power law. Without headers and labels on the axes, the mere data do not reveal their origin but rather the underlying principle in common.

In communication, the quanta, i.e., the received data, changes the recipient's status. But as there are alternatives for the flows of quanta, it is impossible to predict what communication will cause.[26] And there is no shortage of options in the vast neuronal network of the central nervous system. For the same reason traffic affects traffic, the performance of a road network, a telecommunications network, or any other network is ultimately unpredictable. So it is not strange but natural that often the recipient understands the message, if not entirely wrongly, at least not in exactly the same way that the sender meant.

It can provocatively be said that communication is only meaningful when there is a misunderstanding. The recipient can understand the message exactly as the sender does only when there is no difference between the two. In that case, there cannot be anything new in the message for the recipient, and nothing will happen. That is why judicious decisions result from considering contrary views rather than only homogenous opinions.

Like other courses of events, the flow of information cannot be calculated, but still, we wish to know the future. In this quest for certainty, physics narrowed Maupertuis' principle to the exactly computable Lagrange's equation. Lack of computability is often seen as an obstacle to understanding, but the real obstacle is the lack of understanding of computability.[61]

BREAK ONE'S WORD

It is no coincidence that biological information is encoded in the energetically expensive DNA molecule. Information as free energy is valuable. It includes tremendous opportunities not only to survive but also to thrive. The biological functions coded in our chromosomal archive can be recombined into novel genetic variants. Similarly, the information in our memory can be combined into new cross-disciplinary concepts.

Making a new connection is often more relevant than producing just another snippet of information. To that end, we can employ Bacon's method, for instance, in identifying time as the quantum's attribute when encountering a concept of time in various contexts, such as in the expansion of the universe, biological evolution, and in a period of quantum, yet always associated with energy. This leads to straightforward thoughts about thinking.

The physical character of information was painfully evident to me as well as to my colleagues Mahesh Karnani and Kimmo Pääkkönen, as we had a hard time dealing with large amounts of data produced by protein structure determination. We saw that machines also had a lot of work to do in sequencing a whole genome. Nonetheless, Shannon's information theory gives us an incorrect impression of information, as if it were immutable and immaterial. Our everyday experience is quite the opposite. For instance, notes get lost; DNA is vulnerable to damage. 'Breaking one's word' states the destructible character of information pretty much as the thermodynamic theory does.[57]

Knowledge is power. In-depth education is among the most efficient means to evaluate and exploit data. While it is easy to suspect that information is unreliable when email, Facebook, or Tinder offer too-good-to-be-true temptations, it is more difficult to distinguish whether scientific truths authorized by textbooks, such as Shannon's definition of information, comply with reality. The survival and success of an individual, community, and humankind rely on a realistic understanding rather than on the prevailing perceptions.

WHY DO WE SLEEP?

We sleep for the same reason as we are awake – to gain balance.

We sleep for about one-third of our lives. Much is known about sleeping, but the ultimate purpose of sleep remains unknown. However, we can recognize in the sleep the same characteristics as in other processes. The resemblance is illuminating but blinding as well, for the thermodynamic theory cannot explain phenomena in any other way than by the least-time principle. In the end, the essential question is, does the offered explanation make sense?

When we suffer from insomnia, we behave much the same as we do under severe stress. We forget things; we make mistakes. When the whole is in tatters, the most effective mechanisms are out of order, and hence the fastest flows of energy are blocked. An ecosystem, too, is not what it used to be when fragmented; society is in trouble when disintegrated. Conversely, the balance is restored when the neuronal network, food web, or telecommunication network reintegrates by recovering connections.

During the day, the neuronal network moves out of balance, as intense activity generates numerous connections in specialized cortical areas while long-range links degenerate, causing the whole to lose coherence. During deep sleep, the balance is reclaimed by the brain's low-frequency electrical activity renewing long connections. While sweeping throughout the brain, these long waves solidify the whole; pieces unite into enduring memories and holistic insight. Conversely, mid-frequency activity straightens out medium-range connections, and high-frequency dreaming processes prune short-range connections, eroding our more trifling memories. Also, changes in sleep cycles as we age from childhood to adulthood suggest increasing neuronal integration.

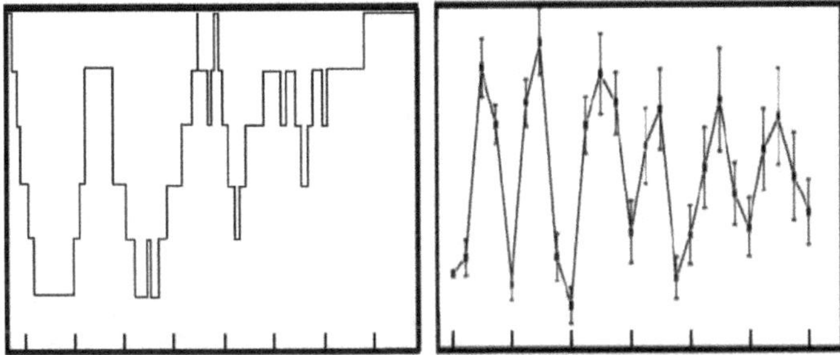

Over a well-slept night (from left to right), deep sleep (low readings) alternates with dreaming (high readings) in a lessening manner toward the morning.[69] The concentration of the genetic information carrier fluctuates in an attenuating manner over the divisions of a mammalian embryo (from left to right).[70] Similar courses suggest that the processes follow the same law.

Through a well-slept night, the balance is approached by repeating and varying the basic theme, where slow periods of deep sleep follow one after the other in a waning manner.[71] This pattern is a characteristic of nondeterminate processes. The sleep cycle changes the neuronal network, after which the need for sleep changes, and so on. There is no efficient algorithm for knowing before bedtime how to sleep,[13,14] because the need for sleep and the sleep itself depend upon one another. Even though the unique sleep-wake cycles cannot be predicted, they still follow the same natural law as other processes. In like manner, ecosystems and exchange rates recover from devastations through damping oscillations.[72]

When conscious reasoning cannot find a way to our goal, it makes sense to extend the search further away from what we can imagine when awake. In particular, long-distance connections add diversity and robustness to the network.[73] Music, too, is known to activate the central nervous system as a whole with variations on a theme.[74] Creativity requires freedom and security, such as good sleep, where the nerve signals may vary their activities freely without harm, for our movements are paralyzed and our reactions are restrained. When unrelated and remote thoughts mingle, exceptional visions of reality open up and prepare us to cope with unanticipated situations. Similarly, search engines screen the world-wide information network for plausible queries. The road network, too, is monitored, maintained, and supplemented to meet conceivable needs. At times, the effects of a new connection are surprising for drivers and thinkers alike. For example, in what is known as Braess' paradox, a new connection to a network may even end up impeding overall flow through it. It is not so easy to put a new idea in operation when it messes up existing practices.

The structure and phases of sleep are also typical of other creative processes. Work communities organize festivities or other seemingly secondary activities that may open unexpectedly rewarding opportunities despite their seemingly unprofitable character. Multidisciplinary and multicultural interaction fulfills the same end. The meaning of art can be understood similarly. It frees our imagination for explorations of reality. Without alternatives, there would be no pool to choose a course of events, ultimately a way of life.

Many a specific problem can be resolved in the general context by apprehending in a similar fashion supposedly opposing phenomena, such as sleeping and being awake. According to thermodynamic theory, all processes are headed toward balance. The principle is the same; only the mechanisms are different. That's why the patterns reiterate *mutatis mutandis* through Nature's hierarchy.

Is free will an illusion?

Having free will means having opportunities to make choices. As the future seems neither predetermined nor altogether open, to what extent is our will free?

Minority Report (2002), directed by Steven Spielberg, addresses this theme. The lead character, Tom Cruise, is running away from minions of the law, having committed a murder – in the future, according to a foresight system of psychics. However, as it is getting closer, it turns out that the future is not so fixed after all.

The plot is more convoluted than the reality we know. We experience neither the future nor the past, only the present in transition due to forces. Those who can sense the forces can see the future. Those who have forces in their power can influence the future. The more that capacity is demanded of you, the more powerful a position you have, for the future is genuinely open yet capped by resources.

In the form of a physics theory, this commonplace understanding is more convincing than mere philosophizing. Evidence for free will is enormous because making choices results in the ubiquitous characteristics of history, namely, skewed distributions, sigmoid curves, spirals, oscillations, and even chaos. The future is neither determined nor indetermined but nondetermined because forces change as they transform today into tomorrow. As everything flows and nothing stands still, there is no ground for predestination.

Freedom of choice

Since antiquity, the existence of free will has continued to be debated in philosophy, psychology, and neuroscience, sometimes also in physics. Free will threatens the traditional worldview of physics, where the motions of particles are entirely determined in principle. It also makes one question the view of modern physics, where motions are thought to be indetermined, that is, defined in probabilistic terms. The thing is, these beliefs hold true only in a state of balance, whereas free will exhibits itself in a state of imbalance. The thermodynamics of time addresses the question of free will in a practical manner. We do not ask whether a person has free will. Instead, we ask what free will *is*.

It is common sense that you have free will as much as you can make things happen. You can direct what will happen with the forces under your command. Your basic metabolism provides enough energy for contemplation, but still, you cannot think freely without any bounds because even thinking is constrained by resources and structures. Granted, few have sought to be free from their own minds. The story goes that Brian Wilson, the linchpin of the Beach Boys, expanded his mind so that it no longer fitted into his head.[75] Freedom of expression requires substantially more force, not to mention the freedom to act. In the same way as other forces, freedom will manifest itself as flows of quanta that channel along open paths through available mechanisms.

We recognize the concrete nature of free will in violations of rules. When there are no alternatives, the will is, in fact, not free. A desperate person has no choice; they are exempt from liability, just like a stone with no choice but to fall straight down. Its course, dominated by gravity, seems entirely predestined, even though it is, in fact, dictated by many additional forces. Still, we do not realize their presence unless we see them in action. Caught by the breeze, the exact trajectory of a falling leaf toward the ground is unpredictable. The course of life goes likewise. Turns affect future turns. This is nondeterminism.

Paraphrasing Leibniz, the cause of an event is an imbalance.[76] When you make a choice, you will consume resources. On the one hand, the decision affects the amount of power at your immediate disposal and, on the other hand, how much of it will be available to you after the choice. The outcome is not solely dictated by the initial state but by all states along the path of events. In other words, although the equation of the least-time law is known precisely, it cannot be solved exactly because the causes and effects are inextricable.[47,77] No chain of events is predetermined, for even a slight change can invite irresistible forces. Just a single quantum may trigger a nuclear chain reaction.

The quest for balance in the least time leads to unique yet regular evolutionary trajectories. If the path were predetermined right from the beginning, the data would not display the variation that we see. If the events were random, the data would distribute symmetrically and not be skewed, as we observe. Of our own accord, we do the things that result in the universal patterns. Conversely, if free will entailed something special, it should stand out in the data and contrast with the *Grand Regularity*.

The practical perception of free will differs from that of the deep thinkers but, as explained, not from observations. The utterance "I have no time or energy," in other words, I have no quanta to do the task, is the accurate answer, in terms of physics, to the philosophical question about the nature of free will.

IN THE LIGHT OF TIME

The questions of free will and consciousness are intertwined. In the 1980s, the neuroscientist Benjamin Libet showed by measuring the brain's electrical activity that, we prepare to do something before we have actually decided to do it.[78] So, are our deeds, after all, predetermined?

Daniel Dennett criticized such an interpretation of Libet's experiments.[79] The American philosopher maintains that it is impossible to figure out the unambiguous order of events from the recordings. It is unclear how the final decision could be distinguished from the preceding focusing of attention.[80] In turn, Roger Penrose, a world-renowned professor of mathematical physics, argues that the processes of consciousness cannot be calculated because of the intrinsic indeterminism of Nature, as claimed by quantum mechanics.[81] But what is the evidence for indeterminism?

According to the thermodynamic theory, the subconscious signals preceding the conscious decision do not imply determinism or indeterminism but nondeterminism. Libet's experiments can be compared to other courses of events. For example, our reflexive reactions tell us that our actions are not all conscious. Advanced systems react autonomously and assemble an integrated response later. The readiness to raise a hand precedes the conscious decision in the same way as the readiness to draft a law precedes a parliamentary decision. It is perhaps impossible to distinguish unambiguously between conscious and unconscious actions because of events preceding events. There is no outcome without some cause since causes stem from previous consequences. Once again, drawing a line where there is none engenders a nonexistent problem.

Interpretations of Libet's results and purported implications against free will necessitate a prior perception of what freedom of will is. The American philosopher Alfred Mele examines this dilemma in his book *A Dialogue on Free Will and Science* (2013). We cannot view a subject without having a viewpoint, that is to say, a theory.

Forces shape free will just as they affect other degrees of freedom. Circumstantial effects, such as a social situation, modify people's thinking and behavior. In any case, we do the right thing when acting responsibly toward ourselves, our community, society, and humanity, for the greatest forces are associated with the most significant values.

The correspondence between free will and free energy does not undermine the importance of the freedom of will. On the contrary, we are fully accountable for our deeds to the extent that we have the power to execute them. After exhausting all our options, there is nothing we can do. Only then are we free from the responsibility.

While the thermodynamic theory is unable to advise us on how we should behave in a given situation, it emphasizes the value of broadening our perspectives. To act wisely, we should gaze far into the future as well as way back into the past, weighing manifold factors in the light of time.

The revelation of free will, consciousness, and mind at large could not be other than materialistic, for, within the thermodynamic theory of time, it is not possible to comprehend anything in terms other than the quanta and their least-time flows. It is hard to grasp anything outside what is known because there are no concepts within the theory for these unknowns. That which light leaves in the shadows is difficult to discern. We need different angles of illumination to see clearly.

KEY POINTS

- Thinking is a natural process to attain a balance.
- Consciousness is a means to make things happen faster.
- Free will is free energy.

8. WHAT IS OUR DESTINY?

> The better we understand reality,
> the longer we live.

"Where do we come from? What are we? Where are we going?" are not only elusive philosophical questions, summed up by French artist Paul Gauguin,[xv] but addressable in practice by thermodynamics. This holistic perspective on our past, presence, and prospects opens up when everything is understood to comprise quanta and all processes are understood as flows of quanta.

From this vantage point, we see, for example, that the ongoing great extinction of animal and plant species will be etched in the history of Earth alongside earlier extinctions. The fallout from nuclear tests, the radioactive streak will be found in geological strata far away into the future as the signature of an iridium-rich meteorite that struck the Earth long ago.[1] Given that our undertakings are now written in the Tellurian chronicles like previous disasters, it has been suggested that a new geological era, the Anthropocene, has begun. It is characterized by human impact on the Earth's geosphere, hydrosphere, atmosphere, and biosphere.[2]

However, isn't it all the same whether we label the most recent activities of man as a new epoch or only as a part of the Holocene? Over this era, which began ten thousand years ago, humankind transformed from hunter-gatherers into farmers and eventually into urbanites. The large-scale extinction already started tens of thousands of years ago. Climate change, the spreading of invasive species, and the fragmentation of habitats are now only more evident than ever. Even as we become more aware of the global ramifications of our way of life, it seems as if we are in for only new adversities.

[xv] The three questions title Gauguin's painting (1897).

Since changes seem hard to master, let us ask why the world is changing in the first place. According to thermodynamics, forces cause changes. We have freed forces from those deposited in fossil fuels all the way down to those bound in the atomic nucleus. This release of free energy has set our way of life on a collision course with the grander reality, where the natural forces on a planetary scale are counteracting our global impact.

NOT A MOMENT TOO SOON

Human activity is unprecedented but not extraordinary. While the scale of our influence is global,[3] our pursuits follow the same principle as other processes ([Chapter 1](#)). Urban areas expand[4] in the same way that species spread.[5] Transport and energy transmission networks and information networks disperse, just as trees and shrubs sprout new branches.[6] Our organizations assemble themselves as other hierarchical systems do.[7]

Even our ominous endeavors, in all their enormity, display the *Grand Regularity*. Wars break out like forest fires. Skirmishes and flare-ups burst out from time to time, while world wars are as rare as walloping wildfires. Data reveal nothing unique in human undertakings. After all, how could humans possibly bypass the universal law of nature?

The magnitude of human impact is not exceptional, either. Natural disasters, too, such as earthquakes, large meteorites, and volcanic eruptions, bring about global changes abruptly. The biggest forces cause the widest and swiftest changes. The massive power in the hands of modern people makes things happen fast. That is not a problem in itself. For example, when the ozone layer depletion was noticed, its causes were investigated and mitigated. Likewise, the acidification of lakes caused by emissions of sulfur dioxide and nitrogen oxide was brought to a halt, and today the waters are recovering. However, now plastics pollute the oceans, some raw materials are running short, the planet is warming up. What is going to wax and wane next?

We tackle the problems we cause. That says a lot about our short-sightedness. Our measures ought to be proactive rather than reactive. Not a moment too soon, we need a worldview that covers more than the world of human affairs to take a course consonant with the planetary forces of nature toward a sustainable balance. To this end, Parmenides' idea of the fundamental and eternal element that makes up everything seems a sufficiently broad and solid ground upon which to erect a truthful worldview.

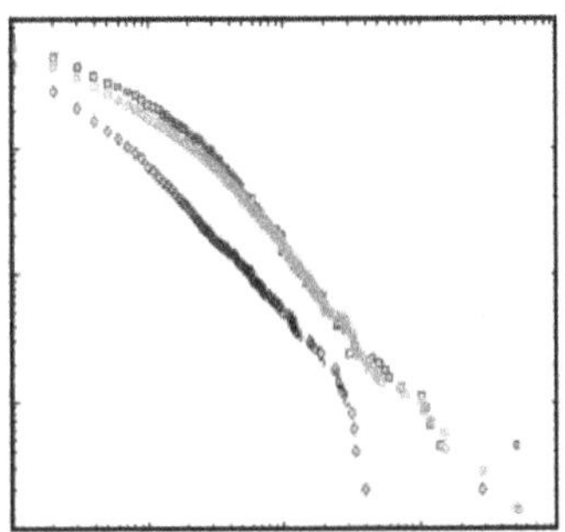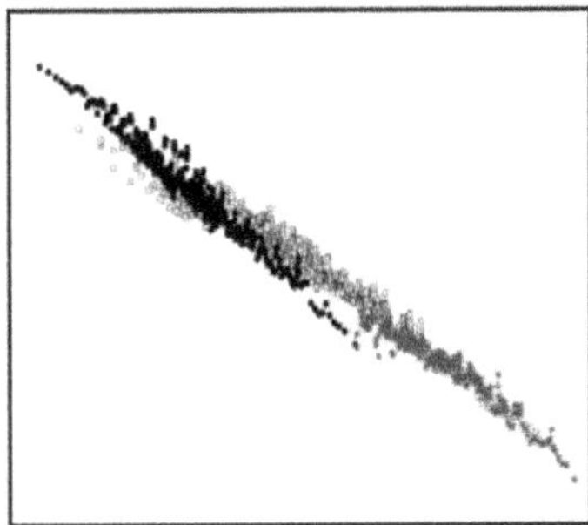

The amounts of wounded and dead in terrorist attacks (left: dead ◊, injured ○, total □)[8] distribute similarly to forest fires (right: Japan ●, United States ●, China ◦)[9]. Massive wars are as few as immense infernos. The same form implies the same law.

We tend to consider ourselves unique. We imagine living in a world of our own because our colossal consumption of fuels detaches us from the whole. So we interpret economic activity as the behavior of agents with their preferences and biases. In the data, however, nothing in our doings seems out of the ordinary. We merely aim for a balance with surrounding resources in one way or another.

Similarly, cormorants and shags, those widespread aquatic birds, fish wherever they can. The flock grows until the environment's carrying capacity is reached, and if exceeded, the population collapses. So why would human activity follow some other law or eventually follow none?

Rather than looking for the causes of the destruction of natural habitats in ourselves, let us examine our activity as a natural process, recognize the universal law in ourselves, and appreciate the need to do things differently. Our blindness to reality is apparent in our theories about behavior and economics, as well as in our outlook on culture and values. To live more meaningfully, we ought to understand what values are in their essence.

CODES OF CONDUCT

The thermodynamic theory is, in character, a worldview rather than, say, a mathematical model of society for political interpretation that explains what society is all about rather than what it should be like.

Instead of arguing for a wide or narrow distribution of income, we should ask why societal distributions are skewed to find the answer in understanding society as a system that does not evolve toward a high or low degree of skew as such but toward balance. Likewise, rather than seeking growth or eventually arguing for downsizing, we should inquire why the economy grows or declines to find the answer in understanding that society does not develop or downshift as such but diminishes imbalance in one way or another. Moreover, rather than presumptively striving for equality or perhaps seconding inequality, we should examine how equality or inequality lessens the imbalance because rights are not values as such but mechanisms that serve motion toward balance. Furthermore, rather than arguing that economic success stems from, for instance, secularism[10] or traditional labor ethics, we should focus on forces that drive the economy in one way or another because a way of life is in the service of forces. The thermodynamics of time allows us to see the root cause of things.

The least-time imperative explains our quest for effectiveness, but it does not mean consuming resources as quickly as possible. On the contrary, drawing in a farsighted manner from wide-ranging and long-lasting sources will get closer to a balance as a whole. For example, while deforestation for farmland decreases the energy difference between a community and its environment locally, it increases the imbalance between sunlight and matter on our planet, eventually with global aftermath. We already knew that our depletion of nonrenewable reserves is shortsighted; now, we know this also in the form of an impeccable natural law.

WHY DO CULTURES FLOURISH AND FADE?

The American anthropologist Marvin Harris saw habits, traditions, and taboos as means for a community to survive in its environment.[11]

Materialism views culture as a response to the natural environment, as well as to technological and economic circumstances. At the time of its birth in the 1960s, this stance focusing on external factors in cultural expressions challenged idealism, emphasizing the importance of the mind in expressions of culture. Harris recognized the constructive role of controversy: "I don't see how you can write anything of value if you don't offend someone."[12]

Today, just as in the past, cultures adapt to circumstances. Lifestyle is changing as both the natural environment and synthetic settings continue in flux. For example, vegetarianism is becoming more popular, as the ecological effects of eating meat have raised concerns for many people. We are adapting to the shifting sands of time to survive. That has happened over and again. For instance, the traditional Nordic Christmas table treats remind us of the leaner olden times.

Although the thermodynamic theory arrives at similar conclusions to those of cultural materialism,[13] only now can we unambiguously see that culture sources from the foundations of our existence. Culture is a way of making a living. It is a mechanism that conveys flows of quanta. It makes things happen.

A WAY OF LIFE

In his book Collapse (2005), Jared Diamond emphasized that a community's way of life paves the way toward its destiny. He recounts the Norse people's brave but ultimately disastrous Greenland settlement, which lasted around 450 years. Even though the Norse, led by Erik the Red, landed first on the big island, the Inuit eventually outlived them. The Norse almost gained a foothold on the American continent too, but indigenous people repelled them. The Norse kept up their high cultural traditions in the face of the harsh arctic circumstances, going so far as using precious resources to institute and maintain a cathedral and a bishop. Generations later, from the 14th century onwards, the Inuit moved further south into Norse regions. The Norse saw the Inuit thrive by whale hunting and harvesting other fruits of the sea but refused to adopt the pagans' unfamiliar practices. They even shunned fish, Greenland's most abundant food. At some point over the minor ice age, as overused pastures became exhausted, the last of these obdurate folk died of hunger.

The British writer Nafeez Ahmed reasons that the deterioration of living conditions will cause contemporary cultures to collapse just as past civilizations perished.[14] While we recognize, for instance, migration as a consequence of changing circumstances, we also need insight into the root causes. As Edvard Westermarck declared in his inaugural speech at the London School of Economics in 1907, "The object of every science is not only to describe but also to explain the facts with which it is concerned; and the object of sociology is to explain the social phenomena, to find their causes, to show how and why they have come into existence."[15] The Finnish philosopher and sociologist warned not to allow personal feelings to bias interpretations. A unifying theory is needed, especially when only little is seen of the whole. To this end, the *Grand Regularity* justifies the drawing of parallels. For example, the means for a community to persist are encoded in its culture in the same way as an organism's phenotype represents the features it needs to survive.

A human is what they do. Depriving them of their identity is a tragedy. To change from a farmer and seafarer into a whale hunter seems to us a no-brainer, but the Norse considered the idea ludicrous. Although their livelihoods did not correspond to the living conditions, the Norse only recognized the resources accessible to their culture as valid. For the same reason, lack of perspective, contemporary communities may be unable to revise their conceptions to correspond to reality.

The Inuit adapted to the harsh region by tracking and catching the available prey, while the Norse did not adjust closely enough to reality. The circumstances became even more pressing as the climate cooled and connections to Europe were disrupted. Presumably, some Vikings must have tried whale hunting, but the community hardly approved of such a pagan practice. A deviant would have been devastated by excommunication. In any case, catching prey would have been tough enough. Even an Inuit hunter had to practice harpoon throwing from youth.

Like the stubborn settlers in Greenland, we are in for trouble by holding on to our narrow-minded perceptions. Many a transformation of society or disastrous economic turmoil has only been seen after a catastrophic event, even though a broader viewpoint would have revealed it already during the buildup. However, by understanding our culture as a manifestation of natural law, we might still be able to revise our worldview and survive.

SOCIAL PHYSICS

Just as the Norse were once holding out in their cottages as their traditional means of living were fading away, contemporary communities are fighting for their existence as standard jobs are going away. Flows of quanta shape society in the same way as rays of sunlight structure an ecosystem. Networks of energy-intensive cities are dense as food webs rich in tropical rainforests. Conversely, job opportunities are few and far apart in the outback, just as vegetation is scant in the Arctic and Antarctic. Only by steadily drawing from vast resources can we attain an enduring high-level balance. Eventually, the whole of humankind will wane away without power, just as a plant wilts without light.

We usually think that energy is produced for us, which is true, but we tend to overlook the fact that energy differences drive us to use fossil fuels and harvest sunlight in the first place. By expressing our ultimate motives through the universal principle, we can understand ourselves better. This portrayal of humankind in its habitat in terms of energy and time is essentially the same as the anthropological narrative about the individual in society. People live both in the natural world and in the worlds of their own making, their own imagination.

When our concepts are down-to-Earth, we are in contact with the bedrock of our existence. For instance, Harris put it plainly that native people in Central America were cannibals due to the scarcity of other kinds of meat. However, cannibalism was not seen as appalling, as folks did not eat their own people but their enemies. Using plain rather than convoluted concepts makes it harder for us to defend unsubstantiated attitudes or cover up pejorative prejudices.

Harris argued that in the olden days, taboos were efficient means of guiding people to live sustainably.[11] In our time, worldviews are no longer passed on as religious beliefs and cultural bans but perhaps as scientific canons. Either way, when our worldview is consistent with our experience, we are able to reason instead of just believing.

The 17th-century English philosopher Thomas Hobbes already saw society as a manifestation of natural laws. Mechanical philosophy[16] can be related to thermodynamics such that different ideologies are seen to represent various forces. At its best, our culture can consume a variety of forces. Then society functions in a versatile manner, like an ecosystem with a diversity of species.

Social physics, the tenet that regards society as a physical system, paved the way for statistical physics – not the other way around.[17a] Formulating the velocity distribution of gas molecules, Maxwell adopted the random, i.e., symmetrical, rather than the causally skewed distribution from Adolphe Quetelet, one-time supervisor of Pierre François Verhulst. The Belgian astronomer, statistician, and sociologist had modeled distributions in societies with the bell curve. Boltzmann, too, drew a parallel between the motions of particles and people. From this historical perspective, we see that the bold 19th-century attempt to derive thermodynamics from core principles failed because statistical mechanics sprang from modeling data as if it had resulted from stochastic rather than causal processes.

ETHNOGRAPHY OF ENERGY

While the richness of our culture seems a blessing, it occasionally conceals the austere causality. We are often beguiled by vibrant language without getting the message. We are readily enchanted by the beauty of Nature without comprehending its law-like character. And we are easily enthralled with new technology without understanding that it distances us from our natural environs. We are wont to focus more on the means of making a living than on the natural principle of life. This is understandable, as we may manipulate mechanisms, but there is nothing we can do about the law of nature.

When we recognize that culture complies with thermodynamics, we come to see, for example, that a community may retain as much of its original culture as is permitted by self-sufficiency and isolation. Habits are the means to make things happen. For instance, the handshake is a common way to start communicating, just as a data transfer protocol opens communication between machines. These and other conventions change when circumstances change; species evolve in response to a changing environment, the supply of goods shifts in response to demand, and practices change in response to an epidemic.

Comparing opens the door to understanding. An advanced society is a symbiotic system like our own body. Our cells accommodate powerful metabolic units, known as mitochondria, and plant cells likewise harbor chloroplasts. Once upon a time, these organelles were independent organisms, which gave up some freedom as they took up residence in bacteria. The resulting symbiosis turned out to be a productive redistribution of work.[18] Eukaryotes, the complex organisms descending from these primeval marriages, effectively consume the imbalance between substance and sunlight.[19] First plants, then animals spread across the oceans and the continents. Today, the same drive to consume free energy in the least time gives rise to organization and cooperation that spread across the world.[3]

When people first harnessed flows of solar energy captured by plants for their use, their identity changed along with the transformed way of life. A farmer depends on the surroundings in a different way than a hunter-gatherer. When fossil fuels were tapped, humans moved yet further away from their natural habitat. When ties loosen up, agents of any kind run amok, wreaking havoc on their surroundings. For instance, cancer spreads when control of cellular growth is lost. The banking business gets out of hand when it is not regulated. Without counterforce, power is used one-sidedly, senselessly. The virtue of modesty is akin to the ideal of balance.

The brightest stars shine for the shortest time. This parallel does not foretell a very promising future for us. Still, perhaps we can avoid the most disastrous scenarios if we embrace our environment sustainably and aim for a global balance.

The anthroposphere is seeking a new planetary balance as the world economy contests the power of the biosphere. However, in the long run, we can hardly compete with the powers of the atmosphere, geosphere, and hydrosphere. A changing climate exhausts our endeavors, as a rising sea level drowns our cities and rain clouds fail to form over our fields. Today, in the words of Rachel Carson, "mankind is challenged, as it has never been challenged before, to prove its maturity and its mastery – not of nature, but of itself."

Which cultural mechanisms serve us best, which modestly, and which only ostensibly? The answers depend on our values. Even if social physics does not advise us exactly what to do and what not to do in a given situation, it gives us grounds to choose. At any rate, evolution is on its way toward balance. So it is not about whether we can benefit from the changes but whether we can bear them.

As Wallace projected in *The Wonderful Century* (1898), revising our way of life does not mean returning to the Stone Age but evading a bleak future. That has happened before. "Hindu vegetarianism was a victory not of spirit over matter, but of reproductive over productive,"[11] reminds Harris. Back then, butchering cows would have caused starvation because they were needed to plow drought-hardened fields. How is it with us? Are we already consuming provisions needed by future generations? Instead of another selfie, the next shot should be a group photograph of reformed people, a new ethnograph.

WHAT IS THE PURPOSE OF THE ECONOMY?

Long ago, the biosphere emerged from the quest of matter on Earth to gain a balance with sunlight. Today, the anthroposphere emerges from the same ingredients, and it should be consciously aimed at the same sustainable balance. The grander the realm we identify ourselves with, the further out we aim to draw power for our activities. For instance, we would not limitlessly clear forests for agriculture if we realized that thereby we are damaging our own home's cooling. We would not limitlessly scoop up the treasures of the sea if we recognized that thereby we were looting our own property. We would not pump oil from the seabed and dig metals from the Earth's crust without any limit if we understood that we are thereby littering the lands and seas of our own backyard. Environmental degradation does not follow so much from our failure to know what we do but more from our lack of comprehension as to who we are.

Since the data show no difference between our endeavors and other processes of the universe, we have no opportunity to free ourselves from the discipline of time. Instead, our opportunity lies in understanding that we are part of the Earth rather than just populating its face.

SO THEY SAY

Economic activity meets human needs – so they say. Economics focuses on humankind[20] by claiming that our insatiable needs boost economic growth. However, there are hardly any grounds to imagine that we are unique in our aspirations, as our accomplishments also display the Grand Regularity's universal patterns.

It is said, for example, that investments support economic growth. However, it should be kept in mind that the sizes of enterprises distribute similarly to the sizes of mammals.[3] Companies and mammals flourish under favorable circumstances, yet the distributions are skewed under all circumstances. It is also said that spending on schooling contributes to economic growth – no doubt about it. Still, education is distributed in a population in the same way as income. When society develops, the distributions of wealth and education change in the same way as the distribution of the speeds of gas molecules changes when the temperature rises. In all cases, the distributions are skewed. By all appearances, the ubiquitous patterns imply that the economy complies with a more general law than the specific laws of economics. The conclusion could not be any different, given that everything comprises quanta.

The recurrent patterns can be copied by mathematical modeling, but it is more important for us to comprehend the universal underpinnings of that regularity. It may well be necessary for us to fathom the complete stack of phenomena from the quantum to the cosmos for us to see ourselves as part of the whole. For example, suppose particles appear to be entangled in incomprehensible ways, and the universe seems to house some inexplicable darkness. In that case, it might be difficult for us to bring into focus the universal law that human beings are also expressing.

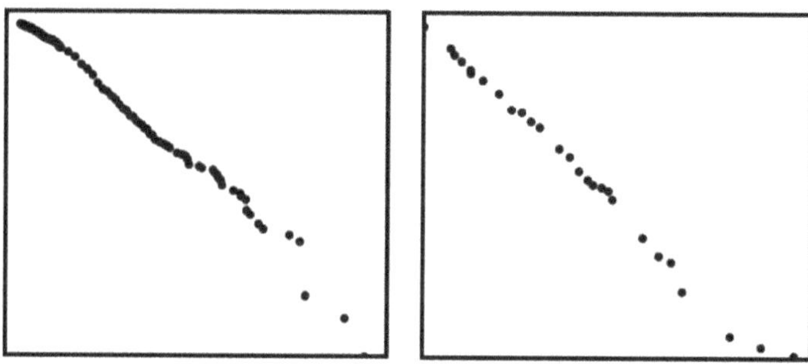

Income (left) is distributed like craters on the Moon (right).[21] There are as many poor people in a country as there are pockmarks on the face of the Moon, whereas the super-rich are as few as gigantic craters. Since the axes are not labeled, you would not know what data you are looking at. All kinds of processes, such as becoming prosperous and facing meteorite bombardment, result in similar data.

The fundamental scientific problems point to problems in our doctrines, which lay the foundations for our worldview, delineating our way of life. However, physics can be renewed now that we know what time is. Biology can be revived now that we comprehend what evolution is. Economics can be revolutionized now that we realize that an economy is not only about human activity but also part of natural processes from photons and atoms onward. It takes a unified worldview to draw parallels, such as likening transistors to neurons, as both computers and humans trade stocks, issue loans, and write the news.

THE INVISIBLE HAND

What is the invisible hand, coined by the economist Adam Smith? What, if anything, guides an individual's vested interest for the benefit of the common good? Do competitors organize themselves into the synergistic world economy in the same way as rivaling species structure into a symbiotic system? Is economic development weeding out agents by the same principle of natural selection as evolution is pruning out species? Does the process ultimately lead to the cooperative system of the whole Earth, known as Gaia?[22]

The similarity between ecology and economic activity is encapsulated in the words themselves; 'eco' comes from the Greek *oikos*, meaning family or household. The idea of their being undergirded by one and the same process is old. Darwin himself got the idea of survival when reading the tome *An Essay on the Principle of Population* (1798) by Thomas Malthus, an English demographer and economist.

Contemporary economics has, in turn, inherited many ideas and concepts from biology. Innovations and competition are seen to propel economic growth in the same way as viable mutations and the struggle for existence are seen to drive the evolution of species. Diversity is desirable in the economy, like in the ecosystem, because an array of agents, just like a spectrum of species, can seize opportunities and prevent setbacks. By contrast, monocultures in both commerce and agriculture are nearly incompetent at progressing and predisposed to disasters. In thermodynamic terms, a wide distribution of mechanisms can consume various free energy sources, whereas a narrow one can tap into only a few resources.

The Enlightenment philosopher Anders Chydenius is known as a sharp-eyed visionary from his main work *National Gain* (1765). The thesis shares the hallmarks of the thermodynamics of time. A prime example is that the course of events cannot be dictated from above because causal relationships are nondeterministic. In other words, the command economy does not work, as it is impossible to know in advance how to command. It is thus better to increase the freedom of choice. That paradoxically requires restrictions. Regulations prohibiting, among other things, monopolies, cartels, and the misuse of insider information, ensure diversity in making things happen. Laws protecting property, infrastructure, and capital secure means and mechanisms that make things happen. A restriction for a few is thus a construction for most, just as a city wall protects numerous transactions and a cell wall various reactions.

Chydenius regarded the freedom of the press as one of his greatest achievements. However, his attack on mercantilism, conservatism, protectionism, and privileges eventually caused him to be excluded from the Swedish Riksdag of the Estates by his own political party.

Following Chydenius, David Ricardo and Adam Smith wondered what the process of early industrialization in England was all about. Since the late 18th century, economic growth has produced products in the most varied forms, but by now, many of them have fallen into disuse, been dumped, or ended up on the shelves of museums. All kinds of goods were produced only a short while ago, but the integration of the global economy narrowed the diversity into cost-effective, standardized product lines. Similarly, as documented by Cambrian fossils found in Burgess' shale, a great variety of early organisms suddenly emerged about half a billion years ago but soon narrowed into lineages of thriving species.[23] Is this universal least-time consumption of free energy by ever more efficient manufacturing what we mistake for our unique rationality?

Natural capital

It is easy to demonstrate that the most well-known laws of economics, such as the law of supply and demand and the law of diminishing returns, are manifestations of evolution toward a balance. It just requires us to state the economic factors, raw materials, products, machines, workforce, etc., in the universal units of energy and time, quanta.[24] In this way, the thermodynamic theory gives a more comprehensive view of economic activity than economics. The dynamics of a society is like that of any other system; market forces are forces like any other. A cent is the quantum of the US economy. This is how Matti Estola, a lecturer in economics at the University of Eastern Finland, relates the laws of economics to those of physics.[25]

Expressing economics with the terminology of physics is nothing new. In the late 19th century, Léon Walras explained that demand and supply meet at a balance.[26] The French economist described the course of events aptly as groping (in French, *tâtonnement*). The dynamic balance between supply and demand is sought somewhat blindly, as it cannot be known beforehand or predicted accurately. Indeed, economies are capricious.[27]

Frederick Soddy, in turn, offered thermodynamics as the foundation for economics in the 1920s and 1930s, but the Nobel Prize-winning chemist was roundly dismissed as a crank. At the time, Soddy's ideas were radical; today, they are routine. The gold standard is gone; currencies are floating; economic cycles are balanced by government surpluses and deficits; consumer price indices serve to measure the state of the economy; banks are required to keep minimum reserves. Also, the mathematician and economist Nicholas Georgescu-Roegen offered ideas from physics and ecology to economics in his classic book *The Entropy Law and the Economic Process* (1971).

An economy evolves as resources from the surroundings flow into it. The global economy extracts most of its power from fossil fuels rather than solar radiation, just as a deep-sea ecosystem draws from springs of chemical energy. In contrast, most of the biosphere lives off sunlight. So we are beginning to understand that we could act in a more meaningful way by considering the global economy as part of Earth's natural economy. That is why we have begun to talk about ecosystem services and natural capital. Even better would be to price everything in energy and time instead of money. The budget would be realistic, for such a calculation would make evident the forces of nature on the planetary scale.

The economy stalls when quanta flow from the system to the surroundings, reflected in the weakening of the currency, loss of capital, and emigration. When infrastructure erodes, cohesion dissolves. At worst, means of growth, such as education and healthcare, are lost. When copper railings, lead roofing, electric wires, and equipment are also being stolen, railroads are robbed of their tracks, and waste management is failing, the disastrous state is indisputable. Similarly, a big carcass, a system, too, is a plentiful supply of food for many scavengers in an ecosystem.

Simply put, a system thrives when it gets more than it loses. Otherwise, the surrounding system flourishes at its expense. Both sequences of events follow the same principle; only the interpretations of success and failure are opposite.

Among other aspects, economics is about bookkeeping, the counting of flows. It is theoretically possible to use flows of quanta instead of money, even to the precision of a single quantum. Surplus and deficit are physical; this we intuit. Yet it is new to think that the world economy, national economy, and household economy are, in principle, no different from the biosphere, an ecosystem, and an organism. All evolutionary courses are unique; even so, they are manifestations of the same law.

I still remember that the worldview implied by the thermodynamic theory of time initially seemed strange to me, even creepy. I asked myself whether everything was just quanta. That is indeed the case, nothing less than everything. It took me some time before I understood that the comprehensive view is not distressing but rewarding. The issues of economics can be explored in the same vein as the questions of biology and physics because thermodynamics explains an economy and an ecosystem similarly and a human being and a particle likewise. This equivalence was the very vision of Quetelet, Boltzmann, and von Neumann.

FALSE PREDICTIONS

Economists are berated for making false predictions. Indeed, it seems strange that even the most developed economies experience unexpected swings. Paradoxically, the high degree of development is the very reason for the pronouncedly capricious character of modernity. Efficient mechanisms, such as stock exchanges, and the means of free communication, mediate changes in value almost instantaneously. When a change is abrupt, the driving forces also change suddenly, which affects the ensuing change. Changes beget changes. In a similar manner, the forces and counterforces compete when a fishing line float bounces for a while or when a chemical reaction oscillates for some time, or when the light flashes for a split second after a laser has been turned on. The data are similar. They can be modeled with simple equations. The cause of oscillations is also trivial: the system follows a great force with high speed,[28] and thus, the reaction to the action overshoots the balance point.[24,29]

While oscillations can be dramatic, they are not random. The event at hand follows from all the previous ones, not deterministically only from the initial state. In all its turbulence, the intricate course of events does not follow from the system's hair-trigger sensitivity to the initial conditions, as is assumed in chaos theory. Slight initial differences do not bloom into an enormous diversity of possible outcomes.[30] The flap of a butterfly's wings in Brazil does not cause a tornado in Texas, but the temperature difference between the warm ocean and the cold upper atmosphere.

Although the logistic map of the S-curve is a good model of chaos,[28] the dramatic effects do not follow from subtle differences at the onset but from the tremendous forces engaged along the way. There is no butterfly whose wing flap could create a tornado in Lapland. Cyclones cannot thrive up there.

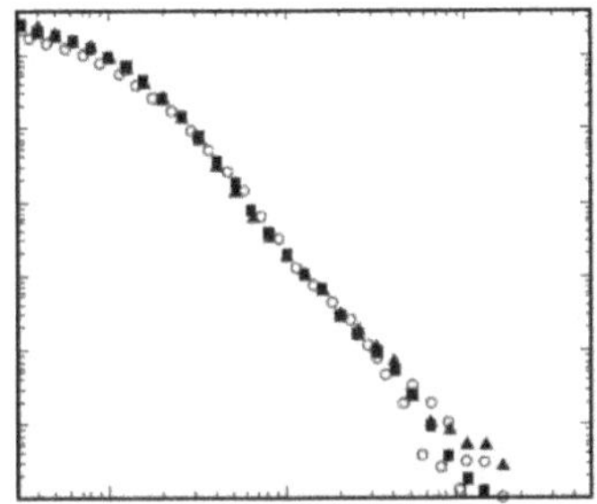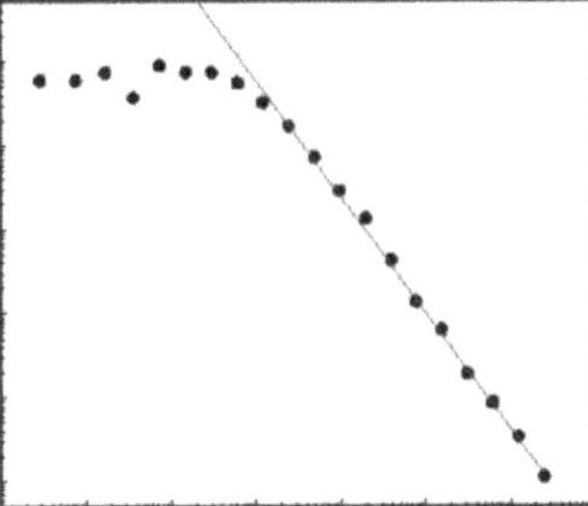

Daily exchanges on a stock market (left)[31] accumulate like rain (right)[32]. Small-time business is as abundant as drizzle, whereas only occasionally does the market crash like a cloudburst. The events are unpredictable, yet not all arbitrary.

Complexity is not per se the reason why predictions do not pan out, but the interdependence of causes and consequences. There is hardly any point in social sciences striving for the same exactness as contemporary physics because the perfect precision is achieved only where the discipline proves pointless – in a balance of stagnation. Often the most fitting, but perhaps also the most useless, 'forecast' is that everything continues as before. Trends in business follow S-shaped curves, just like other sequences of events. Innovations, just like mutations, may lead from one unpredictable growth phase to another, but eventually, there is nothing to predict when a branch of business has matured.

Economies react to changes in the same way as ecosystems do. Developed economies have resources for investments and the capacity to reform their structure. People relocate and take up new jobs. A vital faculty can, however, be wrecked by reckless oversteering in response to severe swings. In a rich ecosystem, too, some species can adapt to new circumstances, but however well-endowed they might be, they cannot recover from extinction.

In a recession, we should secure the means of future growth and maintain a level of unity. Similarly, a starved cell retains its crucial machinery of regeneration and economizes on its functions. And the same wisdom is on display on the level of the whole human body. The vital organs are protected, and the connections between them are diligently maintained; with them, we will be able to strike back as the occasion allows. Such thinking is nothing new. What is new is the understanding represented in the form of a natural law, whence it is more substantiated than before.

REALITY CHECK

Morgenstern contributed to but also criticized the mathematization of economics.[27] Although optimization tasks are presumed to be computable, time-dependent phenomena are, in fact, nondeterministic. For instance, while it is possible to calculate that in the so-called Pareto optimal situation, the position of one person cannot be improved without worsening the position of someone else, the pathway to that balance is impossible to figure out. The optimum is approached by weakening the positions of some, although by less in magnitude than the group as a whole is strengthened. Say the interests of a community overrule those of an individual. Society, in turn, exercises power over the community, and humankind falls victim to the forces of nature on a global scale.

The nondeterminate character of optimization is also exposed in the question of whether purchasing reveals preferences.[33] Sure, but purchasing also changes the preferences. As a person follows forces, those forces are modified by the person's actions. This means that the outcomes will be different when the same commodities are bought in a different order. Again, this accords with our own experience. At any time, our means to buy depend on our past purchases.

In a scientific enterprise, the standard procedure is to reduce the number of variables in order to find the significance of the remaining ones. But a reduction taken to the extreme yields insignificant results, as the object of study has then been disconnected from all else that matters. For example, herd behavior does not reduce to the behavior of individuals, whether those individuals be atoms or human beings.[17c] Morgenstern deplored modern science more broadly than mere economics by maintaining that it is often easier to mathematize an incorrect image than to face reality.[27]

Although the future is fundamentally unpredictable, the course of events can be simulated. In practice, however, such scenarios do not cover everything, as it is impossible to take into account the numerous factors that reside in the long tails of the skewed distributions. Those factors that are furthest away from the mean can have an exceptionally high impact. That is the main message of the book *Black Swan* (2007) by the Lebanon-born philosopher and risk analyst Nassim Taleb. For example, a dreadful epidemic may derail an economy.[34]

In any case, timely actions and effective mechanisms can both suppress and enhance the consequences. Similarly, chemical reactions can be both catalyzed and inhibited. To influence the course of events is, of course, nothing extraordinary; the only novelty is the extent to which parallels across disciplines can now be substantiated.

As with any scientific theory, the defects of economic theory are exposed when observations depart from expectations. For instance, economists wonder why exchange rates fluctuate appreciably over short periods while the average rate remains quite steady over a long time.[35] This is the selfsame phenomenon seen in species populations, which may vary much from year to year but tend to be stable over generations. Although the mean is about right, the variation in the skewed distribution is too wide to fit into the narrow mold of the normal distribution. So we understand that a versatile economy remains on an even keel because it can respond to diverse forces, while the currency of a one-sided economy can lose its value markedly. At worst, the value plummets permanently when the economic structure is wrecked. Diversity of organisms, too, guarantees rapid responses and steady survival, whereas a species-poor ecosystem is vulnerable to stress and may, at worst, be ruined for good.

Economic models are thought to be too simple to explain complex reality; however, is it merely the simple principle of least time that is yet to be understood?

WHAT IS GOING ON IN THE ECONOMY?

The resource-depleting flows of the world economy have swollen by now to immense dimensions and departed dramatically from the organic and resource-depositing flows of the Earth. Our goal should be a sustainable economy where synthetic materials circulate in the same way as natural products circulate in the biosphere by the power of sunlight. Rather than extracting resources from the Earth's crust, we should deposit supplies for the future in the same manner that organic material fossilized into petroleum a long time ago.

This change toward a sustainable way of life is underway, whether we want it or not. It is driven by many factors, particularly by natural forces. At best, we will adapt to the reuse of resources; at worst, we are heading for a collapse. The course of humanity rests not only on economic issues but on our culture, collective identity, and values; in short, on our worldview in its entirety.

THE IDEA OF A FIRM

Why do companies exist? Ronald Coase was already exploring this profound question of economics in 1937. In his publication, *The Nature of the Firm*, the British economist argued that labor division in and among companies improves overall efficiency. The optimal form of cooperation depends on the nature of the business.

The economic theory speaks in terms of economics, the thermodynamic theory in holistic terms. The economy is, in essence, an energy transduction network where workers, firms, and alliances play the same roles as cells, organisms, and biotic communities in an ecosystem. This hierarchy of systems is machinery that mediates flows of quanta, irrespective of whether the imbalance is in the form of money, natural resources, sunlight, or other assets.

The best players make the most of limited resources in the least-time sense. The flows of quanta *naturally select* them as mediators of free energy consumption. Therefore, business moves to where the demand is strongest; a species spreads to where food is most readily available. And a suppressed demand, like a dammed brook, will ultimately find its outlet.

The entry of a new agent into a market is comparable to the appearance of an alien species in an ecosystem. It will prosper, provided that flows of cash and other forms of quanta are diverted from the established agents to the newcomer. While it is hard to get a foothold in advanced markets, opening a closed system offers a window of opportunity for many. Similarly, the ecosystems of relatively isolated islands are highly susceptible to invasions. When the brown tree snake (*Boiga irregularis*) landed on Guam Island onboard American warships in the 1950s, the island's abundant birdlife was greatly impoverished.[36] Similarly, a transnational corporation eats away traditional companies. It is next to impossible to foil the invasion of alien species. Still, the balance can be regained when the original players differentiate into other tasks or develop into viable competitors.

We do not necessarily oppose changes; we might even welcome them. But only when we relate an unknown process to familiar ones will we understand what could happen, for example, from the opening up of markets and their deregulation, from technological innovations, as well as from a revision of the worldview. The *Grand Regularity* of data justifies parallels across scales and disciplines.

In a versatile economy, just as in a vibrant ecosystem, there are many alternative flows of quanta. Such a resilient system not only adapts to changes but can influence and even oppose them for a while. In contrast, an economy based essentially on a single industry, just like a monoculture, is without options and will not last for long in an economic squeeze.

The tough question is, of course, how the economy can be diversified to make it more robust. It is a chicken-and-egg situation. A system that holds ingredients and mechanisms can garner more ingredients and assemble more mechanisms.[37] Even so, investments in education and research, for instance, are only superficial if there is no genuine striving toward novelty.

REASONS TO AGREE

The thermodynamic theory of time maintains that we make agreements to make things happen. For example, a subcontractor works for a contractor based on mutual understanding. Similarly, the cleaner wrasse relies on a symbiotic relationship when picking parasites off of a predatory fish's skin, gills, and even jaws. Just as damming a river consumes resources, contracts are associated with costs. When enough forces are behind the deal, we expect the ensuing string of events to stay on course.

Not all forces, the incentives of an agreement, are visible to all parties. This state is referred to as information asymmetry. Insufficient information is thought to hamper the operation of markets, but ignorance and insecurity, just like knowledge and confidence, motivate agreements. If everyone were fully informed, the parties would constitute one and the same subject, and so the deal would be unnecessary. The reasons to agree cannot but be subjective.

When the ultimate rationale behind deals and agreements dawns on us, we will not fail to honor those we make. Breaking contracts is comparable to breaking infrastructure. It may well be the purpose of reforms, but the end does not justify the means. Structures, albeit different ones, are needed in the future, too, to make things happen. Now that we know that every action causes some reactions, we can at least avoid worsening the problem.

BIRDS OF A FEATHER

According to economics, we aim at maximizing utility. Although it is quite difficult to define utility, this may not be an issue in our new worldview, since the outcomes of economic activity are like those of other activities: economic growth follows the S-curve, and the distributions of assets are skewed. By all accounts, decision-making is motivated by plain forces, just as other events are driven by free energy rather than by the ambiguous utility.[24,38]

The thermodynamic theory gives perspective to Kahneman's point about making decisions: "And we think that we make our decisions because we have good reasons to make them. Even when it's the other way around. We believe in the reasons because we've already made the decision."[39] Francis Bacon said the same: "Man prefers to believe what he prefers to be true."[40] This prudence can be understood as our behavior being dominated by the drive to achieve balance in the least time. So we amplify the first impressions, especially when we have no experience, no firm ground beneath us. Similarly, water quickly carves its path into soft soil.

In his book *Skin in the Game: Hidden Asymmetries in Daily Life* (2018), Taleb argues that you make better decisions when you are made to incur the consequences. On the other hand, rash penalties may only prevent us from pursuing truly worthy goals and preclude us from wide-ranging benefits. It would be wise to nominate persons with a sense of responsibility for positions of trust, but that quality might be hard to judge before it is needed.

When a conflict of forces besets decision-making, it might be difficult to sense which force will be the most significant in the long run. To avoid paralyzing indecisiveness, we tend to concentrate on some aspects and ignore others. The sheer focus on a single issue distorts our image of the world, as functional imaging studies of the brain have revealed.[41] In behavioral economics, this subjective view is called cognitive bias. The very word bias tells us that others have a different opinion or that our subsequent assessment differs from the earlier impression. Therefore, a group of people experiencing many forces tends to end up with a more balanced decision. Even then, birds of a feather flock together; many follow an authority blindly instead of seeing the greater reality, for we humans, social beings, are geared up to mimic masters right from the birth.

Feynman recollected an instance of this multifaceted assessment of forces. The Manhattan Project members Compton, Tolman, Smyth, Urey, Rabi, and Oppenheimer were evaluating how uranium should be enriched.[42] When one of the participants raised a viewpoint, Compton followed up with another, and then a third fellow raised yet another, and so on. The young Feynman was surprised that Compton did not repeat and reinforce his point, which seemed the best. At the end of the round, it appeared as if everyone disagreed. Then Tolman, who was the chairman, said: "Well, we've heard all these arguments, so I think Compton's argument is the best, and now we have to move on." Feynman realized that not the participants themselves were being evaluated but the things they said.

Remarkably, when supported by a community of fellow believers, we are wont to take authorized allegations at face value, however absurd they may be. "People want an authority to tell them how to value things. But they choose this authority not based on facts or results. They choose it because it seems authoritative and familiar," wrote Dr. Michael Burry (Christian Bale) to investors at the end of the film *The Big Short* (2015). In changes of worldview, as in financial crises, beliefs are likewise exposed as groundless.

Simplistic thinking is not always a drawback, though. On the contrary, its prevalence suggests that it is conducive. When only one force stands out, the goal is clear, and objections are pointless. A stone falls straight down – no question about it. It would nonetheless be wise to draw from diverse strands of thinking to discern the various forces at play in each situation. The one who adopts ideas from contemporaries and predecessors transcends oneself and one's moment in time.

Sarah Kaplan emphasizes our need to be aware of the limits of our thinking, not remain captive to them.[43] The professor of strategic management encourages us to stand our ground but justify our stance from a viewpoint other than the one from which the conclusions were initially drawn.

Kaplan also expands the concept of creative destruction,[44] coined by the economist Joseph Schumpeter, from economic reform to the renewal of thinking. It may be easier for us to revise our thinking when recognizing that the structures of thought resemble other structures. Adopting a new way of thinking corresponds to opening a new branch of business.

WHAT ARE VALUES?

Our values are reflected in our beliefs of right and wrong, good and evil, beautiful and ugly, desirable and detestable. The essence of values and changes in values have traditionally engaged philosophers, anthropologists, psychologists, and sociologists, but seldom physicists. The thermodynamic theory allows us to examine the essence of values, just like other things. As Thomas Nagel says, "the development of value and moral understanding... forms part of what a general conception of the cosmos must explain."[45]

Values are subjective. One person can see the same thing as worthy, while another as worthless, as the two differ in their relations to it. By spelling out the nature of subjectivity, it might seem as if the thermodynamic theory would imply that values are relative. This is not the case; values are not of equal value.

One person's values can be compared with those of another, even precisely, in the universal terms of energy. In practice, we look for a more comprehensive evaluation by bringing the matter to the public or putting it in the hands of experts. In the same way, the value of a product is rated on the market and the significance of a mutation is tested in Nature. However, there is no way to get an undistorted evaluation, for the evaluation itself leaves neither the matter nor the evaluator unaffected. But being aware of the subjective nature of our veil of perception, we at least know to ask for versatile and veritable accounts to weigh various forces.

Nagel considers values to have appeared together with consciousness in the late stages of evolution.[45] Our values evolved as we became aware of new forces. So today, democracy is valued as an effective mechanism in pooling and channeling resources, which all are ultimately free energy. After all, democracy displays *Grand Regularity* in its voting statistics, which follow power laws.[17b] Martti Koskenniemi, a professor of international law at the University of Helsinki, argues that democracy means equality, the search for the truth, respect for the opposing party, cooperation with political opponents, as well as respect for pluralism. These are powerful mechanisms altogether to draw means of living from diverse resources.

A VALUE JUDGMENT

Especially in a crisis, values have value. From history, we know leaders who, in the hour of need, respected the values of the community and thereby recruited all means to defend the common good. On the other hand, there have also been those who caused considerable damage by abandoning mutual values for personal gain. We are most likely to act in a responsible way, regardless of our intentions, by respecting core values, even if we do not yet understand why.

The hierarchy of a set of values can be described by enlarging spheres of interactions. Health, integrity, security, hedonism, etc., lie at the core of the individual. Many of us also cling to power, money, and other resources. When individuals identify with their community, they honor social values, say, honesty, fairness, and obedience, and respect ethical values, say, human dignity and human rights. Ecological and environmental values sum up evolutionary history, which has granted us our living conditions. And the values of science express the wisdom needed to comprehend reality. The wider a world of values is, the greater the forces that are respected therein. All in all, in defending values, we defend the capital of history.

CHANGES IN VALUES

Values enrich through interactions in the same way that the economy grows in transactions. The growth opens growth opportunities. It is exemplary of a self-sustaining growth process where motion influences motive forces. As the unpredictable chain of events unfolds, people are exposed to new interactions. Morality matures as they get to know others' perspectives and relate them to their own. Correspondingly, when an individual or a community turns inward and shuts itself off, the values become impoverished. As everything affects everything else, the values explain the circumstances, and the circumstances can explain the values.

The thermodynamic theory enables us to examine values in the same way as other matters. Although the tenet may not be ideal for the analysis of values, it may allow us to see something that has been overlooked as self-evident; after all, the data that reflect our values resemble other data.

Maupertuis already explained various matters with the same principle. Nevertheless, it may still take quite some time before we are poised to see ourselves the way we see other things. In that worldview, every one of us, the whole society and culture, are just mechanisms for making things happen. This naturalistic interpretation essentially expresses functionalism, familiar from social science, in terms of energy and time. The French sociologist Émile Durkheim saw society as an organism, like Lovelock saw the whole Earth as an organism.

Today, we can extend this all-inclusive reasoning to the foundations of existence down to the single quantum.

We may find it hard to believe that human rights, democracy, equality, and freedom of speech, as well as respect for pluralism, property, and contracts, are just effective mechanisms to gain balance with our surroundings. But reverence for values does guarantee that individuals, communities, and societies function so that things will continue to happen. Housing, construction, transport, commerce, research and development, healthcare, education, science, and art are merely processes when expressed in their basic terms of physics. While each traditional tenet has already stated all of this, the thermodynamic theory states the same across the disciplines.

Comprehension of the destiny of humankind can be drawn from answers to Gauguin's three questions:

Where do we come from?
> We are products of the past, irreversible evolution along
> the lines of force.

What are we?
> We are agents, among other agents, materializations of
> the universal law of nature.

Where are we going?
> We orient along the lines of force to explore the unknown, intensify our activities, and eventually realize that
> we are part of the Earth rather than living on its face.

Nicholas Maxwell reasons that providing answers to questions about us and our place in this world is the whole purpose of science. But today, the meaning is lost, as the unity of Nature is reduced to detached details. Blaise Pascal saw likewise that "Small minds are concerned with the extraordinary, great minds with the ordinary." By contrast, the ever more specialized research of our age fails to process knowledge into wisdom.[46] In line with these philosophers' points, the mathematization of our own experience of time and existence with

Galileo's method marshaled physics into a unified view of space, time, matter, and life, and all of that within us.

Comprehending causality will usher us to act wisely and truthfully. Nonetheless, it should be kept in mind that the general principle is only a principle. Its implementation asks for ingenuity. And we already have taken up the challenge, for example, by planning how we might prevent a massive meteorite from hitting the planet.[47] If such a cataclysm were looming, it would be easy for us to see that we are all in the same boat. In all circumstances, we sense reality as forces.

Our lot is not to struggle against the least-time law of nature but to efficiently abide by it. It is difficult for us to resist the enchantment of technology. But by comprehending the whole and considering ourselves to be part of it, we behave sensibly and act responsibly. What we feel unity to determines how we use our resources. Rather than wasting resources on shortsighted mutual rivalries, let us use them unitedly for the future of the Earth.

End point: Our aim, as well as that of the ancient philosophers, has been achieved. The general principle that brings both our everyday experience and scientific experiments into a logical order has been found. We can describe atoms, people, societies, and galaxies in the same manner. This understanding does not devalue us but places us within the full spectrum of existence.

KEY POINTS

- Culture is a way of living.
- Like any other system, the economy is driven by the quest for balance.
- We cannot but comply with the universal law of nature, yet we ought to do so in a comprehensive, planetary manner.
- Values express forces.

9. Why Will the Worldview Change?

New views are not welcomed
but forced by reality.

The revision of a longstanding worldview is particularly impressive when that vision capsizes. How world-shattering it was when the Copernican Sun-centric model of the solar system eclipsed the Ptolemaic Earth-centric one. How revolutionary it was when the Newtonian inertial notion, motion without a force, displaced the Aristotelian idea of forced motion. How radical it was when the absorption of oxygen at combustion substituted for the postulated emission of phlogiston. How epoch-making it was when the universe turned out to be expanding instead of remaining stationary. And how perplexing it was when the modern image of particles as ephemeral quantum fields replaced the classic concept of solid corpuscles. Is it possible that physics is now about to turn the corner and return to the old atomistic idea, where the photon can be seen as the fundamental element of everything and where every process can be understood merely as the evolution of this universal substance?

Scientific progress in its most spectacular form does not inch but leaps. Thomas Kuhn put forward this capricious character of science in his brilliant book *The Structure of Scientific Revolutions* (1962). The American physicist, historian, and philosopher of science was influenced by Ludwik Fleck. The Polish physician's and biologist's sharp-sighted analysis *Entstehung und Entwicklung einer wissenschaftlichen Tatsache* (1935) focused on the most critical factor of science, namely the human factor. Of course, philosophers had pondered the nature of knowledge and interpretation of information much earlier, but Kuhn nonetheless stirred up quite a controversy with his book. He revealed that science could not merely progress step by step. If that were to happen, the best science of any period would never prove to be heretical; conversely, a one-time dissidence could never develop into a contemporary doctrine. That is to say, we have never been liberated from universal truths with ease but only through revolutions.

Kuhn compared scientific progress to biological evolution in the same way as Stephen J. Gould and Niles Eldridge pointed out that evolution punctuates from one equilibrium to the next rather than progressing gradually. The conceptions remain pretty stable throughout the eras that precede and succeed bursts of progress. The specialization of science into subjects is similar to the differentiation of a root form into species. An omnipotent base form keeps developing into finer specialties until observations fail to be explained; only then does the stem sprout powerful general principles again. The revised perception, in turn, changes what can be seen. Scientific progress is, by nature, as unpredictable as other courses of events.

Kuhn was not only celebrated but also criticized.[1] His portrayal of the dominant mindset, the paradigm, was deemed inaccurate. The impression that a new view would be incommensurate with a standard doctrine was met with disapproval, although Fleck had already likewise noted that the scientific community expresses its truth embedded in the very concepts it deploys. One school of thought cannot understand another to the extent that it lacks the concepts of the other stance. Terry Winograd, an American professor of computer science, and Fernando Flores, a Chilean engineer, entrepreneur, and politician, reached the same conclusion: "Reflective thought is impossible with the kind of abstraction that produces the blindness."[2]

Along those lines, a contemporary physicist may find the key concept of causality, the quantum of light as the carrier of time, all incomprehensible. There is neither need nor room for the notion of time in a discipline that is bogged down with calculating only equilibrium properties but blind to the obvious observation that all processes will, in time, terminate at balance.

Like an orthodox physicist, a mainstream biologist who has adopted the modern evolutionary synthesis may find it impossible to conceive that evolution is just a sequence of events. For them, the *Grand Regularity*, the ubiquity of patterns in nature, does not signify anything, let alone would they consider it as a sign of a universal law of nature. Moreover, the concept of time does not explicitly belong to biology, even though life unmistakably evolves with time.

The difference between the conventional and the controversial is relative, nonetheless significant. Kuhn stressed that traditional and revolutionary tenets do not share the same set of meanings. For example, quantum mechanics does not use the concept of time in the sense of causality. The more closed and specialized the discipline, the smaller a deviation from the doctrine is likely to be deemed subversive, even condemned as heresy. A revision is not possible because a theory would thereby abandon its axiom and disrate to a malleable model of the data. A new way of thinking, to be genuinely new, must thus deviate from the prevailing viewpoint in terms of its underpinnings.[3a]

Despite having irreconcilable differences, disruptive and dominant paradigms can be compared. The common measure is how well their perceptions of reality correspond to the observations. Kuhn did not disavow this. On the contrary, he examined the issues involved in juxtaposing one theory with another.

As the evolution of the worldview displays the same characteristics as other processes (Chapter 1), the thermodynamic theory describes the shift from one paradigm to another in terms of physics (Chapter 2) but essentially in the same way as Kuhn and Fleck did. The fundamental open questions signify the forces of change, possibly projecting even a revolution. The prevailing paradigm gives way when the new viewpoint turns out to be more productive.

Instead of an encyclopedic reprise of science history, it suffices to compare present-day events to some notable past episodes to get a perspective on how changes in the worldview take place. In ancient times the discovery of irrational numbers showed that mathematics does not substantiate a natural law but merely expresses or approximates it. It is unclear, though, whether the distraught Pythagoreans actually threw Hippasus into the sea for telling the truth. Later, more verifiable sentences were given to heretics.

Today's purists would hardly contemplate the elimination of a renegade. Then again, our eminently specialized science can be more dogmatic than ever. Conformists may even stifle scientists promoting a holistic perspective.[4] In his remarkable book, Thomas Neil Neubert crisply identifies this weakness of our time: "Physicists are often skeptics, but this is a century of extreme specialization and hence consensus within physics specialties. Specialists, whether in modern poetry or modern physics, become incestuous in their use of journals, processes of peer review, definitions of value, and opportunities for career advancement."[5b]

On a par with reality

Kuhn's outline of scientific progress and his understanding of its nature are today as relevant as ever. The age-old fundamental questions are still at hand, and new puzzles are piling up. Our current view of reality is not on a par with reality; only our models match the data. Likewise, the geocentric model, with its epicycles, seemed once precise but, in the end, did not tally with reality.

Regarding the character of conventional science, Kuhn wrote: "Normal science does not aim at novelties of fact or theory and, when successful, finds none."[6a] Even when observations depart from expectations, as they do today, most scientists blindly patch up mathematical models rather than opening their eyes to the fact that the seemingly infallible doctrines are thereby proven false.

Likening paradigm shifts to other changes, we realize that these extraordinary events should not be too surprising. For example, consider that the smartphone was preceded by generations of cell phones in parallel with the introduction of palm computers. This innovation also took advantage of the increase in mobile telecommunications network capacity. Similarly, the understanding of electromagnetism accumulated first gradually over several centuries, then accelerated in the 18th century, finally culminating in the mid-1800s with Maxwell's equations and his revelation that electromagnetic radiation is light.[7] The increasing vigor and range of activities often presage the advent of a new paradigm.

Kuhn reminded us that the present lacks perspective: "Unless he has personally experienced a revolution in his own lifetime, the historical sense either of the working scientist or of the lay reader of textbook literature extends only to the outcome of the most recent revolutions in the field."[6d] Discoveries are thus serendipitous. For example, when Planck found the quantum as a mathematical necessity in the law of radiation, he did not yet grasp what its meaning was,[8] but became alert. A few years later, as the editor of *Annalen der Physik*, Planck received an extraordinary manuscript from a Swiss patent officer (Einstein). Planck understood that Einstein's work concerned the quantum as the corpuscle of light and published the paper.

Rather than the history of science, the recent success stories and downfalls of companies may be more effective at teaching us how science is done – or, rather, how it should be done. For example, through the gales of the technology market, Nokia's navigation has been both bliss and distress for many Finns. Whoever can sense the forces in the present can foresee the impending change already in their own time.

Science is no different. Breakthroughs stem more often from conflicts than from consensus. Seemingly strange contemporary studies, akin to experiments with electric phenomena in the past, might disrupt the dominant worldview and open breathtaking new vistas. Would we today recognize such a revolutionary viewpoint amid all our hustle and bustle, or would it most likely be found somewhere on the fringes? Can we challenge the prevailing paradigms or even put up with those who question them?

ARE WE IN FOR A CHANGE?

Many a change in worldview has begun within the scope of physics,[3] because there are opportunities to quantify even minute discrepancies between conceptions and observations. Other disciplines follow suit by mathematizing their content in the realization that a lack of precision points nowhere.

Researchers hardly ever endeavor to make a paradigm shift but refine consensus even when discovering discrepancies. The confidence in the prevailing doctrine flaunts itself in the frequently quoted passage from Michelson's book *Light Waves and Their Uses* (1903): "The more important fundamental laws and facts of physical science have all been discovered, and these are now so firmly established that the possibility of their ever being supplanted in consequence of new discoveries is exceedingly remote."[9] In fact, at the time of this statement, modern physics was about to break through due to new observations. So are there now unfathomable data that could foretell yet another revision of the worldview?

The search for dark matter, which began in the 1930s, covers increasingly larger swathes of the universe today. Ever more sensitive instruments are tuned to detect particles of dark matter, but there are no direct signs of them yet. Instead, the speculations about the nature of dark matter have led physics into a strange zone.

In truth, dark matter is an *ad hoc* response to the observations that *de facto* falsify the established theory.[10] Since dark matter does not belong to the postulates of general relativity, it can be tuned to exist in just the right amounts to match the data.[11] Do not these undertakings resemble the search for the imaginary phlogiston to explain combustion at the end of the 18th century? Phlogiston turned out to be a big blunder as soon as it was proven that oxygen was the real stuff. We should thus ask: what is the real substance whose effects in the sky we interpret as dark matter?

In view of history, it seems questionable that terms are added one after another to the quantum field model of a particle to match the increasing precision of the data.[12] The success scored in this way may prove to be no different from the past practices when epicycles were added one after another to the geocentric model of celestial orbs to keep the calculations up with more and more precise observations.

The esoteric dark energy, the incomprehensibly homogeneous horizon of the universe, the inexplicable expansion of the cosmos, and the excess of matter over antimatter also appear as profound misunderstandings. This spectrum of peculiarities reminds us of the confusion among electromagnetic phenomena that reigned in the early 1800s. Are contemporary physicists likewise trying to force pieces of the puzzle to fit without seeing the whole picture?

Physics is in dire straits. "We are not simply confronted with experimental data excluding a model or a class of models," observes Gian-Francesco Giudice. "We are confronted with the need to reconsider the guiding principles that have been used for decades to address the most fundamental questions about the physical world. These are symptoms of a phase of crisis." As the head of the CERN theory group reminds us, the word crisis by origin had no connotation to a catastrophe but meant a turning point.[13]

THE LOST UNITY

Today, just as in the past, physicists are trying to find a theory that describes fundamental forces and elementary particles in a comprehensive manner. The attempt involves uniting general relativity and quantum mechanics, but are these pieces the ones that should be put together? After all, these effective theories only model data, whereas a true theory should explain the causal relationships that underlie the data. For example, we should understand why a distant galaxy is receding, why a stone is falling, why time is elapsing. When such profound "why" questions go unanswered, our view of the world cannot be accurate, and the course we steer in the world cannot be right.

Although the unification is striven after, it is not obvious, once found, that it will be welcomed. For example, when the British chemist John Newlands presented the *Law of Octaves* (1865), the achievement was ridiculed. The Chemical Society did not even consider it for publication. However, only a few years later, Newlands' proposed system became central to the periodic table assembled by the Russian chemist Dmitry Mendeleyev. Likewise, Murray Gell-Mann first sorted elementary particles into octets in the system, known as the *Eightfold Way*, leading to the Standard Model. However, still, that theory does not explain why we have this panoply of particles. Since the key concept of the periodic table was the *period*, shouldn't we now ask what the fundamental period is in a quark, an electron, and a W boson?

Two hundred years ago, chemists asked why some atomic combinations are molecules while others are not. But only a few of them wondered if the atoms have an internal structure that determines chemical bonds and explains the molecules. Nowadays, physicists ask why some quark combinations are particles while others are not. It seems that only a few of them wonder if the quarks have an internal structure that dictates the strong bonds and explains the particles.

In the bygone changes of worldview, perceptions of time and space have been revised. Most notably, when Einstein's works led to modern physics, it was thought that Newtonian absolute time and static space did not correspond to reality. But even today, not everyone is satisfied with timeless equations and the empty geometry of space. Modern physics gave us models that reproduce atomic spectra and astronomical data, but the multiverse and wormholes were thrown into the bargain. Other free gifts to the modern worldview include strange wave-particle dualism, supernatural entanglement, and spooky action at a distance. These mysteries still haunt us. Suppose we declare that these surreal elements of modern physics do not ring true. In that case, we should not take the theory to be real in any other aspect, either, for mathematical identity ensures sameness in both sensibility and absurdity.

Although the ether evaporated from the minds of many after the Michelson–Morley experiment, we sense the void in the form of inertial and gravitational effects. Space seems real not only in practice but also in loop quantum gravity, a theory that renders the vacuum quite tangibly.[14] So, it does not look as if we would be just collecting bits of information at the harvest time of normal science. Instead, it seems that we already have the pieces at hand, but we have yet to figure out how to put them together. Those who see the guiding picture will assemble the jigsaw for a comprehensive worldview. In the words of Michel de Montaigne: "having experimentally found that, wherein one has failed, the other has hit, and that what was unknown to one age, the age following has explained...."[15]

Perhaps the tenet of our time, in which a particle is seen as a quivering quantum field, will transform into a solid understanding of the particle being immersed in a vacuum of quanta. Long ago, but likewise, the Aristotelian idea of a chain that constrains a falling stone swung in Galileo's mind like a pendulum. Similarly, the Ptolemaic paradigm of locked planetary orbits opened up in Newton's mind to a dynamic balance of forces. And having cataloged the weights of numerous compounds, Dalton got the idea of relating the regularity in them to their atomic constituents. Perhaps the ongoing grouping of particles foresees the watershed moment when masses will be related to the curvature of the concrete quanta rather than to the curvature of abstract space-time.

We know the strengths of the fundamental forces and the values of natural constants, but so far, we have failed to figure out their meanings. We need a leading concept to connect the forces and constants of nature to the structures of the universe. Weyl, Dirac, and Eddington reasoned this way but not all the way to the answer. Have we regressed since then, now that anthropic ideas have surfaced from the irrational depths of the mind?[16] Such thoughts do not just elevate humankind to the pinnacle of creation but all the way up to being the purpose of the whole universe.

The impressive ability to collect data and the problematic inability to carry out a dialog are contradictory characteristics of our era. A naturalist asks how the natural constants get their numerical values[17] but categorically denies a purpose. Consequently, they do not study what a process *is* and what it leads to, thereby effectively cutting off inquiry into their own question. The one in need of a teleological explanation is essentially asking what time *is* and why things happen.

THE WINDS OF CHANGE

In the past, seemingly insignificant deviations from expectations have foreshadowed major corrections to convictions. Current anomalies could foretell the same. For instance, the cosmic microwave background has an extraordinarily cold spot. Unexpectedly, the deep sky's largest hot and cold segments are not randomly oriented but aligned relative to one another.[18] It is also unclear where the space probes got a little extra boost as they flew by the Earth.[19] Similarly, the Pioneer 10 and 11 probes acquired an additional unaccountable acceleration toward the Sun on their long voyages.[20] Furthermore, the surprising tiny thrust of an electromagnetic rocket engine seems to question the conservation of momentum, the cornerstone of physics.[21]

These abnormalities, per se, do not force us to revise our theories, as the earlier anomalies did not necessitate revisions either. It is, therefore, expected that some oddities could be explained in an ordinary way while others will only be cleared up with a new perspective. For example, Einstein did not propose general relativity to explain

Mercury's anomalous precession, but the theory nevertheless explained it. Oxygen binding was not suggested to explain the peculiar gain in the weight of metals in some circumstances, but the oxidation hypothesis explained it. The current anomalies could likewise be breaths of the winds of change.

THE STATE OF KNOWLEDGE

At the time of its introduction, Darwin's theory distinguished itself from its counterparts by claiming that evolution has no purpose.[22] But even today, this thesis has not been conceded without reservations. Our current concept of evolution fails to ring true because randomness is no explanation, nor is design. Thus, a paradigm shift seems to be inevitable.[23]

Carl von Linné's systematization of species is continuing today in the form of sequencing genes. Linnaeus used himself as an archetypal specimen of *Homo sapiens*; Craig Venter, who led the human genome sequencing, used his genus as a typical specimen. The endeavor was motivated by a promise to find out the causes of diseases, not to shake up the worldview of biology. However, the results now in hand force us to reconsider the paradigm. Modern evolutionary theory does not explain why our chromosomes house a whole heap more matter than mere genes and why genes alone do not dictate the phenotype. It seems that when research specializes in spotting details and even gets mired in certain ones, the overall picture loses its focus.

While normal science proceeds by spotting ever subtler differences, past scientific revolutions have revealed many seemingly distinct phenomena to be the same. So should we not talk about life, consciousness, and economy using the same concepts? After all, mere data exposes no differences.

When we build comprehension from arguments rooted in the very basics, we are not just complying with the current consensus. The dialog between contemporary science and the history of science both justifies and questions beliefs. As Fleck noted: "The current state of knowledge remains vague when history is not considered, just as history remains vague without substantive knowledge about the current state."[24]

IS THE OLD PRINCIPLE A NEW PARADIGM?

The obvious prerequisite for a scientific revolution is that there is in place a well-defined paradigm to be overthrown. For example, the consensus about immobile continents had to exist first before it could be negated. Already in 1596, when the Flemish Abraham Ortelius had compiled the first world atlas, he wrote that the Americas had been torn away from Africa and Europe. In the following centuries, the same idea was offered time after time, but it is only Alfred Wegener, a German geophysicist and meteorologist, who is seen as the hero of this particular scientific revolution because he faced fierce opposition. The earlier thinkers did not, as the contrary stance had not yet established itself. The orthodoxy of fixed continents solidified only in the 19th century, especially in America, thanks to James Dwight Dana's works. Wegener's arguments for continental drift from 1912 forced others to reconsider the established belief as well. Finally, in the 1960s, decades after Wegener's death, the theory of plate tectonics was adopted, as the mighty forces and mechanisms of continental drift were at last understood. In general, the fledgling ideas are the more fiercely resisted, the more entrenched the established doctrine.

A favorable reception may also be problematic, for it does not bring the idea to the forefront through a necessary struggle of forces; instead, the thesis is relegated to oblivion. This distressed Maupertuis: "[Other philosophers] have certainly established that nature must act by the simplest means, but none of them has really determined what these simplest means are, nor the fund that nature saves in the production of her phenomena."[25] Ever since then, no true controversy has arisen over the principle of least action that instituted itself as Lagrange's instrumental rather than Maupertuis' teleological form. While physicists do know the principle of least time, most of them only recognize it in the special case of balance, where trajectories can be readily calculated to high precision as nothing truly changes.[26]

Only now that physics has locked itself into these specific stationary-state formulas does the old principle of least time, where change is the general rule, distinguish itself clearly as a novelty. Only now that physics posits that particles are fluctuating quantum fields does the old idea of everything comprising photons stand out as unorthodox. Only now that the vacuum is theorized as having no substance does a vacuum embodying real quantum pairs identify itself as a distinctly different position. And only now that general relativity has become the standard way of describing the evolving cosmos can it be questioned as an accurate account of reality and rather regarded as a mere model of gravitation. There is a world of difference between the abstract mathematical models of matter, space, and time, and the concrete worldview where everything is quanta.

Similar to the dogmas of physics, it is necessary that the synthetic evolutionary theory has firmly established itself, for it to be challenged. Only now that genome-wide association studies (GWAS) fail to uncover causes of diseases can the dogma of evolution without direction be refuted with the thermodynamic theory of evolution as motion toward balance. And only now that epigenetic phenomena have demonstrated the role of circumstances can the doctrine of genetic determinism be contested with the thermodynamic tenet that everything depends on everything else.

While the thermodynamic theory can be viewed as reinstating naïve realism, the fact remains that we may only detect something that exists. An electron, for example, is not just a concept of quantum field theory but a torus made of photons. The vacuum is likewise not just sheer geometry or virtual particles but the quintessence of actual photons in pairs. Paradoxically, a reality check is needed most where it is least wanted.[xvi] As Nicholas Maxwell remarked, "Acceptance of instrumentalism by physicists adversely affects physics itself."[27]

Modern physics has abandoned the ideals of modern science, as Paolo Rossi, an emeritus professor at the University of Florence, defined them in *The Birth of Modern Science* (2001). The concepts of quantum mechanics are interpreted in various ways,[28] for the multifaceted language of magic portrays objects beyond substance and existence.[29] So, there is no point in interpreting mathematical models for something real because modern physics was never meant to be taken as anything real, let alone supernatural, say, entanglement, or

---

[xvi] "That which you most need, will be found in the place where you least want to look," is Carl Jung's wisdom.

transempirical, say, the multiverse. Its only object was to deliver numbers that dovetailed with the data. There is nothing to be interpreted in a true theory, either, because its founding axiom alone fixes the view of reality, that is, the interpretation of observations.

In physics, thanks to quantitative data, the contrast between a conformist and a dissident is stark with consequences, whereas in cognitive science, economics, and the social sciences, and especially in philosophy, there are next to no locked-in doctrines. In these disciplines, lacking unambiguously quantifiable objects and phenomena, such as particles and reactions, the thermodynamic theory is not seen as subversive as in physics. Instead, it is just a different theory about consciousness, behavior, free will, economy, culture, and values. Its claims that we think along the lines of force and that free will corresponds to free energy can be seen as refreshing rather than revolutionary.

THE INEXPLICABLE AXIOM

According to Kuhn, for a theory to be subversive, it must contest the established stance by presenting a solution to at least one central problem. This can be done most convincingly in physics, although more significant implications might be seen elsewhere. When there are only a few variables, if any, there is not much room for speculation. An error will immediately be spotted if the calculated mass of a particle, the redshift of light, or the clock's ticking deviates from the recorded data. By contrast, it would seem hopeless, for instance, to unravel precisely how the brain constructs the conscious self or how free access to information boosts economic growth. Presumably, the phenomena and factors relate to each other, but the complexity of issues leaves plenty of room for guesswork.

The thermodynamic theory contrasts conventional stances by providing solutions to several problems, most notably:

- There is no dark matter; instead, galaxies move and spin in the gravity of the expanding universe.

- There is no dark energy; instead, it should be taken into account that the light originating from a receding object shifts toward red not only due to the velocity of recession but also due to the weakening gravity of the expanding universe.

- The flatness of the universe is no coincidence but an inevitable consequence of reactions where matter-bound quanta transform into the quanta of the void.

- The universe's homogeneous horizon does not follow from cosmic inflation but the universal least-time consumption of free energy.

- The vacuum does not comprise virtual particles but real photons in pairs that make up the substance of inertia and gravity.

- Gravity is not the geometry of curved space-time but a manifestation of the vacuum in motion.

- The excess of matter over antimatter does not imply an imbalance but follows from the least-time quest for balance.

- The mass of a particle is not a manifestation of the coupling to the Higgs field but the void comprising paired quanta of light.

- The handedness of natural compounds is not a primordial coincidence; instead, the standard stems from the least-time free energy consumption.

- The genome is not so much of a repository of instructions as a system of its own consuming free energy in the least time.

A truly cross-disciplinary tenet differs from meek multidisciplinary collaboration, where different viewpoints merely mirror one another but do not merge into a unified worldview. According to Fleck, a revolutionary theory interprets the basic concepts in a new way that opens a more unified view of reality than offered by the prevailing thought style.[24] In this fashion, the thermodynamic theory interprets basic concepts as follows:

- Planck's constant is the measure of quantum, not just a constant.

- Time and energy are complementary attributes of the quantum, not just concepts without substance.

- Time advances as energy changes, not all by itself.

- Causality is embodied in the flow of quanta, not merely an impenetrable abstraction.

- The fundamental forces are manifestations of the structures of substance rather than the other way around.

- The vacuum is not sheer geometry but photons in out-of-phase pairs.

- Particles are not quantum fields but consist of quanta surrounded by the field of vacuum quanta.

- New properties do not emerge from the ingredients of a system but from the quanta of the surroundings.

- Information should be measured as free energy for its recipient rather than non-redundancy in a sequence of symbols.

- Consciousness is one means, among others, for consuming free energy.

- Free will is free energy at an individual's disposal rather than an unimpeded ability to choose between courses of action.

It may seem fanciful that many fundamental questions could be answered by one and the same theory. But a narrow stance could not possibly accommodate the catchall paradigm, such as everything comprising quanta of light.[3b] With this in mind, we should not be content with little but instead be receptive to a lot when facing the unity of everything.[31] The holistic worldview may nonetheless strike an expert as pseudoscience because it includes material new to their discipline. Taleb points out that "the problem with experts is that they do not know what they do not know."[30]

The resolutions provided by the comprehensive foundation are inevitably impressive.[6c] The far-reaching implications are what they are, rather than in command of a researcher, since the theory has no parameter to tinker with; everything is either composed of photons or does not exist. The theory's explanatory power is thus by no means unreasonable. On the contrary, it would be a suspicious scientific activity if one did not explore the limits of knowledge.

A theory can explain a lot, but a theory of everything is logically impossible. A candidate does not qualify as a theory unless traceable back to its inexplicable founding assumption. The axiom of the thermodynamic theory, the quantum, is not explained; it just exists.

UNQUESTIONABLE TRUTHS

A paradigm starts shifting when the facts that are thought to be true are found to be unproven. Case in point, the normal distribution is held as the norm for statistics even today, although it has been known for hundreds of years that natural distributions are skewed. Instead of merely modeling them mathematically, these long-tailed distributions can finally be properly understood as energetically optimal partitions. Furthermore, the sum of probabilities is presumed to be invariant (100%), even though it has been evident since Bayes' theorem of 1763 that the norm does not hold, except at balance. After ages of being regarded as a mere number, probability can, at last, be understood as what it truly is, a measure of energetic status (Appendix A).

Moreover, it is illogical to associate the increase in entropy with an increase of disorder because both order and disorder are consequences rather than causes of events. Quantization of energy, too, is a common but inconsistent belief, except in the special case of a stationary system, because photon energy does change in a continuous manner. Furthermore, many laws of physics and constants of nature are considered to be independent of time's arrow, although nothing escapes the sands of time. As a result, many a puzzle stems from a groundless premise.

Evolution is customarily ascribed exclusively to living beings, although the evidence suggests the opposite: everything is evolving. The conventional stance has not been competently challenged insofar as the equation of evolution has not been known. Another common but fallacious belief is that the handedness of natural substances was set at the origin of life. Likewise, the victory of matter over antimatter was set at the Big Bang. These standards emerged from a sequence of events rather than having been settled at any one event. Flaws of thought become apparent now that the ancient logical unity is at hand, the photon being the *atomos*.

WARRANTIES OF THE THEORY

Theories are ways of perceiving reality. No matter how persuasive the mental images it gives rise to, a scientific theory must first and foremost be consistent with the observations. No theory can be proven right but only wrong because of inconsistencies yet to be unearthed. Similarly, a person must be held innocent until evidence proves them guilty. A theory must be predicated on falsifiable axioms; otherwise, it can be modified to be in line with just about anything.

Specifically, the thermodynamic theory would be wrong if:

- even single quantum were found to be divisible.

- even single quantum were to vanish into nothing or pop into existence from nothingness.

- something were found that is not made of light quanta.

- a phenomenon were discovered where the system moved spontaneously away from the balance rather than moving toward it.

A scientific theory is also expected to predict what should be observed. In this regard, as everything hinges on everything else, the more accurate the prediction is, the less there is to predict, and the less useful the prediction is. Such forecasts do not concern a real future, only the properties of a well-balanced system. For example, the properties of particles can be calculated from their wireframe models (Appendix B). Combinations of elementary structures also suggest what kind of particles could exist that have not yet been produced.

Although quantitative analysis is sought in science, Kuhn pointed out paradoxically that the goal of science is not actually achieved by measuring.[3b] In practice, it is difficult to quantify complex systems with high precision beyond elementary particles, as they keep changing all the time. Obviously, numeric values that keep shifting cannot be precise. Nevertheless, quantitative results can be obtained when the course of events follows a trend. For example, the redshift of light from the early universe to the present can be calculated quite accurately because the energy density of the universe is declining gradually.[32] When the environment remains virtually unchanged, the system is also almost stationary; hence, its average properties can be calculated precisely. For example, the gravitational contribution of the expanding universe to the motions of galaxies can be computed fairly accurately.[33]

Besides predictions, a scientific theory is expected to provide conclusions about specific cases from its general grounding. This goal is not only motivated by the observed *Grand Regularity* but also accomplished by the thermodynamic theory.

Even though scientists know the rules of science, they tend not to follow them, choosing time and again to evaluate a new way of reasoning against established beliefs than facts. There is no earthly reason to insist that the novel thinking must side with the prevailing thoughts, for there are always different ways to explain the same data; instead, the new theory should be challenged by the same criteria as the old.

"When we say something is unreasonable," remarked Robert B. Laughlin, "we usually mean it is not suitably analogous to things we already know. Pure logic is a superstructure built on top of this more primitive reasoning facility and is thus inherently fallible. Unfortunately, we need to be most logical precisely when it is most difficult — when confronted with something new that is not analogous to anything we already know."[34] There is thus no good reason to oppose an alternative theory but every reason to find out its worthiness.

Time will tell to what extent the theory of time meets Kuhn's list of the five qualities of a sound scientific theory:[35]

- The theory must be accurate within its domain. The results should correspond to what is observed and measured.

- The theory must be consistent not only in itself but in relation to the prevailing theories.

- The theory should have a broad scope. It should explain much more than its first objective.

- The theory should be simple and bring order to phenomena that otherwise would be individually isolated and, as a set, confused.

- The theory must be fruitful in revealing new phenomena and unknown connections among those already known.

Kuhn stressed that evaluating accuracy, consistency, scope, simplicity, and fruitfulness is demanding. For instance, the Sun-centered model was at odds with what was once thought about why stones fall, how water pumps operate, and how clouds move. Likewise, the thermodynamic theory seems to belie what is nowadays thought of the vacuum, elementary particles, expanding universe, and evolution.

In the foreword of *Novum Organum*, Bacon points the finger of responsibility at those who dare to be complacent and arrogant and present truths as incontestable. They silence and interrupt research, causing the greatest damage to philosophy and science.[35] Today, over half a century since the publication of Kuhn's well-known work, scientists certainly know the qualifications of valid scientific theory,[3c] yet they do not act according to them. Instead of investigating the matter, they are inclined just to label the unusual as fake and cherish the conventional as genuine. Such superficiality is asking for trouble.

HOW TO FACE THE CHANGE?

The changes in the worldview are the gems of science. It is thus good to know how people in the past have reacted to the mind-boggling insights to understand how we should meet future revisions. "In science," as Kuhn phrased it, "novelty emerges only with difficulty, manifested by resistance, against a background provided by expectation."[6b] In its simplicity, the all-encompassing principle that underlies the *Grand Regularity* is strange to our sophisticated epoch, where the disciplines have detached from one another. On the other hand, modern physics is a very odd tenet judged by common sense.

SCANDALOUS IDEAS

No matter the degree of irrationality, theories are held in high esteem. As Yuval Harari, the historian of our time, verbalizes baldly: "The theory of relativity and quantum mechanics argue that you can twist time and space, that something can appear out of nothing, and that a cat can be both alive and dead at the same time. This makes a mockery of our common sense, yet nobody seeks to protect innocent school-children from these scandalous ideas. Why?"[36]

Everyone has noticed the necessity of comprehending – in one way or another – when being knocked over all of a sudden. Then it will take a split second before understanding what happened and in which way one happened to land. At that moment, when comprehending nothing, one does nothing. We need perceptions, even wrong ones, to do at least something. So, even an outlandish explanation is preferred over having none because being perplexed is paralyzing.

Most of the phenomena that modern physics entertains are ones about which we lack personal experience involving very high or low temperatures and exceptionally short or long distances. Apart from the photon, we do not see the elementary particles that quantum field theory imagines. Relativistic effects reveal themselves only at extremely high speeds and cosmic expanses. So we are at the mercy of the interpretations of experts, not only because our senses are limited but also because getting acquainted with modern physics is a demanding task.

We doubt an explanation that we cannot relate to our own experience. A recondite account does not *feel* right. While unorganized feelings are not science, an incomprehensible interpretation isn't either. Instead, the universal patterns displayed by data of all kinds suggest that the world is one, hence comprehensible by a universal law of nature that also complies with our experience.

Caught by Thoughts

The adoption of incomprehensible concepts is indoctrination when having no way of knowing whether they are true or not.[37] Specifically, we cannot liken entanglement or action at a distance to anything that we have experienced. Similarly, curved space-time is a cryptic notion. We do not sense abstract geometry; we feel gravity and inertia.

When a doctrine has been imbibed uncritically, the consciousness may have become closed to the possibility that the teaching might not be true.[38] As teachers pass on only the paradigm that they blindly believe in, students do not develop critical thinking, accepting fiction for facts and beliefs for truth. In a generation or two, nobody will know anymore how to think differently. Scholars imagine themselves as critical thinkers when labeling results that deviate from the adopted doctrine as pseudoscience. So it is that skepticism and irrationality displace sense and experience.[39]

As much as we cannot relate modern physics to our own experience, its weight is entirely a matter of believing in scientific authority. According to Winograd and Flores, our thinking is largely based on prejudices, and scientific knowledge does not protect us from this — in fact, to the contrary.[2]

Through their education, students of liberal arts know that knowledge is uncertain and collective by nature, whereas a seasoned scientist may not even know how to doubt a doctrine. Thus, the idea of questioning my learned beliefs did not cross my mind before I became curious about how to formulate evolution as a law of nature.

The limitations of the conventional mindset become apparent first when a nonconforming approach resolves the longstanding problems. In our case, rather than zooming further in, we should open the focus of a telescope to encompass much more than a galaxy now that the gravity of hypothetical dark matter can be understood as the gravity of all ordinary matter in the expanding universe. Likewise, rather than zeroing further in, we should raise our eyes from the microscope and take a look at the whole now that evolution can be comprehended without any demarcation between the animate and inanimate as a mere manifestation of the universal quest for balance.

Although the *Grand Regularity* is perceptible in the data, and its explanation has already been provided on several occasions over the centuries, a scholar may find it even today impossible to think outside the box. A doctrine may be so indelibly imprinted on one's mind that questioning it is no longer seen as science. Nonconformity is regarded as abnormal rather than exceptional. It is swept aside and shut out. Smolin is outspoken on this matter of science being no longer self-correcting but self-protecting: "Science requires a balance between rebellion and respect.... research universities do not tolerate revolutionary thinkers. It is no wonder, therefore, that there is no change, even though problems clearly call for it."[40]

On the brink of a scientific revolution, accepted theories do not work, but working theories are not readily accepted. Sabine Hossenfelder says the same: "To me, our inability – or maybe even unwillingness – to limit the influence of social and cognitive biases in scientific communities is a serious systemic failure. We don't protect the values of our discipline."[41] The hard-core theoretical physicist continues: "This means, for example, that we shouldn't punish researchers for working in unpopular fields, filter information using friends' recommendations or allow marketing tactics, and should counteract loss aversion with incentives to switch fields and give more space to knowledge not already widely shared (to prevent the 'shared information bias'). Above all, we should start taking the problem seriously." It is brave to point out the disdain of the scientists themselves for scientific values and virtues. In her book *Lost in Math* (2018), Hossenfelder proposes guidelines for scientists, administrators, and editors on how to correct the skew state of contemporary science. Only by acknowledging reality as it is can science find the truth, whereas the incomprehensible truths are merely the insipid fruits of a tussle of opinions short on sound reasoning.

The specialization of science is itself a natural course of events, no different from a species differentiating into an ecological niche until facing extinction due to changing circumstances. Just as a branch of business comes to an end, a faculty of technical virtuosity, spending its time refining calculations, will end up in a blind alley. Sometimes it is more important to question the question itself than to seek an answer. Those who comprehend something profoundly new, far from meeting their end in dwindling demand, go on to create new kinds of demand. Lucid thinking renders many problems moot.

On a pedestal

While a pioneer's job is to open up the vista, it is each person's own responsibility to dare to see it. Smolin calls for candor when confronted with new thinking.[40] But as Neubert says, "Such well-intentioned advice is difficult to implement, not because we are hypocrites; but because one man's revolutionary is another man's fool."[5a] The true nature of profound scientific problems is indeed human nature.

We favor what works, even if we do not understand why. It is easy for us physicists to spot this aspect in ourselves but hard to fix it. Even though the accounts by quantum mechanics of the double-slit experiment and action at a distance are clearly supernatural, they have been thrust through the decades so deeply into the gullible social consciousness that the very act of questioning modern physics is seen today as incredulous. We abide by the same formula, almost irrespective of how irrational it might be, because collaboration requires collective conceptions. Human beings are disposed to form a superorganism.

We also tolerate inconsistencies in explanations surprisingly well, as we tend to believe, not in what is believable, but in what is commonly believed. For example, we know from our own experience that rays bend when crossing from air into water. Light similarly slows down when going through the gravitational field of the Sun.[42] And it likewise takes light a little extra time for light to propagate through a momentarily increased vacuum density, that is, a gravitational wave. Instead of this consistent explanation, physicists have accepted the tale given by general relativity that the gauging gadget itself shortens when the gravitational wave passes through it.[43] Einstein's theory refers to observations such as the slowing of time and the shortening of length as relativistic phenomena. By contrast, all phenomena can be grasped in ordinary terms, comprehending time as the period of the photon and length as its wavelength.

Interpretations by modern physics are inexplicable yet not exactly reproachable, aware as we are of why we ended up with them. It is, however, reprehensible to put modern physics on a pedestal. Only a few people know what it takes to question. Einstein did and seized it tragically: "To punish me for my contempt for authority, fate made me an authority myself."[44]

One hundred years ago, the essence of the vacuum, the ether, free space, was as vague a substance as it had been since antiquity. Newton did not solve this problem but instead found the law of gravity. Neither did Ampère, Faraday, and Maxwell solve it but established the laws of electromagnetism. The Michelson–Morley experiment did not clarify the issue; its outcome was interpreted as if the vacuum were nothing, however, not unambiguously. Although Einstein's space-time geometry is in line with this famous experiment, general relativity describes the void as unreal without any substance. The vacuum lacks essence in quantum mechanics, too, where it is modeled as virtual, random, and ephemeral. Be that as it may, the vacuum *feels* real. The feeling can be discredited as a lack of learning. However, there is no gainsaying the failure of conceited expertise to explain our experiences of falling, accelerating, decelerating, and spinning.

In the early 20th century, physics got no grip on the vacuum and began to wander off from the concrete to the abstract. Weinberg confesses that this instrumentalism does not feel good: "But I admit to some discomfort in working my whole life in the theoretical framework that no one fully understands."[45] It is alarming that many professionals think that the task of physics is not to explain phenomena but to model data. Analyses, derivations, and simulations lead to no new insights. The mathematical notation expressly keeps the reasoning within the fixed mindset, whereas a new view calls for a reinterpretation of the basic concepts.

At one time, natural philosophy branched into various disciplines. We tend to see this specialization as progress. Consider, however, the words of Nicholas Maxwell: "Our understanding of our place in the universe is obscured. Our ability to see what is of value in life, and our ability to achieve what is of value, are undermined."[46] A scientific worldview is flawed if it fails to explain how meaning, values, and purpose arise from substance, existence, being.

We fancy scholarly sophistication but would need philosophy the way Wittgenstein defined it: "Philosophy is a battle against the bewitchment of our intelligence by means of our language."[47] For over a hundred years, we have been entranced by concepts of wave-particle dualism, entanglement, action at a distance, and space-time. Breaking free from the conceptual straitjacket of modern physics is hard for those who do not even realize that they are prisoners. You cannot figure out how things should be based on how they are. Only when you view the world from another angle does it look different. Galileo was able to cut loose from Aristotle's doctrine of motion because he also knew the ideas of impetus theory. Still, no question about it. Expertise is valuable, especially when it is broad.

IDENTITY CRISIS

The reassessment of the core concepts is painstaking precisely because they are central. At one time, Lord Kelvin, a big name in science in the late 1800s, apparently suspected that X-rays were just a sham.[48] Their discovery questioned established perceptions and practices. At first, it seemed unbelievable that all work with cathode rays needed to be re-evaluated. Well, so it was. Hardly anyone remembers that concern anymore, but many are aware that Wilhelm Röntgen, an engineer and a physicist, opened a new window onto the world.

The light quantum is a key concept of physics. Abandoning the flickering virtual photon for the concrete one as the permanent primary constituent of everything would therefore entail substantial revisions. At first, it may seem unbelievable that all the work based on quantum mechanics and general relativity would need to be re-evaluated. This may well happen. Sometime in the future, hardly anyone will recall this task, but many might well know that when we broke with our preconceptions, we progressed.

Revision of the basic concepts brings discipline to an identity crisis. Once, Dalton defined by his atomistic proposition that chemistry *is* the reactions of atoms.[49] Now, Lewis' vision of the photon as *atomos*[50] is challenging the identity of physics with the notion that physics *is* the reactions of quanta.

In his Nobel lecture in 1954, Max Born, the founder of the standard interpretation of quantum mechanics, explained taking from Einstein the idea of photon density but, instead of sticking to its concreteness, he interpreted the wave function, introduced by Schrödinger, as the probability density of a particle.[51] Had he taken the wave function only as a mathematical model of the vacuum's photon density around the particle, we would have been on the right track all along.

Although the abstract wave function and curved space-time still qualify as mathematical models of the void, physics becomes concrete and comprehensible regarding everything in terms of quanta. Today, physicists calculate the properties of particles using quantum field theory; in the future, they could think of calculating the photons around the particle. Although the implementation is the same, the meaning is worlds away.

Physicist identifies themselves with their work as any other professional. I embrace such an identity too. Back in those days, I revered the world of abstract formulations and belittled concrete facts — groundlessly, for it did not even occur to me that modern physics is only a mathematical model of data, not an account of reality. Neither did I realize that having a preconception of what physics is suppresses one's passion for and blocks one's passage to truth, as demonstrated by Ernst Mach's disposition toward Max Planck's faith in instrumentalism.[52] It seems to me that the very pursuit of mathematical unification of quantum theory and general relativity has prevented contemporary physicists from deriving a unified worldview in its mathematical form from the atomistic axiom.

The holistic worldview at hand challenges the identity of a bioscientist just as that of a physicist. If the living does not differ from the inanimate, what is the study of life's origin all about? A neuroscientist, too, may ponder what specific issues they should study if cognition is nothing more than a sequence of events aiming at balance. Experts in other fields may ask similar questions. The answer is the same. The goal of science is the wholeness of thought. So it is still sensible to figure out, say, how behavior and cognition relate to each other through their discernible counterparts, for example, by considering how economic activity and governance are connected.

INTELLECTUAL CHALLENGES

Our time is troubled by sophisticated theorizing. Especially in those fields of science where specialization has advanced furthest or even ended up in a blind alley, it is hardest to recognize the value of elementary reasoning and general principles. While complex theories furnish us with mental gymnastics, in the end, reality commands us to comprehend the world physically, the way we experience it.

Contemporary physicists seem so confident about modern physics that they may find it only a frivolous waste of time to explore other ways of understanding Nature. But as Karl Popper pointed out, "Whenever a theory appears to you as the only possible one, take this as a sign that you have neither understood the theory nor the problem which it was intended to solve."[53]

The philosopher's piece of advice may seem irrational. What is the value of expertise if it, in fact, impedes rather than enables progress? More germane than some specific knowledge is to see evolution from a new perspective to realize that it is nothing but the least-time consumption of free energy. The conclusion is trivial, and that's why it is revolutionary.

While biology revolves around evolution by natural selection, the study of life cannot define itself. Evolution cannot be understood through populations, genes, or molecules but through a more thorough perception. Focusing only on one facet makes it hard to see that evolution entails everything.

Just as Kuhn did, it is worth emphasizing that expertise is essential and a paradigm is practical for a scientific community to identify problems as soon as possible. Of course, some problems turn out to be harder than others, and a few have yet to be cracked. No timeframe, in principle, is too long. Then again, many a puzzle becomes with time so byzantine that the simple answer, clear as crystal, is no longer regarded as the answer.

Normal science does not look for new views. It is even blind to them. Naturally, an expert assumes that the solution to a problem encountered within their métier is available by their means. Based on past success, it seems only a matter of deepening the specialty, but the more science specializes, the fewer incendiary ideas fit into the narrowing capacity of the field.[3b]

Those who are young, or have come to the field only recently, are still free of dominant thinking and think in their own way or in the way adopted elsewhere. Dalton, as a meteorologist, intruded into the grounds of chemists. Wegener, also a meteorologist, encroached onto the field of geologists, and Soddy, as a physicist, onto the territory of economists. Gatecrashers are not welcomed, but eventually, their work, where it renews disciplines, is recognized.

The achievements of many a generalist speak for the value of cross-fertilization. Gamow understood radioactive decay, the nuclear synthesis of the Big Bang, the temperature of the cosmic background radiation, and the nature of the genetic code.[54] Von Neumann founded game theory and self-organized cellular automata, not to mention his achievements in mathematics and physics.[55]

Thinking that revises one discipline is often just normal science in another because, by now, the disciplines have grown so far apart that what is normal in one is revolutionary in another. In physics, time is the problem; therefore, it is radical to reason free from conventions that time comprises photon periods, in fact, heretical. In truth, the starting point of the inquiry is just far enough from the goal for the view to cover more than the conventional stance. For example, the evolution of animal species is superficially far from the reactions of elementary particles, but comprehending both phenomena requires associating the arrow of time with the flow of quanta.

THE HARSH REALITIES

An expert leads the scientific community to a problem quickly. However, the expert does not swiftly solve a problem that is beyond their expertise. In the history of science, the professionals are often the last to recognize the solution because it is not in line with their expectations regarding the gist, author, use of concepts, references, or publishing forum. The specialists evaluate holistic thinking only within the scope of their specialty.

The reception was curt in 1845 when the unknown Scottish physicist John Waterston, a lecturer at Bombay's Grant College, sent a manuscript on the kinetic theory of gases to the Royal Society. Sir John William Lubbock judged: "The paper is nothing but nonsense, unfit even for reading before the Society." Moreover, Baden Powell, a professor of geometry at Oxford University, found Waterston's idea that the pressure of gas would result from molecules bouncing on the walls of the container very difficult to accept.[56] The theory was approved a decade later when proposed by Rudolf Clausius and James Clerk Maxwell.

Lord Rayleigh pointed out typical, if less noble, features of scientific practice by "The history of [Waterston's] paper," which, "suggests that highly speculative investigations, especially by an unknown author, are best brought before the world through some other channel than a scientific society, which naturally hesitates to admit into its printed records matter of uncertain value. Perhaps one may go further, and say that a young author who believes himself capable of great things would usually do well to secure the favorable recognition of the scientific world by work whose scope is limited, and whose value is easily judged, before embarking upon higher flights."[56]

J. S. Haldane, a famous Scottish physiologist, also took notice of Waterston's case with a rebuke. In the long and glorious history of the Royal Society, nothing had more devastating effects on British science than the rejection of Waterston's paper contrary to the vaunted principles of science. Nevertheless, Sir Rayleigh saved a lot of the institution's credibility by seeing to it that the Society published Waterston's work, albeit posthumously in 1892, and acknowledging the Society's failure in its most important task: promoting science.

Even today, the traditional permanent position of a professor barely provides the necessary protection to face the harsh realities. Prejudices just refuse to die. Smolin gives an example of the disturbing attitude toward the new: "In fact, professors with tenure who lose their grant funding because of having switched to a riskier area can quickly find themselves in hot water. They cannot be fired, but they can be pressured with threats of heavy teaching and salary cuts either to go back to their low-risk, well-funded work or to take early retirement."[40] It is telling that many scholars engage with the fundamental questions only after retiring.

PEERLESS IN PEER REVIEW

Never in the history of science has it been easy to get revolutionary results published. Distinguished journals publish mainstream science and avoid the merest chance of printing nonsense. On the other hand, many traditional publishers have also set up omnivorous series to increase their earnings.[57] Everything technically correct in light of peer review and contains something new is published. Only when a paper differs from conventional beliefs do suspicions awaken; have peers actually reviewed the work? However, verity rather than verdict should be the object of science.

Peer review works adequately on average, but the exceptional remains problematic, as the familiar is regarded as factual but the original berated as groundless. While a highly networked researcher becomes oft-cited and credited,[58] obviously only a few people are behind a new idea. The peerless tend to have a hard time with peer reviews.

As an anecdote along these lines, the renowned cancer researcher Kari Alitalo, when getting a manuscript rejected at an early stage of his impressive career, countered: "I have nothing against peer review, but these reviewers are not my peers." The same sense of self-esteem is evident in Einstein's indignant retort to the Editor of *Physical Review* in 1936: "We (Mr. Rosen and I) had sent you our manuscript for publication and had not authorized you to show it to specialists before it is printed. I see no reason to address the – in any case erroneous – comments of your anonymous expert. On the basis of this incident, I prefer to publish the paper elsewhere."[59]

Einstein was not used to peer reviews, yet he recognized his peers. In 1924, he received a paper from an unknown Indian physicist Satyendra Nath Bose on the distribution law for photons in the vacuum (Appendix H) and, without delay, translated the study into German for publication in *Zeitschrift für Physik*.

Only a small fraction of submissions are peer-reviewed in the most prestigious journals because the editors reject most manuscripts straight away. So the value of work cannot be inferred from where it was published. For example, an obscure journal called *Transactions of the Connecticut Academy* published Willard Gibbs' groundbreaking works on thermodynamics from 1873 onwards. In turn, Wilhelm Röntgen reported the extraordinary rays he discovered first in a low-key journal of the Physical-Medical Society of Würzburg. John Bell's theorem of hidden variables concerning action at a distance was published in 1964 in the newly established but soon discontinued journal *Physics*. The paper became more famous than the journal itself.

When the voluminous open digital archives became feasible, it was hastily concluded that peer review would soon be obsolete since anybody could evaluate any paper. Not quite. The moderators of the traditional open archive, *arXiv*, have been railed against for excluding unorthodox papers.[60] And even when freely available, it is not guaranteed that pioneering ideas will surface from the immense depths of scientific production, doubling every ten years or so.[61] As Yogi Berra said: "It was impossible to get a conversation going, everybody was talking too much."[62]

Today, in the era of media hype, highlighting research results may be seen as a scientist seeking exposure. But to hold back results would be against the professional ethics of a scientist, even irresponsible. The pursuit of publicity is indeed correlated with scientific deceit; the most prominent journals publish comparatively many unreliable results.[63] Between 1998 and 2001, five *Nature* and seven *Science* papers by Jan Hendrik Schön turned out to be fabricated.[64] The works of Hwang Woo-suk on the cloning of human embryonic stem cells published by *Science* in 2004 and 2005 were based on ethically ruthless fraud.[65] In a way, cheating meets expectations, whereas the originality is truly unexpected.

Science expresses the zeitgeist: publicity is valued more than comprehension. Taleb also recognizes these forces that are in charge of science today. "When you are rewarded for perception, not results, you need to show sophistication."[66] When those in power regard fame, throughput, rankings, publicity, etc., as more important than the truth, their management is distorted accordingly. In the end, it is easy for the scientific community to spot a fraudulent result. By contrast, it may take an unusual episode for society to wake up to the reality that an academic institution itself has become estranged from its values and, in fact, dismisses scientific results.

Valid currency

As it is unusual for a scientist to test the fundamentals of a discipline, results will readily be misinterpreted as something else. Questioning the Copenhagen interpretation by a simple, ingenious experiment, the young Iranian-American physicist Shahriar Afshar was accused of dishonesty and seeking publicity rather than being refuted by sensible arguments.[67] The *New Scientist* journalist Marcus Chown asked: "Why the extreme reaction? Perhaps people thought that such a simple experiment as Afshar's must be wrong. Perhaps those who interpret the quantum theory are not accustomed to their argument being tested. No matter what the reason, there is no excuse for such treatment."[68]

Normal science does not doubt its doctrines but counts on them. Even questioning is misjudged as heretical, even though it tends to illuminate and often enrich the prevalent perception. Questioning, as such, does not undo anything; only meaningful answers may change the world. Normal science defines the norms; revolutionary science revises the definitions. At best, the tension between conservatism and reformism nourishes; at its worst, it tears the academic community apart.[3b] What will happen depends on scientists' capacity to confer with one another.

A researcher who has made a strange observation, like Röntgen, is skeptical of their own results. As much as they try to rule out spurious factors, not all errors can be eliminated until the phenomenon is fully understood. At one time, the result of the Michelson–Morley experiment surprised physicists: is the ether not some substance as expected but none at all? One hundred years later, the supernova findings amazed scientists: is the expansion of the universe not slowing down as expected but speeding up? The subsequently adopted stand for accelerated expansion is not truly revolutionary, for it is only opposite to the expectations of decelerating expansion.

In view of the history and philosophy of science, the stance that a researcher should not venture outside their field of expertise is untenable. On the contrary, it seems necessary to break down the disciplinary barriers to make a discovery. "It is often the case that the kinds of people who originate ideas are not without faults when measured against the criteria that normal scientists use to judge excellence. They can be too bold. They may be sloppy about the details and unimpressive technically. These criticisms often apply to original thinkers whose curiosity and independence led them to fields they were not trained in."[40] Smolin goes on, arguing that novelty is spurned because the conventional ensures funding. Comprehension is no longer valid currency.

IS THE SCIENTIFIC COMMUNITY UP TO DATE?

Today's iconoclast will hardly be condemned like Galileo one time, although it is not all unimaginable either. Nevertheless, the forces allied to doctrine are formidable, even downright crushing, unless the core values of academic institutions defend the original thinker. The Canadian physicist Paul Marmet recalled an astounding incident when the head of the department came in and said, "We do not want your office, your problem is that you keep questioning the fundamental principles of physics."[69] When a researcher who doubts the foundations of a discipline is evicted in the belief that the university would thereby vindicate itself from practicing pseudoscience, the purity of doctrine is being protected by dirty tricks. That risks the credibility of the whole institution.

The ethics of science is violated when it is insinuated that a researcher should not question the foundations of the discipline. Neither can a journalist do their job when restricted in terms of what can be investigated. Progress in science calls for freedom, just as the development of society does. Unless we are conscious of the perils and reality of banning, it will befall us all. Sooner or later, a flawed worldview will guide us into destruction.

Thoughts outside the prevailing paradigms are excluded subtly, as the discourse itself defines the rules for investigating reality and the forms that reality is supposed to have if it is to be real. In this manner, nonconformity will be 'proven' insane, ignorant, irrational, and incorrect, as the French philosopher Michel Foucault argued. Spinoza had already reasoned that the determination of truth is always a matter of power – whether taking the form of an argument or dwelling in the corridors of an institution.[70]

By now, modern physics has attained the status of being self-evidently known. There is no question of particles being quantum fields as observations are interpreted by quantum field theory. There is no question of dark energy and dark matter existing as measurements are interpreted by general relativity. Similarly, when evolution is exclusively associated with changes in the genome, then the genes are everything there is. However, the prevailing perceptions lack absolute legitimacy as long as we do not comprehend everything. Even what we think we know well may turn out to be otherwise. This has played out before. Can we even discuss different views?

FASHIONABLE NONSENSE

Paradoxically, scientists have only little interest in what reality is and how to rationalize it. If you have never seen another interpretation, it is almost unthinkable that the same observations could be understood differently. Truth is not questioned; it is taken for granted. For instance, the Earth-centered model must once have been so sure and self-evident that the mere idea of the spinning globe was quite incredible. We may likewise fail to reconsider the worldview of modern physics and recognize its faults.

This examination of uncertainty in knowledge and the social nature of science might remind many of the squabbles between scientific realism and postmodernism in the late 1990s that Alan Sokal and Jean Bricmont triggered by their book *Fashionable Nonsense* (1998). The two physics professors laid the blame mainly on sociological studies for using mathematical and physical concepts without any intention of understanding what the words mean.[71]

The concern about the state of physics is exactly the same when insisting on knowing what the wave function, entanglement, curved space-time, and other concepts of instrumentalism mean in substance. The answer rests on the perceptions themselves, but Sokal and Bricmont explicitly refuted relativism.[71] It is difficult to figure out what the discourse is all about unless the concepts can be traced back to the elemental processes of experience. I experience, therefore I am.

Kuhn believed that when the revolutionary view is not communicated in an understandable manner, the insights of the philosopher Willard Van Orman Quine and the linguist Eugene Nida might be of help.[6e] A dissident uses an unusual language, so we should translate our concepts to them, and they, their terms to us. But, of course, this requires that we refrain from judging the unfamiliar reasoning as madness. The danger is real; the subversive seems to be standing on their head.

When we do not understand what the other person means, we easily imagine something must be wrong. I remember an incident from my exchange year in the US. I was with my buddies on our way across the Cascades near the Canadian border. It was Friday, and as a habit, I used to call my family back in Finland to tell them the news of the week. I found a phone booth in a mountain village, and I had a lot to tell. The device devoured coins at a rate of knots, so the operator in the background was asking for more. Turning to my pals, I shouted: "I need more coins. Would you please give me some?" They did not move an inch but stared at me as though I was a madman. I repeated my clarion call more forcefully, but they only took a step back. Then I realized I was shouting in Finnish. There is no communication without a common language.

The philosophers Paul Ricœur and Hans-Georg Gadamer reason that language limits one's ability to interpret messages. An expert uses exclusive lingo. Its specific grip prevents them from fully grasping the meaning of the text, even that of their own script. For example, the specialist may find it impossible to see that the electron wave function does not correspond to the particle itself but to the surrounding field of the vacuum quanta. The expert has no other concept for the electron beyond the quantum field, that is, down to the source of the field. Although everyone distinguishes the Earth's solid ground from its gravitational field, the comparison to the corpuscle and its surrounding field means nothing to the professional. "It is possible that at some future point our concept [of the field] will seem as archaic as the 'aether-field' seems now," anticipated the philosopher Nancy Nersessian in her book *Faraday to Einstein: Constructing Meaning in Scientific Theories* (1984).[72]

THROUGH THREE STAGES

Scientific revolutions are typically slow processes despite all their radicalism. It took a whole century to adopt the Copernican view. Newtonian mechanics were absorbed in Continental Europe in a half-century. In turn, Newton's authoritative view of the light particle was overruled first after Foucault measured the speed of light in 1850, to return in the early 20th century in the oblique form of wave-particle dualism.

There are counterexamples, too. In 1982, Stanley Prusiner proposed that scrapie, a disease of sheep, was caused by a protein.[73] Initially, that idea was thought to be entirely out of the question; bacteria and viruses, but not proteins, are pathogens. A few years later, even when the evidence was gleaned, Prusiner was accused of striving for glory at the cost of science, but in the late 1980s, the tide started to turn. When mad cow disease raged in Britain between 1986 and 1998, the American neurologist and biochemist's foresight was generally acknowledged and awarded the Nobel Prize in 1997.

According to the founder of biology and embryology, the Estonian Karl Ernst von Baer, every successful theory undergoes three phases: "First, it is ignored as 'unrealistic'; Then it is rejected as 'anti-religious'; Finally, it is accepted as a dogma, and every scientist claims to have long appreciated its truth." The Baltic German professor's maxim is the same as Arthur Schopenhauer's thesis of all truth passing through three stages: First, it is ridiculed. Second, it is violently opposed. Third, it is accepted as self-evident.[74] When choosing his words, von Baer might have had something particular on his mind as he held the opinion that forces direct evolution. Darwin acknowledged the critique,[75] but von Baer's view that evolution follows its course along the lines of force without a predetermined goal was missed and still goes missing without consideration and comprehension.

Even after over 150 years since the publication of *On the Origins of Species* (1859), despite every piece of evidence to the contrary, the irrational notion of evolution without purpose as an explanation for seemingly meaningful outcomes holds sway. We should, therefore, ask, what is the purpose, even if the question is banned as akin to crimethink. Only then will we see that the goal becomes clearer as evolution draws nearer to a balance.

A new view, such as quantum mechanics or general relativity in the past, does not emerge as a fully-fledged theory but rather with potential. If modern physics were to hold substance, shouldn't the past century have been long enough to realize it? Instead of getting answers to the essential questions of existence, we got perplexing puzzles about existence. The concept of the quantum of action, presented at the beginning of the last century, continues to be incomprehensible to many as the fundamental element of time and energy.

The facade of science

Times and habits have changed since Einstein. He faced a barrage of criticism, but in a positive spirit, among others, from the Finnish physicist Gunnar Nordström, whereas today, a study departing from mainstream science is classified forthwith as pseudoscience or nonsense. Academic institutions no longer seek to make sense of reality through discourse but to cement their authority by defending a doctrine.

The culture of science has become a closed professional activity, despite unanswered questions – or perhaps just because of them, argues David Bell. The professor of cultural geography considers the attitude behind the rational facade of science toward nonconforming reasoning both interesting and revealing.[76] John Horgan, a science journalist, best known for his 1996 book *The End of Science*, puts his finger on the issue: "As modern science has become increasingly institutionalized, it has started to resemble a guild that values self-promotion above truth and the common good."[77] Without logical arguments and solid evidence against dissidence, there is nothing more than a cliquish consensus behind the verdicts of heresy. But a devoted member just turns a blind eye to this revelation.

Bell underscores that rather than brushing aside strangeness, we ought to perceive its value, as we need people who recognize universally accepted assumptions as unjustified and ask new questions. However, even the best ideas have not been acknowledged when the scientific community has evaded dialog with nonconformists. Such aversion is irresponsible even in cases where malice plays no part; it marginalizes people and can even lead to their exclusion from the academic congregation.

While a critical attitude and intrepid thinking are, in principle, lauded in the values of academic institutions and speeches by their leaders, in practice, dissidents are opposed or vilified. According to Brian Martin, a professor of social sciences at the University of Wollongong, most dissidents are merely ignored, some come against the brick wall blocking their career, a few even get sacked. Martin, formerly a theoretical physicist, has explored the dynamics of power. He mentions climate change and genetic modification as themes in which unorthodox views are next to unacceptable. Questioning is not tolerated either when it concerns evolutionary theory, quantum mechanics, and relativity theory.[78]

As a rule, academic communities themselves are incapable of correcting their mistakes, but revision requires public exposure. Those in the media have thus a great responsibility. Comparing a scientific dissident to a political dissident may seem unjustifiable. Nonetheless, it may be precisely the selfsame thing. Mistrusting the power structure corresponds to doubting the groundwork of a discipline. The parallel is substantiated by the fact that science has made significant progress when the foundations have been rectified; vital social reforms have been achieved when the constitution has been revised.

Arcetri

With his new telescope, Galileo was able to take in things that no one else had ever seen before. In his astronomy thesis, *Sidereus Nuncius* (1610), Galileo questioned the age-old view of the cosmos. This launched an intricate controversy.[79]

Jesuit astronomers quickly corroborated Galileo's observations and found his inferences relevant, but the authority took a dim view of the new outlook. The Copernican model was not seen as an exhaustive explanation, so it was not yet obligatory to accept it. Earth's orbital motion about the Sun should have caused apparent movements of the fixed stars. The parallax is indeed visible, just not with the instruments available at the time, because even the nearest stars are extremely far away. However, the phases of Venus and the moons of Jupiter would have been clearly visible to Galileo's antagonists had they only bothered and dared to take a quick glance with their own eyes through the telescope.

The worldview of the time, just like the worldview of today, was the central scientific doctrine, doubts against which the almighty institution could not deal with. Following the proposal of the high-authority theology committee, the Inquisition issued its order on March 5, 1616: the Copernican model is heretical. Galileo was biding his time, but eventually, he was unable to keep silent on the issue of essence. When one has something to say, it must be said, irrespective of what others might say.

The Second Act began in 1632 when *Dialogo dei Due Massimi Sistemi del Mondo* was published. Galileo wrote for the common people. The dialog compared the Earth- and Sun-centered models in formal terms within the permit given by the Inquisition, but he did not conceal his conviction of the superiority of the Copernican theory. The truth became a matter of prestige, as a character who spoke for the stance of Pope Urbanus VIII Galileo named Simplicio. As a result, Galileo was convicted as a heretic, and the publication of his works was banned, including eventual future works.

Galileo facing the Inquisition. Painting by Cristiano Banti (1857).

For the rest of his life, Galileo lived in his villa close to Florence. There he also noted down remarks on the ubiquity of regular patterns in his last book *Discorsi e Dimostrazioni Matematiche Intorno a Due Nuove Scienze* (1638). And from the hills of Arcetri, Galileo's writings fell into the hands of Newton and others. In the end, the arrogant institution's order, detached as it was from reality, failed to rein in the revision of the worldview.

The case of Galileo was not so much a conflict between science and religion but more the case of a prestigious but parochial establishment trying to uphold its flawed worldview. Nowadays, as the holder of the contemporary worldview, a scientific institution can equally ruin its legacy by condemning those who have brought up facts and figures that correspond to reality. Academia, especially when fixated on an aging paradigm, can display all the features of an intolerant authority. It neither promotes nor tolerates the new and may silence dissidents. Rather than tackling criticism with facts, the intelligentsia evades polemics by claiming that the critic is not an expert, so he cannot be right. This arrogant argument only reads that the visionary does not accept the dogma, as it is the very issue under reconsideration.

ASTRAY ONCE AGAIN

Often science is presented as straightforward progress, even though its true course is full of twists and turns. When we do not recall these wretched odysseys, we do not even think of the possibility that we could have gone astray once again. While amused by outdated old-time beliefs, we take the 'truths' of our time seriously, even those that are demonstrably absurd. Ancient mythology was real for the people who lived in the midst of it. Today, we are in thrall to scientific theories. Ultimately, grains of truth lie only in the things that could not be otherwise.

Despite all the accolades contemporary science receives, some are aware of the actual state of science. Thomas Nagel "is willing to bet that the present right-thinking consensus, i.e., reductive materialism, including Neo-Darwinism, will come to be seen as laughable in a generation or two – though, of course, it may be replaced by a new consensus that is just as invalid. The human will to believe is inexhaustible."[80] Beliefs influence science gravely, especially if we are not even aware of them.

Nowadays, ever more is expected from researchers; hence, ever less has become the standard. In the name of intensifying operations, modern universities walk away from their core function by abandoning organizational structures, such as institutional norms, collegial decision-making, and permanent positions. This streamlining of seemingly effective but dead-end-bound normal science is weeding out trailblazing thoughts. "The whole ethos of scientific research – reflected in institutional practices of how scientific research is incentivized, evaluated, and research funding distributed – has changed. While academic values, collegiality, freedom of research and publishing, criticism, and creativity, aim to ensure that science is in the service of society, today, science has to be of use to a well-identified group, or it is deemed to be of no use at all," maintains Michela Massimi, an Italian-born philosopher of science.[81]

Jean-François Lyotard declared that knowledge is no longer related to wisdom or education but valued on the market as a product. The French philosopher saw that knowledge is losing its social and cultural ties and the ability to produce grand narratives for humanity about its meaning, life, and future.[82] Utilitarianism threatens to ignore traditional, tangible thinking as old-fashioned. Since instrumentalism is expressly not realism, defending modern physics as a rational outlook on reality is not only ludicrous but also detrimental.

In 2015 Sari Kivistö and Sami Pihlström reasoned, at a Finnish forum on science and science policy, that a mindset is emerging in which change appears to have a value of its own. Science is no longer directed by old-fashioned values such as truth. A year later, their message was more alarming: academic freedom is in danger because the university cuts in Finland may lead to arbitrary dismissals and structural reforms. The two professors urged the whole scientific community to react against the reforms for the benefit of science and the effectiveness of science.

Among the most important duties of professors is to point out wrongdoing. So, how unforgivable must a violation of academic values be before the community reacts so vigorously enough to recognize the mismanagement and revise its course? "Nothing strengthens authority so much as silence," said Leonardo da Vinci.

We must understand how the scientific community functions, for future generations depend on the achievements of our time. At any given time, the prevailing belief system is taken as self-evident, however alien these conventions might be to outsiders. "The individual within the collective is never, or hardly ever, conscious of the prevailing thought style, which almost always exerts an absolutely compulsive force upon his thinking and with which it is not possible to be at variance,"[24] wrote Fleck. When a scientist questioning the prevailing paradigms is not only derided but not even tolerated, who then will reveal verity?

Despite their distinctive features, Kuhn concluded that scientific communities are like other communities and called for answers to these questions: "What does the group collectively see as its goals? What deviations, individual or collective, will it tolerate? How does it control the impermissible aberration?"[6f] The answers, even when tacit, reveal values and morals. Unless the values are held in common, there is no community and no morality. The epithet *universitas magistrorum et scholarium* shortened with time to its most valuable character, *universitas*, the community. Does it exist anymore?

WHAT IS THE MEANING OF THE REVISION?

Our gaze into the future is reflected incoherently from flickering ripples on the surface flow of time, reaching not the deep dynamics. Yet, in the light of time, we can see that science has leaped forward by drawing on history. Everyone remembers Newton; Newton understood the importance of remembering: "If I have seen further, it is only by standing on the shoulders of giants."

Darwin's theory of evolution changed the way we see ourselves. We are not special, just one among other species, and hence the studies of all living beings have revealed to us a whole lot about ourselves. The impact is immense in agriculture, forestry, health care, and environmental protection. International competition and market mechanisms are seen to parallel the struggle for survival and natural selection. So we press for reforms, however, not quite understanding what they are and why they happen.

Many a revision to the worldview has begun from seemingly insignificant observations but ended up with grand conclusions. When Galileo saw the moons of Jupiter with his telescope, he immediately realized that there are worlds besides ours. The idea of an orbit, recurring in the model of an atom, laid the foundation of many a technology. Our worldview dictates what we target by means of education, research and technology, economic growth, and social reforms. Our worldview defines what we consider to be valuable. The future will evaluate our values.

We have amassed more data than ever, but only now have we explained its universality. We have measured the natural constants more precisely than ever, but only now have we elucidated their meaning.

We have organized our tasks more efficiently than ever, but only now have we realized that our deeds express the universal law. These revelations of profound problems are not found through state-of-the-art research but through a re-examination of age-old foundations of research.

Our focus will change when our viewpoint repositions. When we do not see dark matter as real, we will no longer search for it in vain but tune our receivers to detect true substance. When we do not see anything other than quanta of light in the elementary particles, we will no longer try to generate them but harness our technology for other purposes. When we do not think of evolution only as a hereditary process, we will no longer sieve through insignificant cause and effect relationships perceived to be in our genome but look at the whole. When we embrace the reality that the biosphere, atmosphere, geosphere, and hydrosphere comply with one and the same natural law, perhaps we will also get our econosphere into the same harmony. The revision of the worldview is not merely an academic exercise but a desideratum that drives us toward a sustainable way of living.

The opening of the worldview, first from the geocentric perspective toward the Sun, then encompassing our home galaxy, and over the past decades finally extending to the beginning of the cosmos as well as zooming in on the elemental constituents of matter, gives us a grand view of reality. From this vantage point, we can see remarkably far. It only takes courage to look.

CONCLUDING WORDS

The limits of human knowledge are ever shifting. But there must be an ultimate limit, because we are a part of Nature, and cannot go beyond it. Beyond this limit, the Unknown becomes the Unknowable, which it is of little service to discuss, though it will always be a favorite subject of speculation. But whatever is in this universe can be (or might be) found out, and therefore does not belong to the unknowable. Thus, the constitution of the middle of the Sun, or of the ether, or the ultimate nature of magnetisation, or of universal gravitation, or of life, are not unknowable; and this statement is true, even though they should never be discovered. There are no inscrutables in Nature.[1a]

So wrote the self-taught physicist and mathematician Oliver Heaviside in 1893. Like Heaviside, I reason that reality is one and, therefore, comprehensible. Although our means are not enough to uncover all the details, they are sufficient to understand the laws of nature.

Reality is about causes and consequences. That is why we ask, "why?" But until now, time, the essence of causality, has been a mystery to science. The irreversible flow of time and energy is in substance, the flow of quanta toward balance. This harks back to Heraclitus' famous dictum: "Everything flows, nothing stands still." We can liken various processes to our own experiences with this universal principle. Comprehension brings freedom but also obligations.

Ten years ago, while I rejoiced in discovering the law of time, I was also alarmed by the revelation: humankind itself complies with the universal law of nature. But, unknowingly about it and under the illusion of wielding power, it looks ahead only a step or two, hence stumbles along, hopefully without falling down.

Wisdom is ancient. In *Nicomachean Ethics*, Aristotle dealt with the question of destiny in one's own life. I live my time to the fullest at the very moment when the future transforms into the present. Mathematizing such an experience is the first physics, in Galileo's manner. When the theory is formulated in that way, it serves us. It allows us to inspect (*theoría*) the world in a practical way (*praxis*) and comprehend existence.

Often people think that the researcher produces new information; mostly, the researcher looks for a way to explain the observations. After obtaining it, they may gain access to something novel. When I realized that Newton, Maupertuis, and Carnot had already understood the universal law of nature, my task was to check whether subsequent observations and measurements made since could be explained likewise. Finding not a single exception, I grew curious about why the simple principle is not used more generally, if at all. In the presence of a master key to all the lairs of insight, why do researchers rely on discipline-specific models?

The triumphal march of science also includes long odysseys. We would not have found our way back to the right track unless someone had finally thought differently. It should, however, also be kept in mind that wisdom without disclosure leads nowhere. So, aware of his looming demise, Maupertuis was anguished that the principle of least action would be buried along with him. The deterministic spirit of that era was unreceptive to reality as it is. Today, every one of us is contributing to the spirit of our time. Who is concurring with fashionable truths? Who is closing their eyes to the facts? Who is boldly recognizing reality and acting accordingly?

I am concerned that truths are more pleasing than reality now, as then. We keep on pursuing efficiency shortsightedly and purporting nonexistent things. Instead, we should target a planetary balance where humanity coheres with the global environment. To achieve natural harmony at this scale, we need a holistic worldview.

The essence of time and space, as well as the nature of life and consciousness, are mostly deemed eternal philosophical questions. So I was amazed that the thermodynamic theory explains many of the biggest puzzles in a simple way contrary to common presumptions.

In the light of time, the reality is not strange but comprehensible. Quite the opposite, the mysticism that has gained a foothold in modern physics spoils the sense of science.[2]

We need to know the structures and reactions of substance in a tangible way to understand how energy can be released from matter and bound to it. This is not just academic proficiency but valuable knowledge for all of us to attain a sustainable way of life. We must, above all, be conscious about who we are to comprehend our thinking and behavior. It is time to acknowledge the facts. It is time to appreciate the real thing. It is time to mature from a citizen of the world to a citizen of the Earth. The revision of the worldview does not change the universe; it changes us.

"If a man will begin with certainties, he shall end in doubts; but if he will be content to begin with doubts he shall end in certainties," reminded Francis Bacon.[3] We trust in science since science distrusts beliefs. Doubting is not disruptive skepticism but disciplined questioning. A strong scientific community is critical of its own paradigms, whereas a weak one is hypocritical.

From cover to cover, we have asked questions, not to admire the complexity of Nature, not to get carried away by the achievements of science, not to wonder at the mysteries of the universe, just to understand. Dear reader, I hope that the book, now ending here, has, in your hands, fulfilled the promise of the opening words by giving "a hitherto undreamt of outlook on the whole".

EPILOGUE

The revision of the worldview is not inevitable. It rests in our hands. When the authority in its outdated opposition shoots the messenger instead of tackling the problem, courage is symbolized in the words, "and yet it moves." Four hundred years later, we remember Galileo's mistreatment. The courses of such events should be noted down today as well, although there is no guarantee of being none the wiser in the future, either.

In 2016, to make the cuts demanded by the government, the management of the University of Helsinki chose to defend neither the primary function of the institution nor the necessary preconditions of science but got to grips with the autonomy and freedom of research embodied by the professors. It was possible to discharge staff, for permanent positions had been demoted to contract employment a few years earlier.

To get fired, you must stand out from the crowd. And stand out I did – my research on the holistic worldview distinguished itself from the specialized studies of my colleagues. There were neither financial nor productivity-related grounds for my dismissal. It was not about the money, as the university heads did not even want to discuss my proposal for an unpaid four-year sabbatical, but another professor was hired to carry on the discipline in my stead. The comprehension of reality, summarized in this book, was met with disbelief and suspicion. So, given the opportunity, I was dismissed to free the community of a heretic.

My concrete answers to fundamental questions stem from the tradition of science, hence differing here and there from contemporary hypotheses. Despite tens of peer-reviewed publications, the overarching results derived from the atomistic axiom were condemned as inconceivable. It is deeply discordant with not only academic values but with wisdom itself to dismiss a person working to unite disciplines into an all-inclusive worldview for the benefit of humankind at a time when the world is thrashing in the throes of wicked problems.

Instead of seeking the truth in accord with its age-old ideal, the university defined its own truths with dismissal notice. The lessons from the history of science were lost on it. The uncertain nature of knowledge was not understood. The establishment stuck up for the doctrine troubled with its faults and fables while posing itself as the herald of progress. So the principles were trodden down as soon as they were needed.

I understand it if someone points out that my papers are not generally accepted, albeit not disproven either, for the novelty of science is thus acknowledged. I understand it if someone claims that I have not proven everything, for I have not. Absolute certainty is imaginary; what matters is that I have presented the arguments that underlie my conclusions. And I understand it if someone considers me to have interpreted the key concepts differently from the conventions since I indeed have. As a scientist, I must report results truthfully, seeing that they comply with the observations and measurements.

The truths of today are held high for a time, but in the long run, it is impossible to escape from reality. All this would be inconsequential if a questionable academic game were our comprehension of reality only figurative. The thing is that we behave according to our understanding with far-reaching consequences.

As a scientist, I know how mistakes are made. They, too, pave the way for understanding so long as they are corrected. Science is meant to work that way. But I cannot fathom that the decision-makers of the traditional stronghold of wisdom and open-mindedness should fail in their foremost task, upholding the search for the truth. Regrettably, the most renowned university in Finland will go down in history as a hall of ivy infiltrated with people who despise scientific principles and academic values. Why did the university jeopardize the respect and generosity granted to it by society?

A dissident is, above all, a thinker.

Arto Annila
Professor, Emissus

ACKNOWLEDGMENTS

This study, directed toward a better comprehension of reality, has been, at its best, most rewarding. Unfortunately, at its worst, when encountering groundless skepticism, it has been immensely distressing. During such times, the support of my wife, my children, and my parents has kept my spirits up. I also thank Tarja Niemelä, executive manager of the Association of Professors, for the resolute defense of its members, as well as the Association's trustees Seppo Sainio, Jukka Finne, and Per Saris.

I have received encouraging and constructive feedback from many colleagues, which has motivated me to pursue the authoring of this book. I am grateful in particular to the following: Pentti Alanen, Kimmo Alho, Keith Baverstock, Michael Bergman, Malcolm Dean, Matti Estola, Ray Fleming, Thore Graepel, Gary Goldberg, Pekka Heikkinen, Elliot Holar, Tarja Kallio-Tamminen, Mahesh Karnani, Erkki Kolehmainen, Mikael Koskela, Esa Kuismanen, Antti Laakso, Marcel-Marie LeBel, Lauri Lehmonen, Kimmo Pääkkönen, Eero Rauhala, Tapio Salminen, Stanley Salthe, Teemu Teeri, Pekka Teerikorpi, Esa Vepsäläinen, Stefano Vezzoli, Jeremy Wheeler, and Mårten Wikström.

Constructive critical feedback has been necessary and deeply appreciated. I have found differences in opinions most rewarding and extremely helpful. Dialogs with colleagues have contributed to the emergence of a more consistent worldview. Many refining modifications have also been brought forward by Terhi Annila, Tapani Kärnä, Pia Ketola, Martti Raivola, and Henri Annila.

I also owe a debt of gratitude to Kimmo Jylhämö, CEO of the Vastapaino publishing company, for critical comments and courage in issuing the original book.

Scala Naturae
Length (m)
10^27 The Universe
10^24 The Local Group
10^21 The Milky Way
10^18
10^15 The nearest star
10^12 The Solar System
10^9   The Sun
10^6   The Earth
10^3   The Golden Gate
1   Human being
10^-3 Pinhead
10^-6 Bacterium
       Cell wall
10^-9 Sucrose molecule
       Atom
10^-12 Proton
10^-15 Electron
       Neutrino
10^-18
Earthrise (NASA)

APPENDICES

A worldview does not correspond to reality when observations differ from expectations. Resolutions for such discrepancies, especially solutions to problems in physics, have often led to changes in the worldview with impact beyond physics.

The worldview of thermodynamics dates back to Ludwig Boltzmann. He understood that not only a gas through collisions but everything through various interactions evolves to thermodynamic balance. Compounds reach chemical equilibrium through reactions, as Willard Gibbs theorized. Light, as photon gas, attains thermal equilibrium with matter, as Max Planck formalized by the radiation law. The evolution of any substance can be understood likewise. The quanta, the fundamental elements of everything, redistribute energetically ever more favorably through all kinds of events.

The appendices introduce thermodynamics based on statistical physics and demonstrate its applications. Since the calculations agree with the observations, the theory appears to be an accurate account of reality, whereas differences would expose delusions.

APPENDIX A: EQUATIONS OF THE NATURAL LAW

Data displaying the same patterns irrespective of source and scale suggests that all processes comply with the same law of nature. This law can be derived from the statistical physics of open systems.

THE STATE EQUATION

In a complete theory, every element of reality has a counterpart. Accordingly, any system can be described by the same theory, assuming that the fundamental elements, i.e., the quanta, embody everything. This scale-free description in a mathematical form can be deduced from the energy level diagram of a system.

Let us examine an entity, indexed with j. Its existence is quantified by probability, $_1P_j = \phi_1\phi_2\phi_3\ldots = \Pi_k\phi_k$, in the form of product, Π_k, over ingredients, indexed with k. Thus, if any one component k is missing altogether, $\phi_k = 0$, then also $_1P_j = 0$. For example, an enzyme in a cell could not possibly exist if any one of its ingredients were missing altogether. We can express the probability, $_1P_j$, even if we do not know what components k are in the product, Π_k, provided that quanta make up all entities.

When the system houses several indistinguishable entities, for example, a cell multiple copies of an enzyme, the probability $P_j = [_1P_j][_1P_j][_1P_j]\ldots /N_j! = [_1P_j]^{N_j}/N_j!$ of that population is a product of $_1P_j$ over the size of the population, N_j. Again, the product form ensures that if any entity is missing altogether, $_1P_j = 0$, then $P_j = 0$.

When the entities are identical, their mutual order makes no difference. Hence, the expression is divided by the number of ways, $N_j!$, the entities can be arranged into a sequence. The factorial (!) means, as an example, that three entities can be arranged $3! = 3 \cdot 2 \cdot 1 = 6$ ways.

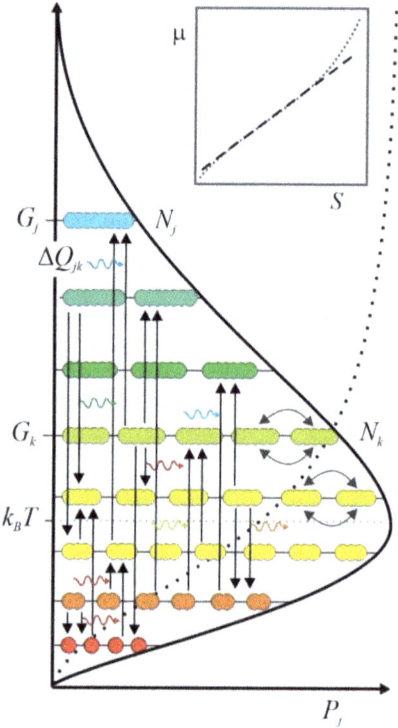

The general energy level diagram can present any system when everything comprises the same fundamental elements, known as the quanta. In a sense, the diagram is the worldview of thermodynamics. The entities of a system, in numbers N_k, with the same energy G_k, are on the same level. The bow arrows portray their mutual exchange, which changing nothing causes no change in the average energy of the system, k_BT, either. The vertical arrows indicate events moving the entities from one level to another. For example, in a chemical reaction, starting materials, N_k, transform into products, N_j. The horizontal wave arrows denote the quanta of light coupling to the transformations by entering the system from the environment or vice versa. Since the quanta carry energy, ΔQ_{jk}, all events, as flows of quanta, move the system and its surroundings toward thermodynamic balance. When the surroundings is higher in energy than the system, the system evolves toward higher average energy and the surrounding systems toward lower average energy, and vice versa. The cumulative probability distribution curve (dotted line) is a sigmoid. When its logarithm, entropy, S, is plotted as a function of (chemical) potential energy, μ, it mainly follows a power law, i.e., a straight line on the logarithm-logarithm scale (inset).

The total probability P of the system is the product Π_j over P_j

$$P = \prod_{j=1} P_j = \prod_{j=1}\left[\prod_{k=1}\phi_k\right]^{N_j} \bigg/ N_j! \tag{A1}$$

where each factor $\phi_k = N_k \exp[(-\Delta G_{jk} + i\Delta Q_{jk})/k_B T]$ denotes the population of starting materials, N_k, and the energy differences, $-\Delta G_{jk} + i\Delta Q_{jk}$, relative to the average energy of the system, $k_B T$. Temperature, a statistical concept, was taken in use long before the concept of energy, and hence multiplying T by Boltzmann's constant, k_B, makes it commensurate with the other terms of energy.

When a single event perturbs $k_B T$ only slightly, the system evolves smoothly. In a statistical system, an energy difference can be expressed as an exponential function (exp).[1] The base of the natural logarithm, the limit of continuous compounding, is a natural choice, as the function $f(x) = e^x$ is a self-similar function under a change, $de^x/dt = e^x$. The gap in energy, ΔG_{jk}, between the starting material, indexed with k, and the product, indexed with j, can be bridged with the flux of energy, $i\Delta Q_{jk}$, between the system and its surroundings that couples to a jk-transformation from the starting material into the product. The label, i, in front of the energy term means that the system is open to the surroundings for the flows of quanta. For example, the flux of photons from the Sun makes photosynthesis happen.

The state equation A1 is the main result of the thermodynamic theory: what follows is a straightforward mathematical derivation. The state of the system is customarily given by an additive, Σ, measure obtained from a product, Π, as the logarithm (ln) of the state equation A1. As an example, the product $10 \cdot 100 \cdot 1000$ can be rewritten as a sum of the logarithms of its factors $\ln(10) + \ln(100) + \ln(1000)$ and characterized by the geometric mean $(10 \cdot 100 \cdot 1000)^{1/3} = 100$. For historical reasons, the logarithm of probability, when multiplied by k_B, is known as entropy

$$S = k_B \ln P = k_B \sum_j \ln P_j \approx \frac{1}{T}\sum_{jk} N_j \left(-\Delta\mu_{jk} + i\Delta Q_{jk} + k_B T\right), \tag{A2}$$

where $\Delta\mu_{jk} = \mu_j - \mu_k$ means the potential energy difference between the populations N_k and N_j. The population of k-entities embodies the (chemical) potential, $\mu_k = k_B T \ln\phi_k$, and that of j-entities the potential, μ_j. The entry $\approx$ stands for the statistical approximation, $\ln N_j! \approx N_j \ln N_j - N_j$, which is excellent for $N_j > 10$.

It is worth emphasizing that entropy A2 is just the logarithm of probability A1. As mathematical manipulations do not change but keep conclusions within the atomistic axiom, the entropy concept adds nothing to the description beyond the energy concept. The total energy of the system, TS, temperature, T, times entropy, S, comprises the system-bound energy, $\Sigma N_j k_B T$, and free energy, $\Sigma N_j(-\Delta\mu_{jk} + i\Delta Q_{jk})$.

The equation of motion

In a statistical system, the quantum-by-quantum changes in N_j can be conveniently denoted by differentials, dN_j, to see that a free energy term, $-\Delta\mu_{jk} + i\Delta Q_{jk}$, drives a forward reaction, where N_j increases, and the opposite force drives the reverse reaction, where N_j decreases. As a result of jk-transformations, the total energy of the system, TS, changes with time, t,

$$T\frac{dS}{dt} = T\sum_j \frac{dS}{dN_j}\frac{dN_j}{dt} = \sum_{jk} \frac{dN_j}{dt}\left(-\Delta\mu_{jk} + i\Delta Q_{jk}\right).$$
(A3)

As the quanta redistribute along the gradients, dS/dN_j, temperature, T, changes as well. Since variation in T follows from variation in S, the average energy, $k_B T$, is not explicitly differentiated with respect to time.

The equation of motion cannot be solved because the changes in a population

$$\frac{dN_j}{dt} = \frac{1}{k_B T}\sum_k \sigma_{jk}\left(-\Delta\mu_{jk} + i\Delta Q_{jk}\right),$$
(A4)

proportional to the free energy terms by mechanism-dependent factors, $\sigma_{jk} > 0$, cannot be separated from their driving forces, i.e., $\Delta\mu_{jk}$ is a function of N_j.

In the scale-free description, a mechanism, such as an enzyme, is a system of its own that facilitates the consumption of free energy by speeding up the conversion of N_k into N_j or vice versa. To attain thermodynamic balance in the least time, the flows of quanta *naturally select* the most efficient mechanisms. Thus, the forces at present point to the future and transform the present into the past through various mechanisms.

When influxes of free energy fuel the growth, $dN_j/dt > 0$, and conversely when effluxes consume N_j, $dN_j/dt < 0$. Thus, at any given time, entropy cannot decrease, $dS \geq 0$, as can be seen by squaring the free energy terms, orthogonal in the jk-basis of equations A3 and A4. Every motion follows its line of force. A quantum that leaves the system will end up in the environment or vice versa because no quanta can come out of nothingness or vanish into nothingness. There is thus no exception to the second law of thermodynamics.

Along an evolutionary path from one state to another, free energy may only decrease. As there are no energy barriers to be crossed, thermodynamics and kinetics are consistent. For example, a flow of water opens when the water level rises over the spillway crest. A chemical reaction proceeds from starting materials to products when the energy of the starting materials, including chemical and kinetic energy, as well as absorbed photons, exceeds the energy of the products. Accordingly, a catalyst does not change the energy level diagram or landscape but only speeds up the conversion of starting materials into products or vice versa. As the quanta flow along the lines of force, energy differences diminish in the least time, and hence entropy does not just increase, it does so in the least time.

While the course of evolution, or any change, cannot be predicted, it can be simulated a step at a time according to equation A4 to demonstrate and model the emergence of standards, skewed divisions, growth curves, oscillations, and chaotic courses.[2]

THE CONTINUOUS EQUATION OF MOTION

Although any system evolves from one state to another quantum by quantum, many phenomena, such as the flow of water, appear as if they were continuous motions.[3] We obtain the continuous equation of motion from equation A3 in terms of continuous potentials U and Q using the definitions $\mu_j = (\partial U / \partial N_j)$ and $Q_j = (\partial Q / \partial N_j)$

$$\frac{d}{dt} 2K = -\mathbf{v} \cdot \frac{dU}{d\mathbf{x}} + i \frac{dQ}{dt}, \tag{A5}$$

where the change $TdS = d2K$, is expressed as the change in kinetic energy, K, and the change in the scalar potential, $dU/dt = \mathbf{v} \cdot dU/d\mathbf{x}$, is a short-hand notation for the change per time, $d/dt = \sum (dx_j/dt)(d/dx_j)$, equal to the change per position, d/dx_j, multiplied by the velocity component, v_j.

When equation A5 is integrated along the path of motion, Maupertuis' least action is obtained. The action adopts the least-time path of quanta, for example, an electron whose quanta form of a torus.[4] Other particles can likewise be described as quantized actions.

We can also work equation A5 from the original form of Newton's second law of motion by writing the kinetic energy, $2K$, in terms of momentum, $\mathbf{p}$, and velocity, $\mathbf{v}$, multiplying by $\mathbf{v}$

$$\mathbf{F} = \frac{d}{dt}\mathbf{p} = m\mathbf{a} + \mathbf{v}\frac{dm}{dt} \mid \mathbf{v}$$

$$\mathbf{v} \cdot \mathbf{F} = \mathbf{v} \cdot \frac{d}{dt}\mathbf{p} = \frac{d\mathbf{x}}{dt} \cdot m\mathbf{a} + \mathbf{v} \cdot \mathbf{v}\frac{dm}{dt} = -\frac{dU}{dt} + i\frac{v^2}{c^2}\frac{dE}{dt} = -\frac{dU}{dt} + i\frac{dQ}{dt}. \tag{A6}$$

Thus, the second law of thermodynamics, Maupertuis' principle of least action, and Newton's second law of motion are one and the same law, the equation of time.[3] Poynting's theorem is also the same law given in electromagnetic terms,[5] in which the work exerted by the electromagnetic forces on charges equals the change in the density of electromagnetic energy. It is noteworthy that the force, $\mathbf{F}$, also contains the change in energy, idQ, either absorbed or emitted when the system is displaced by $d\mathbf{x}$. The concomitant change in mass, dm, is big in nuclear reactions, small in chemical reactions, and always finite. In other words, mass, i.e., the curvature of quanta, changes until the system becomes stationary.

Once the net flow of energy between the system and the surroundings vanishes, the system has attained balance in its surroundings. In the equilibrium, $dQ = 0$, the equation of motion A6 reduces to $2K + U = 0$, known as the virial theorem. In the form of Noether's theorem, $2Kt = nh$, this balance shows that the steady-state system comprises a total of n quanta with kinetic energy, $2K$. They complete their full orbit in the period, t, whether in the form of an electron torus or a planet orbiting the Sun. Such stationary-state trajectories are computable.

A simple system, such as an elementary particle, allows us to weigh the theoretical results against the collected data most convincingly. In principle, we can compare the calculation with the steady-state properties of any system, such as a cell.

Still, in practice, it is challenging to glean sufficiently detailed information about a complex system, ultimately to the precision of a quantum, to make such comparisons. Therefore, we will find it more meaningful to compare the theoretical results with available statistical data. First and foremost, the thermodynamics of time explains the characteristics of the *Grand Regularity*: skewed distributions, cumulative curves, oscillations, and chaotic motions.

THE UNIVERSAL PATTERNS

The characteristic S-shape of a growth curve can be seen from equation A4. In the beginning, there is a wealth of resources, i.e., free energy for growth. So, we can *assume* that mechanisms, $\Sigma_k \sigma_{jk}$, limit the consumption of free energy, $-\Delta\mu_{jk} + i\Delta Q_{jk}$, consumption, and *define*

$$\frac{d}{dt}\frac{1}{k_B T}\sum_{k=1}\left(-\Delta\mu_{jk} + i\Delta Q_{jk}\right) = \frac{dN_j}{dt}\frac{d}{dN_j}\frac{1}{k_B T}\sum_{k=1}\left(-\Delta\mu_{jk} + i\Delta Q_{jk}\right) \approx \sum_{k=1}\sigma_{jk} \quad \text{(A7)}$$

$$\Rightarrow \frac{dN_j}{N_j} = \sum_{k=1}\sigma_{jk}dt$$

using $d\mu_j/dN_j = d(G_j + k_B T\ln N_j)/dN_j = k_B T/N_j$, as μ_k, Q_j and Q_k have no explicit but only a stoichiometric dependence on N_j. The growth by equation A7 is approximately exponential because initially, the amount of free energy seems infinite. In turn, when the free energy dwindles down toward balance, the growth decreases slowly, almost exponentially.

We see that the growth between the initial and final phase follows a power law closely by expressing N_j as the product of its constituent N_k that, in turn, is the product of the basic elements, N_1, and using the atomistic axiom $N_j = \Pi_k \phi_k = \alpha_j N_1^j$ where $\alpha_j = \Pi_{mn}\exp[(-\Delta\mu_{mn} + i\Delta Q_{mn})/k_B T]$ contains the free energy terms that force the assembly of N_j from N_1. So the change

$$\frac{dN_j}{dt} = j\alpha_j N_1^{j-1}\frac{dN_1}{dt} = j\frac{N_j}{N_1}\frac{dN_1}{dt} \quad\Rightarrow\quad \frac{dN_j}{N_j} = j\frac{dN_1}{N_1} \quad \text{(A8)}$$

when integrated, follows a power law $\ln N_j = j\ln N_1 + \text{constant}$. The power law in the continuous form, $d\ln\mathbf{p} = d\ln\mathbf{v} + d\ln m$, is apparent from Newton's second law of motion A6 dividing by $\mathbf{p}$.

When the assumption of a nearly constant change in free energy does not hold, we may *model* the change by adding the term $-\beta N_j$ to equation A7

$$\frac{dN_j}{N_j} \approx \left(\textstyle\sum_{k=1}\sigma_{jk} - \beta N_j\right)dt \Rightarrow N_j(t) = N_j(t_o)\left(\textstyle\sum_{k=1}\sigma_{jk} - \beta N_j(t_o)\right) \quad \text{(A9)}$$

where the population, $N_j(t_o)$, at a time, t_o, determines the population, $N_j(t)$, at a later time, t. According to this model,[3,6] evolution is almost predictable when the change in free energy, $|-\Delta\mu_{jk} + i\Delta Q_{jk}|/k_B T << 1$, is small compared with average energy. And when not, oscillations and chaos occur, for example, when a solid-state laser is turned on, when the animal population proliferates and exceeds the environment's carrying capacity, or when the banks need more money than is available.

Logarithmic, exponential, and truncated distributions and their power-law-like cumulative distribution functions are mathematical models of the physical processes given by equations A3 and A4. The models allow us to describe and categorize various data, but they do not explain how the data came about; that is, causality. It is worth emphasizing that equations A3-A6 cannot be solved, except at balance, because the variables cannot be separated. It means that a chain of events is fundamentally unpredictable, not due to the complexity of a system or ambiguity in its initial conditions, but due to mutual dependencies.

When the system evolves gradually, i.e., the change in energy is small compared with the average energy, $|-\Delta\mu_{jk} + i\Delta Q_{jk}|/k_B T << 1$, the variation, n, is small, $n << j$, around a representative, mean, or an average factor, ϕ_j. So the natural distribution of factors, given in logarithmic terms, $\ln\phi_j = j\ln\phi_1$, of the elemental factor, ϕ_1,

$$\ln\phi_{j-n...j+n} = \ln\phi_j + \sum_n n\ln\phi_1 \tag{A10}$$

is approximately logarithmic. The typical form, j, can be recognized in each member within the distribution $j \pm n$. For example, all sized Northern pike looks like pike and not bream. On the other hand, if the weights of pike and, say, cars were presented in the same graph, we would see two distributions: one about a typical pike and the other about an average car. Spirals, such as shells, cyclones, and galaxies, are also approximately lognormal distributions in polar coordinates, i.e., energetically optimal shapes.[3]

In summary, statistical physics of open systems, the thermodynamic theory of time, explains processes as flows of quanta. As the quanta carry both energy and time, the irreversible consumption of free energy sets the arrow of time. The correspondence between the theory and data implies that every process seeks thermodynamic balance in the least time. The maxim of least-time free energy consumption is inherent in the quantum itself: Planck's constant, $h = Et$, determines the change in energy over time, i.e., the power, $dE/dt = -E/t = -\mathbf{F} \cdot \mathbf{v}$, as motion with $\mathbf{v}$ in the direction of $\mathbf{F}$.

APPENDIX B: STRUCTURES OF SUBSTANCE

Modern physics maintains that particles are quantum fields. By contrast, ancient philosophy posits that the substance and the void are made of basic building blocks. The elemental constituent, the quantum of action in its most familiar form, is the quantum of light, the photon. The quantum is a string with energy on its period of time, momentum on its wavelength, and angular momentum on its revolution.[4,7,8]

The geometry of a string of quanta that makes up an elementary particle manifests itself in three ways: the helical coiling as an electric charge, the curl as a magnetic moment, and the curvature as mass. These particle properties give rise to electromagnetic and gravitational fields in the surrounding vacuum of quanta.

THE VACUUM

When the quanta of a particle transform into the quanta of the vacuum, e.g., in a nuclear reaction, the famous formula $E = mc^2$ relates the vacuum energy, E, with the particle mass, m. The same applies to the opposite reaction where the vacuum quanta transform into the quantized matter. The equation, when multiplied with the period, t, sums the total quantized action of a system, say a particle, from the momentum, $\mathbf{p}$, on the orbit, $\mathbf{x}$,

$$nh = \int E\,dt = \int 2K\,dt = \int \mathbf{p}\cdot d\mathbf{x}, \qquad (B1)$$

where n is the number of quanta, h is Planck's constant, and c is the speed of light. This integrated form of equation A5 is Maupertuis' action. In balance, the total action of the system, i.e., the kinetic energy, $2K$, on the period, t, is constant. The definition of the speed of light, $c = E/p = dx/dt$, can be understood by equation B1 to follow from Planck's constant, the photon measure.

The quanta of the void condensed around a particle constitute the particle's gravitational field. So, the quanta of the free space make the gravitational field of all bodies, the universal gravitation. The balance between all matter and the void means that the vacuum energy density, about $0.55\cdot10^{-9}$ J/m³, is equal to the average energy density of matter, $\rho_M c^2$, in the whole universe.[9] The mass of the expanding universe, M, within its radius, $R = ct$, of about 14 billion light years, defines the average density by the law of gravitation

$$Mc^2 = \frac{GMM}{R} \Leftrightarrow c^2 = \frac{GM}{R} \Leftrightarrow \varrho_M = \frac{1}{4\pi Gt^2}, \qquad (B2)$$

about $0.6\cdot10^{-26}$ kg/m³, as the critical energy density, $\Omega = c^2/4\pi Gt^2$, where G is the gravitational constant. The vacuum characteristics G and c are thus not constants but variables that change as matter transforms into the vacuum.

The photons in out-of-phase pairs make up most of the vacuum.[7] Even though not displaying a net electromagnetic field, they couple to particles. The coupling strength is called mass. We experience the coupling to the local surroundings as the Earth's gravitational field and to the universe's gravitational field as inertia, i.e., a reaction to acceleration and deceleration.

According to the Stefan–Boltzmann law, the fraction of unpaired photons in the void is only about 0.1 per mil, as the energy density of the universe, in the form of light, $u = 4\sigma T^4/c$, is about $5\cdot10^{-14}$ J/m³ at temperature $T = 2.725$ K above absolute zero. Planck's law likewise governs the paired photon energy density, the void's primary substance (Appendix H).

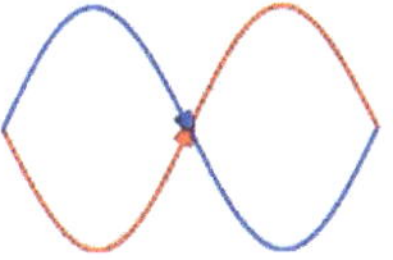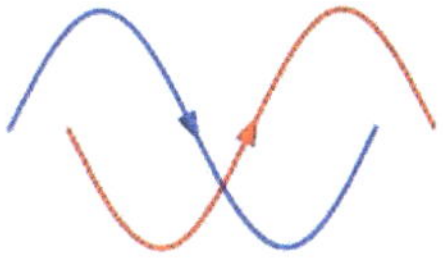

When two photons co-propagate out-of-phase (left), the pair is not visible as light. As the photon pairs do not couple to matter electromagnetically, they manifest themselves only as gravitational and inertial effects and interference phenomena. When the phases are not opposite (right), the photons are detected as electromagnetic radiation, e.g., visible light.

The paired-photon vacuum, embodying the four-potential of the electric scalar potential, φ, and magnetic vector potential, $\mathbf{A}$, satisfies the Lorenz gauge,

$$\nabla\varphi + \frac{\partial\mathbf{A}}{\partial t} = \frac{1}{c}\frac{\partial\varphi}{\partial t} + c\,\nabla\cdot\mathbf{A} = 0 \,, \tag{B3}$$

as a seemingly continuous and indestructible substance whose wave nature gives rise to interference phenomena. For example, in the Aharonov-Bohm experiment, a phase difference,

$$\Delta\phi = \phi_1 - \phi_2 = \frac{e}{\hbar}\int \mathbf{A}_1\cdot d\mathbf{x}_1 - \frac{e}{\hbar}\int \mathbf{A}_2\cdot d\mathbf{x}_2 \,, \tag{B4}$$

develops between waves when charges, e, traverse $\mathbf{A}_1$ and $\mathbf{A}_2$ along paths $\mathbf{x}_1$ and $\mathbf{x}_2$.

Gradients in the vacuum are fields. The electric field,

$$\mathbf{E} = -\nabla\varphi - \frac{\partial\mathbf{A}}{\partial t}\,, \tag{B5}$$

for example, encircling an electron is a gradient in the winding of paired-photon rays that tallies the electron winding number, a topological quantum number. Therefore, it takes the 4π, rather than only 2π, rotation of the electron, e⁻, to return the vacuum to its original state. This SU(2)$_L$ symmetry is disclosed by monodromy of the vacuum around the electron, $e^- = \int\rho dV = -\varepsilon_\circ\int\nabla\cdot\nabla\varphi dV = -\varepsilon_\circ\int\nabla\varphi dS$ enclosed in a volume, V, equivalently covered by an area, S, as denoted by Gauss' law.

As the paired-photon rays wind up to counteract introduced charges, the vacuum polarization is similar to the polarization of a dielectric material. Conversely, when charges neutralize, the unwinding of the paired-photon rays gives rise to a time-varying electric field, i.e., displacement current, $\mathbf{J}_D = \varepsilon_\circ\partial_t\mathbf{E}$.

The magnetic field,

$$\mathbf{B} = \nabla\times\mathbf{A}, \tag{B6}$$

is a vortex of the twists in the paired-photon rays, for example, encircling a line of moving charges. The twist in the vacuum rays due to the electron structure rotates photon polarization by the angle equal to the fine structure constant.[10]

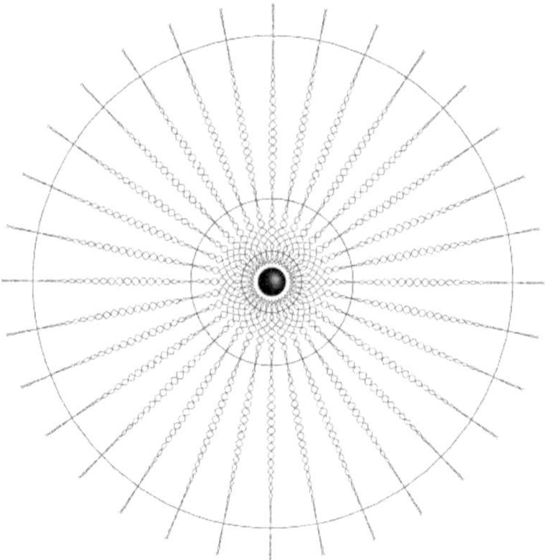

The electric potential diverges from a current-carrying wire (black, pointing straight at) as paired-photon rays whose winding matches the winding number of the charge density along the wire and decays inversely with distance, r. The magnetic field lines correspond to the lines encircling the wire at a constant winding.

The electromagnetic fields (Eqs. B5 and B6) give rise to the Lorentz force,

$$\mathbf{F} = q\mathbf{E} + \mathbf{v} \times \mathbf{B}, \tag{B7}$$

experienced by a charge, q, moving with velocity, $\mathbf{v}$. Thus, it is not that charges themselves would attract or repel each other instead, the charges move as they couple with the vacuum that is leveling off its gradients.

The gradient in the gravitational four-potential of the scalar, Φ, and vector, $\mathbf{D}$, components, is the gravitational field, i.e., acceleration,

$$\mathbf{\Gamma} = -\nabla\Phi - \frac{\partial \mathbf{D}}{\partial t}. \tag{B8}$$

For example, at a distance, r, from a body of mass, M_o, the density gradient of the paired-photon potential, $\Phi = GM_o/r$, tallies the geodesic curvature, the Euler characteristic, χ, proportional to M_o. The density gradient is observed as the bending of light rays, a gravitational time delay, and a gravitational frequency shift, $f_e/f_o = 1 + z = (1 - GM_o/c^2 r_e)^{-1/2}$, between emission, f_e, at the radius, r_e, and absorption, f_o, at the detection. In turn, the leveling of a density gradient is detected as the Doppler shift of light emitted by the moving body coupled to the moving vacuum.

Similar to $\mathbf{B} = \nabla \times \mathbf{A}$ (Eq. B6), the rotational part of the gravitational field,

$$\mathbf{\Omega} = \nabla \times \mathbf{D}, \tag{B9}$$

for example, at a distance, r, from a body with inertia, I, rotating with the angular velocity, ω, revolves at the rate, $\boldsymbol{\Omega} = GI/c^2r^3\nabla \times (\boldsymbol{\omega} \times \mathbf{r})$. This vortex in the paired-photon density displays itself as frame-dragging precession.

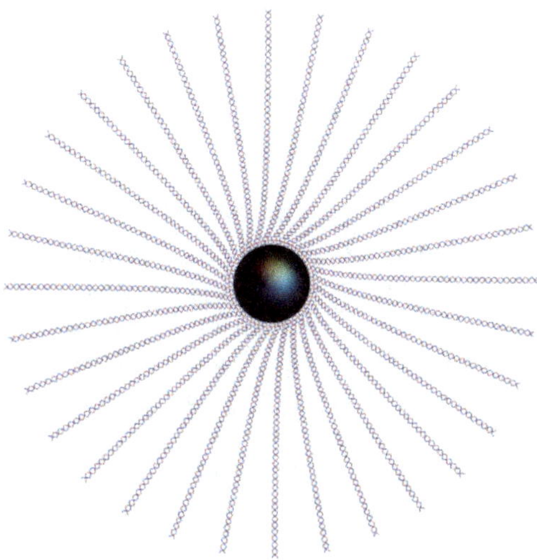

The gravitational potential of a body (black sphere) spreads out as paired-photon rays whose energy density, i.e., frequency, decreases inversely with distance, r. The paired-photon rays are dragged around a rotating body..

Moreover, similar to Gauss' law, the divergence of the density gradient in the vacuum potential relates to the mass density, $\rho_M = -(1/4\pi G)\nabla^2\Phi$. Expressly, $\nabla\cdot\nabla\Phi = \nabla\cdot\mathbf{a}_R = \partial_t GM/R^2c = -1/t^2$, of the all-embracing gravitational potential, Φ, due to all mass, M, of the expanding universe of radius, $R = ct$, relates inversely to the age, t, squared using the mass-energy equivalence, $Mc^2 = GM^2/R$. In other words, the average mass density of the universe, $\rho_M = 1/4\pi Gt^2$, decreasing with time, t, is the source of space embodied by the paired photons.

THE ELECTRON

Ampère proposed that the electron is a circular helix. Parson and Compton thought likewise,[11] but for them, the number of threads in the torus remained a mystery. It can be deduced from the action of the electron and the quantum of action h, whose ratio is known as the fine-structure constant[12]

$$\alpha = e^2 Z_o/4\pi\hbar ,\qquad\qquad (B10)$$

where e is the electric charge and Z_o is the vacuum impedance. The inverse $1/\alpha$, approximately 137.036, expresses the ratio of the length of the torus to one of its loops. The arc length $[x'(t)^2 + y'(t)^2 + z'(t)^2]^{1/2}$, given in terms of differentials $x'(t)$, $y'(t)$, $z'(t)$, can be calculated from the parametric form of the toroidal curve[13]

$$x(t) = \cos(t)[R + r\cos(nt)]$$
$$y(t) = \sin(t)[R + r\cos(nt)] \qquad\qquad (B11)$$
$$z(t) = r\sin(nt)$$

as an (elliptical) integral. The total length, $2\pi rn[1 + ((R/r)/n)^2]^{1/2}$, suggests by the approximation, $1/\alpha \approx 137[1 + (\pi/137)^2]^{1/2} \approx 137.036$, for the perimeter-to-loop ratio $R/r = \pi$, that the electron comprises $137 + 1$ loop quanta, the one accounts for the pitch, dz/dt. In other words, due to its pitch, the torus is slightly longer than an equivalent array of stacked loops. Each quantum joins the next one in a slightly earlier position along the screw thread.

The electron magnetic moment, approximately $\mu_e = \mu_B(1 + \alpha/2\pi)$, is anomalous, i.e., slightly larger than the moment of a plain circumference, Bohr magneton, $\mu_B = e\hbar/2m_e$ because of the thread's gradual rise at a small angle, $360/137$ °. The vacuum quanta curling around the torus manifest themselves as the magnetic field.

The ratio R/r of the large and small radii can be easily determined to a precision of over ten decimal places because R depends mainly on the magnetic moment, i.e., the curl of the curve and r on the fine-structure constant, i.e., the length of the curve. By contrast, the corresponding calculation is demanding with quantum electrodynamics,[14] because the magnetic moment anomaly is obtained as a series expansion of α with a large number of terms.

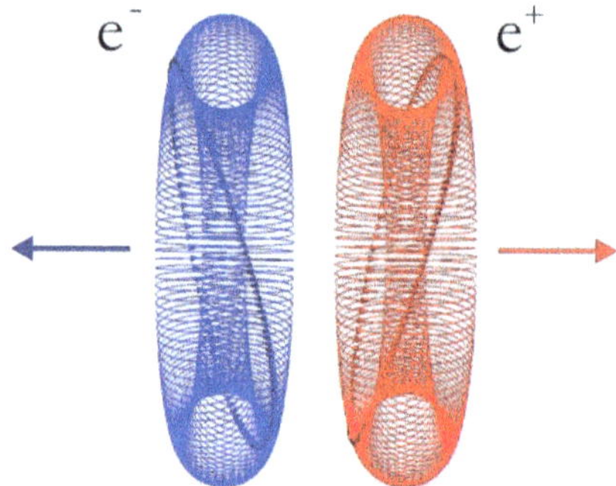

The electron (blue) and the positron (red) are tori comprising 138 quanta (lines). In the drawing, for the sake of clarity, the ratio of the circumference to that of a loop departs from the actual value ($\sim\pi$). The opposite helical windings of the electron (e⁻) and the positron (e⁺) manifest themselves as the opposite electric charges. The opposite circulations generate opposing magnetic moments (arrows). The geodesic curvature of the torus relative to the flat vacuum displays itself as the particle mass (m_e). The electron is matter, the positron antimatter. At annihilation, the two tori open up, unthreading their quanta in pairs among those of the surrounding vacuum. Besides the paired photons, two photons of the opposite phases propagating in opposite directions balance the disappearance of the two opposite charges.

When the electron and positron open up and annihilate each other, the quanta of the tori become the quanta of the vacuum. The open tori can also combine to form the Z boson, a short-lived particle that mediates weak nuclear force. Besides decaying, the Z boson may also transform back into the electron and positron.[15]

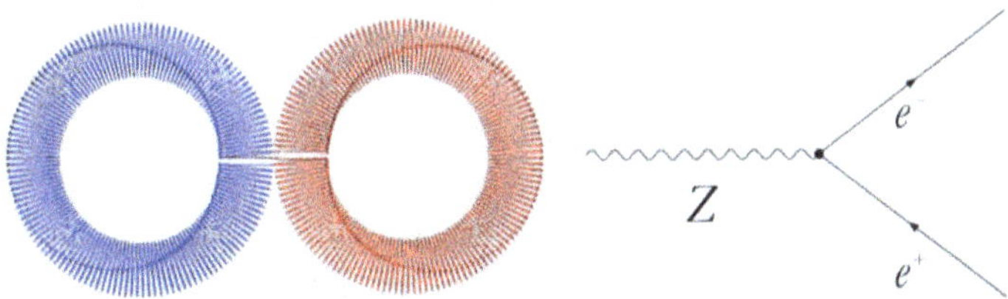

The Z boson consists of two open tori of opposite handedness (blue and red) that connect with the short-wavelength photon (gluon, not shown) (g). The Z boson is thus its own antiparticle that can decompose, for instance, into the electron (e⁻⁻) and the positron (e⁺⁺), as depicted in the Feynman diagram (right).

Since the down quark charge is $1/3$ of the electron charge, we can think of it as $1/3$ of the electron torus, i.e., an arc of 46 quanta. Similarly, we can reason that the up quark consists of 92 quanta because its charge is $2/3$ of the positron charge. Moreover, by considering the anti-down quark as $1/3$ of the positron and the anti-up quark as $2/3$ of the electron, we can deduce the structures of many particles.

Mass, as a measure of the particle-vacuum coupling, can be computed from the particle and the vacuum structures. At its simplest, mass is proportional to the curvature characteristic, $\chi = \int k_g \, d\gamma$, where the geodesic curvature

$$k_g = \mathbf{n} \cdot \frac{\gamma''}{|\gamma'|^2} \times \frac{\gamma'}{|\gamma'|} \tag{B12}$$

is calculated from a string of photons, γ, by projecting the cross-product of tangential velocity, γ', and its change acceleration, γ'', onto the universal surroundings defined by the normal, $\mathbf{n}$. The gentle curvature of the universe shows up only in the electron neutrino's non-zero mass. The curvatures can be put on a scale of a known mass, for example, that of the electron, $m_e = 9.11 \cdot 10^{-31}$ kg. When the mass is small, such as m_e, the particle couples only weakly to the vacuum's paired quanta. The particle therefore readily turns in the direction of the magnetic field.

There is a natural connection between a particle's electric charge, magnetic moment, and mass, as these properties are geometric characteristics of the same curve, i.e., the same particle. As best, the helical winding, curl, and curvature of a model are computed numerically. However, it is good to know an approximate way, too. To this end, the 'bare' quantum mass $m_1 = 137 m_e / 138\alpha = 69.518$ MeV can be inferred from the electron mass, $m_e = 0.511$ MeV, given that one quantum spans $1/138$ part of the 137-loop torus. This line of reasoning was pursued first by Werner Heisenberg and later by Yoichiro Nambu.[16]

With m_1, the mass of a particle can be calculated 'by hand' from a wireframe model quantum-by-quantum. In the simplest case, the two loops of toroidal arcs are parallel, i.e., the angle between them $\theta = 0$, having opposite phases, i.e., $\varphi = \pi$. Then, the two loops on the opposite sides amount to $2m_e/138$. By contrast, the vacuum strongly couples to perpendicular loops. In general, when $\theta \neq 0$, the mass is about $m_1(1 - \cos^2\theta)$, plus $\varphi \neq \pi$ contributing about $m_1^{1/2}[1 - \cos(\varphi - \pi)]$. In this manner, the approximate masses of the ordinary particles can be calculated from the wireframe models and compared to the measured masses.

THE π MESON

A meson consists of a quark and an antiquark connected by a gluon (g). For example, the Pi meson (π^-) consists of down quark (d) and anti-up quark (u*). While its mass, $m_\pi = 139.57$ MeV, can be summed from the curvatures along the quarks and gluons, it is insightful to infer it also approximately from the wireframe model.

Since the quanta of u* in the range from 1 to 23 and from 70 to 92 have opposite curvature, apart from the helical pitch, they contribute to the meson mass only by $^1/_3 m_e$. For the rest of the quanta, too, the toroidal pitch gives $^2/_3 m_e$. Most mass is thus due to offsets in phase, i.e., u* ending a half-turn before the gluon and d starting a half-turn after the gluon. So, while parallel, the 24th quantum of u* and the 1st quantum of d are in the same phase $\varphi \approx 0$, and hence give m_1. Likewise, the 69th quantum of u* and the 23rd quantum of d are in the same phase $\varphi \approx 0$. Thus, the total mass $2m_1 + m_e = 139.55$ MeV explains the measured mass of π^-.

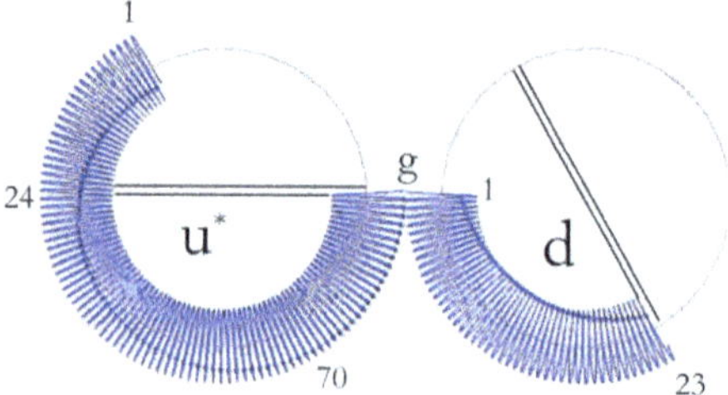

In the π^- meson, the anti-up quark (u*) links via the gluon (g) to the down quark (d). The mass mainly stems from the 24th quantum of u* and the 1st quantum of d and likewise from 69th quantum of u* the 23rd quantum of d, which, while parallel, are not in the opposite but in the same phase.

From the particle structures, we can grasp the reaction $\pi^- \to W^- \to e^- + \nu_e^*$. First, the d quark revolves about the gluon and meets the beginning of the u* quark. Then, this newly formed W^- boson closes as the electron e^-, while g loops out as the antineutrino ν_e^*.

The quarks of π^-, as well as those of other pseudoscalar mesons, are on a plane. By contrast, mesons whose quarks are on the two faces of the tetrahedron are referred to as vector mesons. For example, in the Rho meson (ρ^-), the arcs of the down quark (d) and the anti-up quark (u*) are at a 60 ° angle. Therefore, the total curvature of ρ is a lot larger than that of π; hence, m_ρ is much larger than m_π. Mesons are short-lived, like other particles with open structures.

The vacuum's quest for low-energy states manifests itself in pairwise complementary structures (molecules) of mesons as well as of other particles. For the same reason, the opposite charges and magnetic poles complement each other.

THE PROTON AND NEUTRON

Three quarks can combine via gluons to a closed string only when on the three faces of a tetrahedron. For example, the proton has two up quarks and one down quark, and the neutron has two down quarks and one up quark on the three faces of the tetrahedron. The proton charge, $q_p = {}^{+2}/_3 + {}^{+2}/_3 + {}^{-1}/_3 = +1e$, and the neutron charge, $q_n = {}^{+2}/_3 + {}^{-1}/_3 + {}^{-1}/_3 = 0$, are the sum of the charges of their quarks.

From the proton and neutron structures, we can imagine how the vacuum quanta, whirling around the arcs of quarks, manifest as a magnetic field. The magnetic moment denotes this particle's capacity to generate a magnetic field around it. In units of the nuclear magnetons, μ_N, the proton and neutron magnetic moments are $\mu_p = 2.79\mu_N$ and $\mu_n = -1.91\mu_N$.

The magnetic moment of a particle can be calculated from the geometric model in the same way as the moment of an electric circuit is derivable from the shape of the wire loop with Stokes' theorem. As the proton quarks encircle a larger area than the neutron quarks, the ratio of the two magnetic moments is about -1.46. As the result agrees with the measurements, the models explain the proton and neutron magnetic moments.

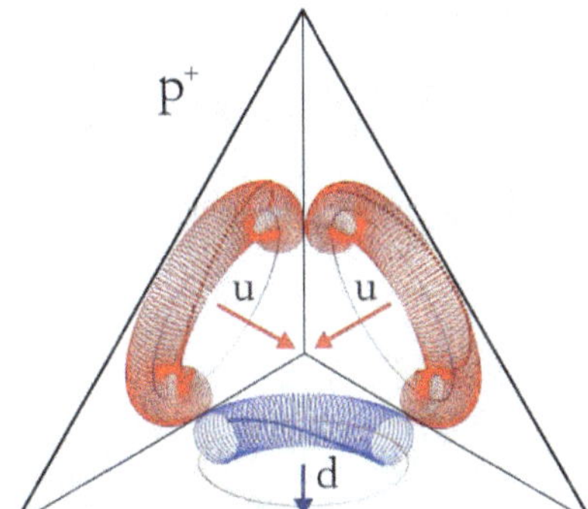

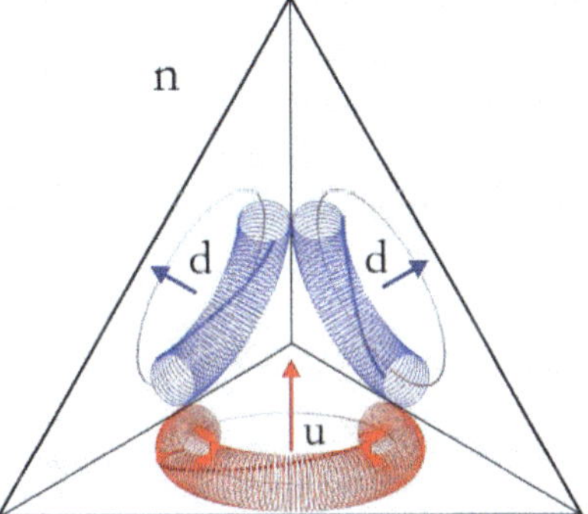

The up (u) and down (d) quarks of the proton (p⁺) and neutron (n), as arcs of the tori, are on the three faces of the tetrahedron. The vacant face points to the viewer's line of sight. The vacuum quanta curl about the quarks. The arrows indicate this net magnetic effect of the vortices. In the picture, the sum of proton quark vorticity points toward the bottom of the page and slightly away from the viewer $(^{-2\sqrt{2}}/_3, 0, ^{-1}/_3)$. In the neutron, the sum points toward the top of the page $(^{2\sqrt{2}}/_3, 0, 0)$. The proton magnetic moment is bigger (by -1.46) than the neutron moment because the area circumscribed by the proton quarks is larger (by 1.38) than that of the neutron. The area can be added up from the quark-enclosed circular segments and the triangle in between. The proton mass and the neutron mass are nearly the same because the geodesic curvatures of the two structures are almost identical.

The proton and neutron masses $m_p = 938.27$ MeV and $m_n = 939.57$ MeV can be calculated as outlined above. In the up quark, the quanta from 1 to 23 and from 70 to 92 are opposite (except for the helical pitch), contributing by $^1/_3 m_e$. In the tetrahedron, the angle between the two quarks spans from 0 ° through 45 ° and back to 0 °. From 0 ° to 45 °, the mass sums up loop by loop approximately to $4.5 m_1$ using the above formula $m_1(1 - \cos^2\theta)$. The proton mass is thus $3 \times 4.5 m_1 + {^5/_3} m_e = 938.39$ MeV.

The curvature of a wireframe model explains the particle mass, but not at once that the neutron is slightly more massive than the proton. Electrical forces between the quarks could cause subtle distortions in the structures, manifesting as small deviations between the measured and calculated masses and magnetic moments.

MODELS OF PARTICLES

While the calculations using quantum field theory agree with the properties of many particles, "there is no theory that adequately explains these numbers. We use the numbers in all our theories, but we don't understand them – what they are or where they come from. I believe that from a fundamental point of view, this is a very interesting and serious problem."[17] At last, this need for understanding, expressed by Feynman, can be satisfied. The wireframe models make sense of the particle properties and their reactions. Quantum field theory aims in the same but backward fashion, trying to deduce the properties of a particle from its field.

It is easy to determine the coil and curl of a particle structure, i.e., charge and magnetic moment. By contrast, the geodesic curvature, i.e., mass, is rather tricky to calculate. For example, two ρ^+ (ud*) vector mesons pair as an energetically favorable 'molecular' structure. Likewise, the psion J/ψ (cc*) dimer is a tetrahedron. Moreover, a varying particle conformation resembles a vibrating molecule. For example, the Δ^{++} (uuu) resonance properties average over the conformations.

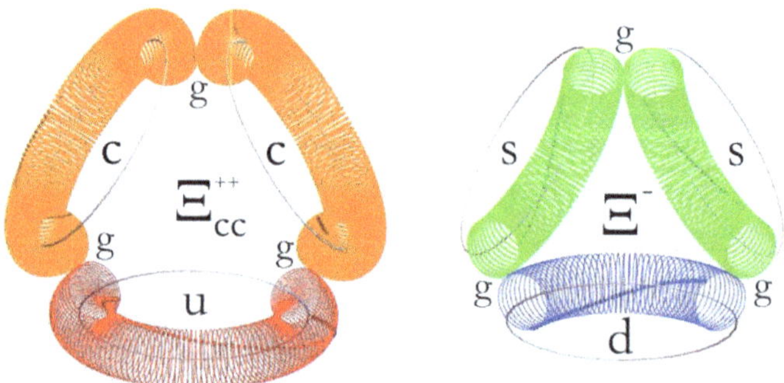

A baryon consists of three quarks and three gluons (g). For example, Ξ_{cc}^{++} comprises two charm (c) quarks and one up (u) quark, and Ξ^- two strange (s) quarks and one down (d) quark. The charm (orange) quark is like the up (red) quark and the strange (green) quark is like the down (blue) quark, but quanta are more curved.

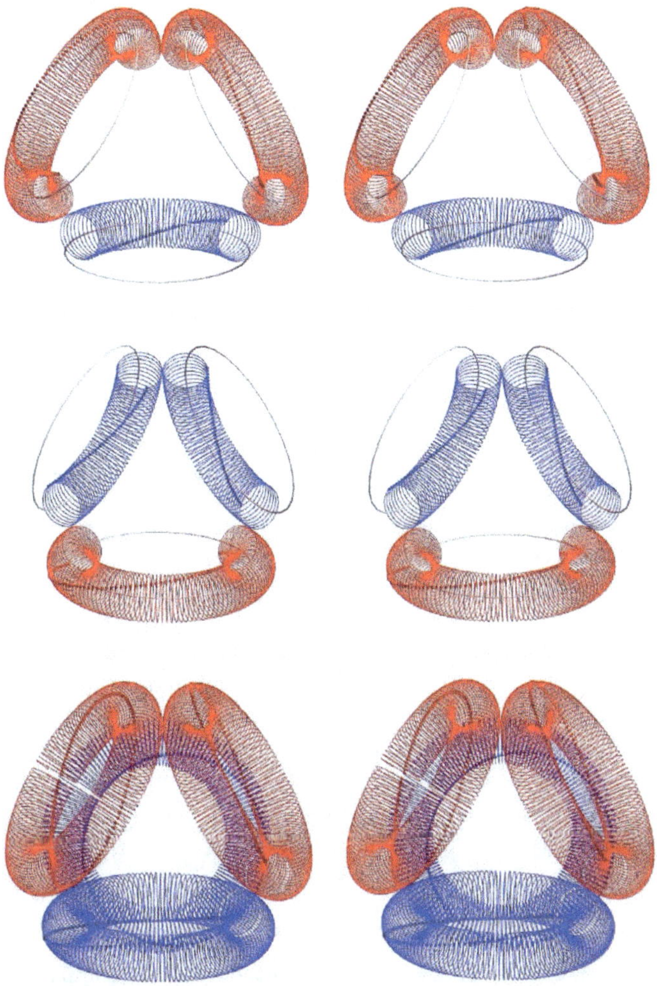

Stereoviews of the proton (top), neutron (middle), and Higgs particle (bottom) models become vivid when the left eye looks at the left image and the right eye looks at the right image. It takes some training to point each one of your eyes on only one image. You can also view the models cross-eyed to see them from the opposite direction. In that case, it helps to focus first on your finger, placed halfway between your eyes and the images.

Regularities in hadron masses stem from constituents and their conformations. Since the elemental constituent of the 2nd flavor is curved, and that of the 3rd flavor even more curved, their quarks are heavier than the 1st flavor quarks. For example, while the up and down quarks without much inner curvature contribute equally, the charm quark (c) adds to the mass a lot and much more than the twice as short strange quark (s). Minor mass differences follow from differing quark conformations about the gluon, i.e., planar, tetrahedral, and resonances between them. The contributions of constituents and conformations to the total curvature can be deduced from the measured masses.

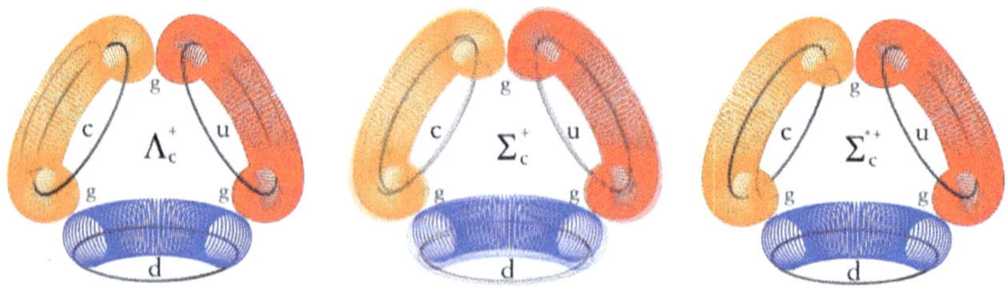

Three baryons Λ_c^+, Σ_c^+, and Σ_c^{*+}, comprising the same three quarks (udc) connected by gluons (g), differ by mass because they differ by conformation. Λ_c^+ (2286) is the ground state with angular momentum, $J = \frac{1}{2}$. Σ_c^+ (2453) is an excited state with $J = \frac{1}{2}$, whose quarks resonate between two modes, above and below the plane of gluons. Mass is higher because the quark curvatures in the transient planar conformation do not cancel each other so well as in the ground state. Σ_c^{*+} (2518 MeV/c^2) is another high-curvature state having one quark (c) in the opposite mode compared to Λ_c^+, hence $J = \frac{3}{2}$.

Subtle structural details cause high masses when the vacuum quanta sharply curve about them. While the 2nd and 3rd flavor 'bare' quantum masses m_2 and m_3 can be deduced from measurements, it is still hard to calculate the W$^\pm$, Z, and Higgs boson masses. Small changes in the narrow slots of their tori have large effects.

In practice, 3D printed particle models help to grasp magnetism, packing, and reactions in the atomic nuclei and the stellar cores, as well as correlations between protons and neutrons.[18]

APPENDIX C: WAVE-PARTICLE DUALITY

The buildup of an interference pattern from independent particle impacts is thought to be impossible to describe in any classical way. This standard interpretation of quantum mechanics presumes that the surrounding vacuum plays no part. However, vacuum waves rising from the particle propagation could reflect back to the particle.

THE DOUBLE-SLIT EXPERIMENT

The double-slit experiment is one of the experiments that led to the extraordinary ideas of modern physics. Yet, the experiment is ordinary enough to be done at home. Shine a laser pointer on three patent lead pencil fillings placed parallel next to one another, and a diffraction pattern shows up. YouTube also holds other instructions on how to carry out the double-slit experiment. For example, Barry Fleagle demonstrates how to question the Copenhagen interpretation.

To challenge the standard interpretation, only one photon at a time should go through the slits. Then it is not obvious why the diffraction pattern emerges. An ordinary laser pointer of one milliwatt power sends about 10^{15} photons per second. That is a lot, but even so, there is not much traffic in the slits because the wavelength of the visible light is short. Curiously, even when only a single photon at a time passes through the slits, the standard calculation of the diffraction pattern expressly assumes that the slits are flooded with light. Feynman pointed out this clash between reality and calculation: "Of course, actually there are no sources at the holes. In fact, that is the only place that there are certainly no sources. Nevertheless, we get the correct diffraction pattern by considering the holes to be the only places where there are sources."[19] The counterintuitive instrumentalism works because the vacuum bristles with photons but in pairs.

We can also calculate the diffraction pattern by drawing a spiral named after the French physicist Alfred Cornu.[19] Through that curve, a straight road or railway track can be led to an arc of a circle so that the centrifugal force does not jerk those on board unpleasantly. The motion gradually changes relative to the vacuum.

The propagating photon causes the vacuum to ripple. After the slit, it becomes subject to the vacuum waves that went through the other slit. Because the motion and the force affect each other, individual events cannot be calculated exactly or predicted deterministically, only modeled and simulated.

While quantum mechanics is an excellent model of a large number of events, its random variable statistics is not a realistic model of any single event. The particle probability distribution, the wave function, collapses suddenly to a single value at the event of observation. Afshar questioned this Copenhagen interpretation by showing that the photons already interfere before the detector.[20] The interference pattern did not disappear by placing metal wires at the points where the light waves through different slits negated each other's effects.

In line with these observations, coherent sources generate a wave pattern across the vacuum, not first at a detector screen. Such a broad structure would be perturbed, for example, with a narrow obstacle placed at a crest halfway to the screen introducing further splittings in the diffraction pattern.

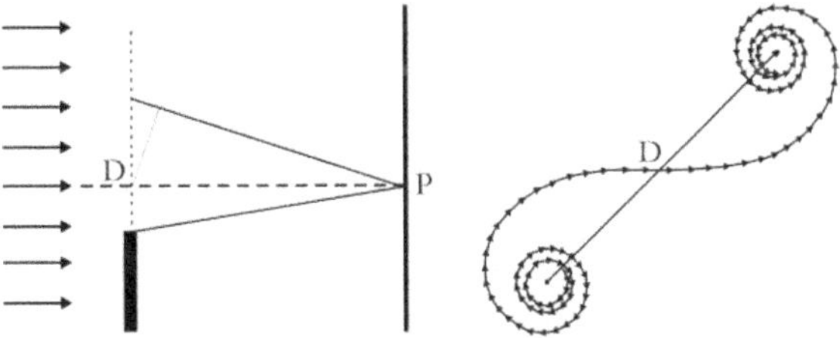

When light rays come from the left and pass by the edge of a block (black bar), a diffraction pattern appears on the screen (in the middle). The sum of all rays that hit, for example, point P can be calculated graphically (right). The ray straight to P, through point D, corresponds to the small horizontal arrow. The ray slightly above D travels a slightly longer path to P; hence, it is slightly off-phase to the straight ray. The more the direction deviates from the horizontal, the larger the phase difference. The sum of rays at P corresponds to the line through D from the lower spiral to the upper one.[19]

TWO-PHOTON INTERFERENCE

In 1987, Chung Ki Hong, Zhe Yu Ou, and Leonard Mandel directed two photons perpendicularly to hit a semi-transparent mirror simultaneously.[21] Remarkably, both photons, one transmitted and the other reflected, ended up on the same side but never on the opposite sides.

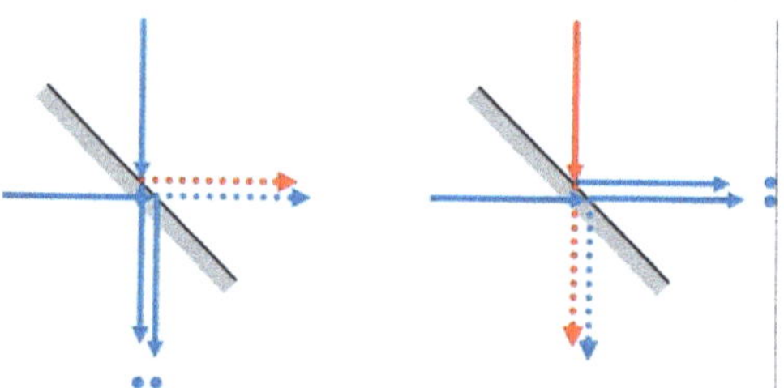

A single photon (solid arrow) can just as well go through a beam splitter (gray bar) as to reflect. However, when two arrive from opposite sides at the same time, either in-phase (blue) or out-of-phase (red and blue), the course of events is different. When one photon is transmitted, the other is reflected, or vice versa. So, the two photons always emerge parallel only on the same side of the beam splitter, never on the opposite sides.

According to quantum mechanics, the four scenarios for the photon passage are equally possible. But as the beam splitter does not store information about what happened, the four possibilities add up to unity giving the observed outcome.

Thermodynamic theory questions such an interpretation.[22] Instead, the surrounding paired-photon vacuum, say, evanescent fields, at the beam splitter forces the two photons to depart on the same side. If not, the photons would be propagating without fields, i.e., associated vacuum waves.

A vacuum wave produced by one photon interfering with another photon would contest wave-particle duality. The vacuum wave and the other photon could be tuned to arrive simultaneously at a beam splitter by observing two photons like Hong, Ou, and Mandel did.

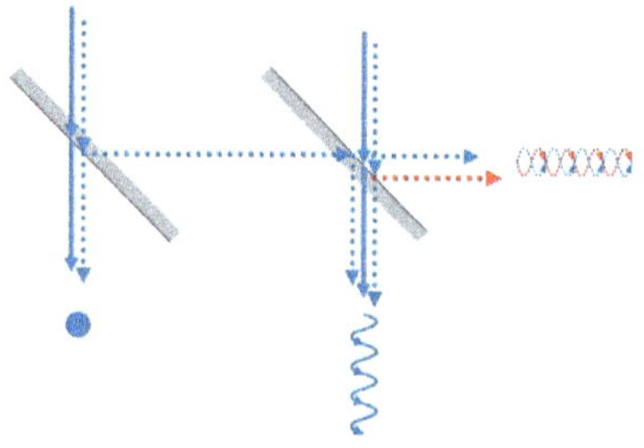

First, a photon (solid arrow on the left) goes straight through a beam splitter (gray bar) to detection (sphere). Then, a second photon (solid arrow on the right) hits a second beam splitter exactly when the first photon, had it been reflected, would have also hit it. An interference pattern emerging from such events suggests that the second photon interferes with vacuum waves (dotted arrow) generated by the first photon.

When an interferometer is perfectly aligned and adjusted, the vacuum waves interfere constructively on one port and destructively on the other. In turn, slightly off-set optical path lengths produce opposite patterns of concentric bright and dark fringes as the waves arrive out of step. Slightly misaligned mirrors reflect patterns with stripes, similar to a slit experiment, where the waves interfere off-parallel and out-of-step.

Quantum mechanics calculations reproduce the patterns. However useful of a model, a single-photon wave function spread on the two paths is not a verifiable concept because the direct measurement reveals the whole photon, not its superposed components. Moreover, the single-photon interference is predicated on Bohr's complementarity principle of two indistinguishable paths. However, momentum on one path is orthogonal to the other. These theoretical troubles come to nothing, realizing that the photon interferes with the vacuum waves it produces. Indeed, the photon-caused waves can be detected in the absence of the photon.[23]

Appendix D: Quantum Entanglement

When the measurement of one particle seems to determine the measurement of the other one, although no interaction between particles is possible, the quantum states of the particles are said to be entangled.

Consider a simple experiment where two photons with opposite phases go off in opposite directions. When the phase of one photon, i.e., the direction of oscillation in the electromagnetic field, is recorded, then the phase of the other one is immediately known to be the opposite. By common sense, there is nothing strange about this because the two photons were expressly out-of-phase to start with. Their phases were just not yet related to the detector phase. By contrast, according to quantum mechanics, the photons have no phase whatsoever before the detection.

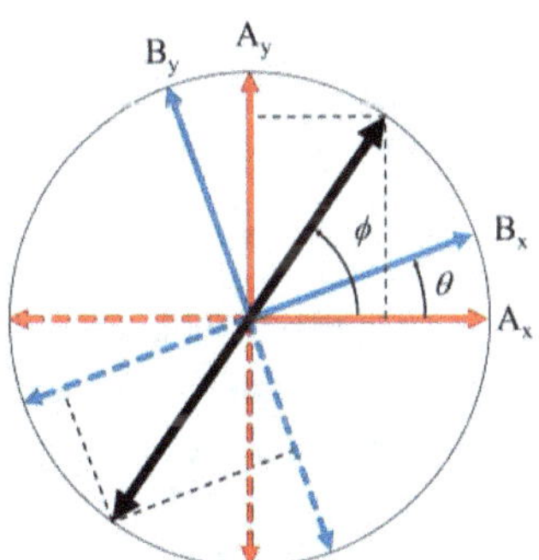

Entanglement is supposedly witnessed in measurement in which each of two photons out-of-phase (black arrows) is captured by a two-channel detector (A and B). The channels are perpendicular ($A_x \perp A_y$ and $B_x \perp B_y$) and the corresponding channels (A_x and B_x, B_y and A_y) at an angle θ. The probability of one photon with phase ϕ entering the channel A_x is $\cos\phi$ and the channel A_y $\sin\phi$, and the probability of the other photon with the opposite phase $\phi + \pi$ entering the channel B_x is $\cos(\phi + \pi - \theta)$ and the channel B_y $\sin(\phi + \pi - \theta)$. When the A detector is rotated relative to the B detector, the two-photon correlation pivots on the angle θ between the two detectors, and the signal varies as $\cos\theta$.

When the photon enters a two-channel detector, the probability of getting into one channel is proportional to the cosine of the direction of oscillation ($\cos\phi$), i.e., the photon phase ϕ, and into the other channel proportional to the sine of the photon phase ($\sin\phi$).

The two-channel detector is like two antennas crossing each other orthogonally. The probability is at maximum when both arms of the cross are at the 45 ° angle to the photon phase because $\cos(45\degree) + \sin(45\degree) = 1.41$. Correspondingly, the probability is at the minimum when the arms are either at the 0 ° or 90 ° angle to the photon oscillation because $\cos(0\degree) + \sin(0\degree) = \cos(90\degree) + \sin(90\degree) = 1.00$. So the *total* probability varies as the detector, say, a polarizer cube, is rotated. When one detector is pivoted relative to the other, the correlation between the two photons varies sinusoidally.

The two-photon correlation, $E = (N_{xx} - N_{xy} - N_{yx} + N_{yy})/(N_{xx} + N_{xy} + N_{yx} + N_{yy})$, is recorded from a large number of coincident photons, N, entering the x or y channel, of the A or B detector. For example, N_{xy} signifies the number of photons recorded pairwise on A_x and B_y. As the angle, ϕ, between the photon phase and the receiver phase varies randomly, the correlation, E, is the integral over all angles

$$
\begin{aligned}
E &= \frac{1}{2\pi} \int_0^{2\pi} \Big[\cos\phi\cos\left(\pi+\phi-\theta\right) - \cos\phi\sin\left(\pi+\phi-\theta\right) \\
&\qquad - \sin\phi\cos\left(\pi+\phi-\theta\right) + \sin\phi\sin\left(\pi+\phi-\theta\right) \Big] d\phi \\
&= \frac{-1}{2\pi} \int_0^{2\pi} \Big[\cos\theta + \sin\left(\theta-2\phi\right) \Big] d\phi = -\cos\theta
\end{aligned}
\tag{D1}
$$

where θ is the angle between the corresponding channels of the two detectors.

The calculation $-\cos\theta$ matches the data.[24] So, there is neither spooky action at a distance nor entanglement. The result is a mere truism. The correlation coefficient denoting covariance in counts between the two detectors is per definition available through the dot product $\mathbf{a}\cdot\mathbf{b} = |a|\,|b|\cos\theta$ between the two polarizer vectors $\mathbf{a}$ (Alice) and $\mathbf{b}$ (Bob). The angle θ is, of course, known right from the beginning. Since the ordinary correlation between the detectors explains the data, quantum entanglement is an unwarranted concept.

One should not mistake correlation for expectation value, i.e., coefficient of determination, $\cos^2\theta$, defining variance in counts of one detector that is predictable from counts of the other. From this perspective, the probability of a random, uncorrelated photon having a 45 ° phase relative to detector A to register on A_x is indeed 0.50. Yet, the probability of a correlated photon is 0.71 on the condition that the other photon in the pair was simultaneously registered on B_x. Finally, when the photon polarization instead of phase ϕ is measured, the correlation is proportional to $\frac{1}{4}(\cos2\theta + 1)$.

The long-lasting miscomprehension, leading to the erroneous expectation of a linear response, stems from perceiving probability as a normalized constant (100%)

rather than a physical state measure. Surely, each detector registers photons with 100% probability irrespective of their phases, i.e., $\cos^2\phi + \sin^2\phi = 1$. But still, the probability for the photon to go through the detector's phase-sensitive area depends on the phase in the usual manner, i.e., $\exp(i\phi)$. While we have learned to hold heads and tails equally probable, the experience of tossing a weighted coin reveals right away that probability is a physical measure.

APPENDIX E: THE PASSAGE OF LIGHT

Since we make observations of the universe primarily by means of light, an accurate account of the passage of light is a prerequisite for a true view.

According to the standard cosmology, i.e., the ΛCDM model interpretation of the color and brightness of Type Ia supernovae, the universe is expanding at an increasing rate. The rate, known as the Hubble parameter, H, is determined from the redshift z of light and the optical distance, D_L, to the exploded star, given by its brightness. In the ΛCDM model, H depends on the energy density of matter and dark matter, to a small extent on the energy density of the radiation and the curvature of the universe, and to a large extent on the amount of dark energy. When the model is tuned to match the data, the portion of dark energy is 68%.

LIGHT IN PROPAGATION

Instead of modeling data, let us inspect the light's passage from an exploded star as a series of events. The energy, hf_e, emitted at the explosion is leveling off toward the surrounding energy density in the least time when the fiery flash is spreading out at the speed of light. After traversing a distance, D_L, in the expanding universe, the light reaches the detector on Earth. The distance, D_L, can be deduced from the flux of light, $F = L/4\pi D_L^2$, i.e., the power, L, per area, $4\pi D_L^2$. The scale is calibrated by measuring the flux, F_r, from the explosion at a known distance, r_r. Such an event could be a supernova in the Andromeda Galaxy at a nearby distance of 2.5 million light years, thus without a marked difference between the emitted and observed frequencies, i.e., $f_e \cong f_o$. Thus,

$$F = \frac{L}{4\pi D_L^2} = F_r \frac{L/4\pi D_L^2}{L_r/4\pi D_r^2} \cong F_r \frac{hf_e f_o r_r^2}{hf_e f_e D_L^2} = F_r \frac{1}{1+z} \frac{r_r^2}{R^2} \frac{(1+z)^2}{z^2} = F_r \frac{r_r^2}{R^2} \frac{1+z}{z^2}, \qquad \text{(E1)}$$

where the frequencies are given in terms of the redshift, $z = (f_e - f_o)/f_o$, and the optical distance, $D_L = Rz/(1 + z)$, relative to the radius of the universe, R.

On its way, light shifts to red due to the supernova's recession velocity and the expanding universe's decreasing energy density. The effect of gravitation on the frequency[25] is not about the hypothetical tired light.[26] However, parametrizing the physical process with a scale factor, as in the standard cosmology, the distance seems longer than it is. That is why the Type Ia supernova data gives a false impression of the universe expanding at an accelerated rate. Likewise, the angular size redshift relation in the ΛCDM model is non-monotonous, i.e., $z > 1.5$, objects appear as if ever larger.

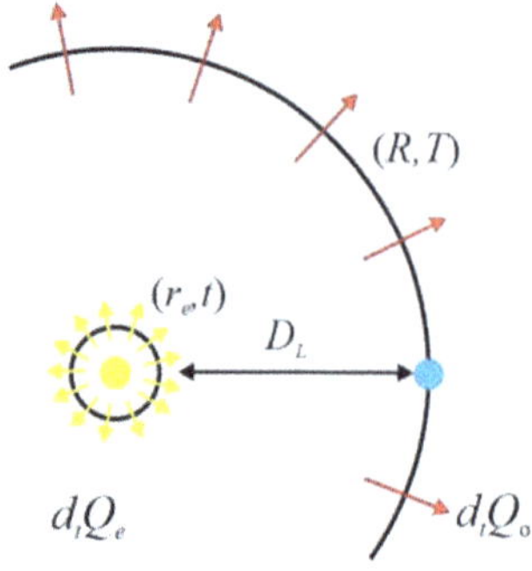

The intensity of light, d_tQ, from a supernova (yellow center) decreases directly proportional to energy and inversely to the optical distance, D_L, squared. Thus, by the time the light arrives at Earth (blue dot), its color has shifted to red (red arrows). The redshift is not only due to the rate of expansion but also due to the energy density of the present (a large arc with radius, R, at the present time, T) being smaller than that of the past (a small perimeter with radius, r_e, at the time of the explosion, t).

The brightness of a star (Eq. E1) on the logarithm scale, the distance modulus

$$\mu = -2.5\log \left(\frac{L/4\pi D_L^2}{L_r/4\pi D_r^2}\right) + K = -2.5\log \left(\frac{R}{r_r}\right)^2 + 2.5\log \left(\frac{z^2}{1+z}\right) + K \quad \text{(E2)}$$

is a function of two terms, $\log(z)$ and $\log(1 + z)$. So, $\mu(z)$ does not follow a straight line but curves at $z \approx 1$. The instrument factor, $K \approx 5\log(1 + z)$ corrects for sensitivity, i.e., the more the light shifts to red, the less is detected.[27]

As the difference between the calculated and observed distance moduli is mostly within the margin of error, the least-time light propagation explains the supernovae data without dark energy. As the amount of matter decreases, the rate of expansion, Hubble's parameter, $H = 1/t$, decreases at the rate, $-d_tH = 1/t^2 = 4\pi G\rho_M$, where G is the gravitational constant and ρ_M density.

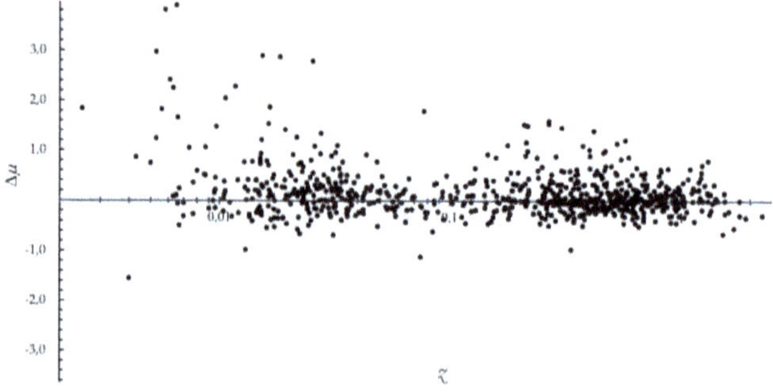

On average, the difference, $\Delta\mu$, vs. redshift, z, between the supernova brightness[27] and the values calculated by the thermodynamic theory do not call for further explanations by dark energy or dark matter.

The thermodynamics of time explains the expansion as an active process where matter transforms into the vacuum and the density decreases according to equation

B2. Therefore, as Hubble's law states, the galaxies in distant groups move away from us while the nearby galaxies move toward us. By contrast, the standard model does not explain the cause of the expansion itself but simulates the course. The initial small irregularities are amassed as galaxies and galaxy groups, while the attraction between the distant groups has not yet disturbed the innate expansion.

BENDING OF LIGHT

General relativity is considered a valid theory of gravity. For example, as calculated, a radio signal passing by the Sun to Venus and reflecting back experiences a gravitational time delay of 195 µs.

According to thermodynamic theory, light selects the least-time path when passing by the Sun. The gravitational field embodied by the photon pairs is mathematically quite like space-time geometry. The stronger the gravitational field, the more curved the geodesic. Since the paired photons at a given distance are denser around the Sun than the Earth, the photons measure a longer orbit about the Sun.

The curvature causes an angular acceleration α that integrates a change in the angular momentum L over time to an angle of precession φ,

$$\frac{dL}{dt} = I\alpha = I\frac{d}{dt}\frac{d\varphi}{dt} = mr^2\omega^2 = (2\pi)^2 mv^2 = (2\pi)^2 \frac{GmM_\mathrm{o}}{r}$$

$$\Rightarrow \varphi = \int\int (2\pi)^2 \frac{GM_\mathrm{o}}{r^3} dt dt = 2\pi^2 \frac{GM_\mathrm{o}}{c^2 r}$$

(E3)

where G is the gravitational constant and $r = ct$ is the distance to the center of the Sun. Since the mass, m, corresponding to light's energy, hf, per the speed of light squared c^2, is negligible compared with the solar mass, $M_\mathrm{o} = 1.99 \cdot 10^{30}$ kg, its effect on the precession φ is insignificant. The equation E3 gives $\varphi = 8.65''$ for the ray that tangents the Sun's surface at its radius, $r_\mathrm{o} \approx 695{,}700$ km. (One radian is $360 \cdot 60 \cdot 60 / 2\pi$ of arcseconds.) In concord with the measurements, it takes an excess time, $\Delta t = 2r_\mathrm{o}\varphi/c = 196$ µs, for the light to make a round trip along the geodetic line than along the straight line.

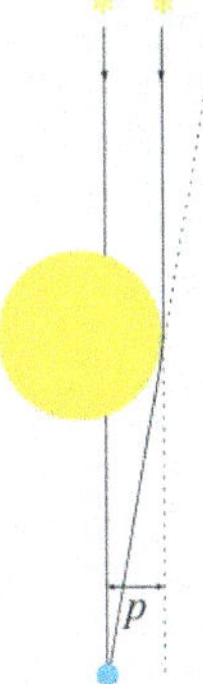

Ray of light from a distant star (solid line) bends in the gravity of the Sun (yellow sphere). In the drawing, the deflection is massively magnified. The amount of bending, as seen from Earth (blue point), can be determined from the star's apparent direction (dotted line extending to the gray star) measured at the time of the total eclipse and the star's actual direction (line extending to the yellow star) six months later seen in the clear night sky. It is essential to note the parallax (p) between the bent and direct rays. Otherwise, the bending will be found to be smaller than it truly is.

The least-time principle reproduces bending and delay with one and the same equation, as the two phenomena are one and the same. In contrast, by general relativity, the delay and bending are calculated using two different equations. By E3, light coming behind a galaxy deflects $2\pi^2/4 \approx 5$ times more than general relativity.[24] Thus, there is no need for dark matter.

The so-called Einstein ring forms when a ray comes straight behind a galaxy. Its bending can be calculated likewise. The thermodynamic theory also gives the time delay between two different images of a pulsating object.[29] Moreover, the results of Gravity Probe B tally with thermodynamic theory.[30]

The curvature is also evident in orbital precession. At the distance of the Mercury half-axis, $r = 57{,}909{,}050$ km from the Sun, the angle φ advances about $0.504 \cdot 10^{-6}$ radians per cycle according to equation E3. Advancing over a century amounts to $43.09''$ from Mercury orbiting the Sun about 415 times in agreement with the data.[31] More precise calculation is demanding since other celestial bodies influence the orbit besides the Sun, especially Venus, Earth, and Jupiter.

Moreover, according to the least-time principle, equation A5, the angle between the tangent and radial line, $\phi = \pi b/2 = \arctan[(-d_t U + id_t Q)/d_t 2K]$, is a constant of motion on a spiral orbit, $r = a\exp(ib\omega t)$. Using Kepler's law, $\omega^2 r^3 = Gm$, for the change in variables, the characteristic change in the angular velocity, the hallmark of emitted gravitational waves, $d_t\omega = (Gm)^{5/3}\omega^{11/3}/2c^5$, is obtained from the kinetic energy of a cycle, $2K\omega = (d_t 2K\, d_t E)^{1/2}$. The geometric mean, familiar from matching for the maximal transmission, includes multiplicative natural processes from those causing the change in the orbiter's kinetic energy, $d_t 2K = 2mr^2\omega d_t\omega$, to all of those generating the power of the universe, $d_t E = GM^2 d_t R/R^2 = c^5/G$, from the energy $E = Mc^2$ bound in all mass M. The chirp $d_t\omega$ of a binary system can be expressed in terms of the reduced and combined mass.

STELLAR ABERRATION

A star straight up in the sky appears off from its true position due to the Earth's orbital velocity about the Sun. The aberration, $\theta = \arctan(c_y/c_x)$, is the ratio of light's velocity y- to x-component, hence independent of the index of refraction, n.[32] However, James Bradley mistook in 1727 the Earth's orbital velocity component, v_x, for the x-component instead of equating c_x with v_x. The error became an issue assuming a light-carrying medium, the luminiferous ether. There was no medium-dependent aberration, as François Arago proved with a water-filled telescope in 1810. The conundrum culminated in Einstein deriving the aberration from the Lorentz transformations between frames of reference. But the simple truth remains that the speed of light is $c/n = [(c_x/n)^2 + (c_y/n)^2]^{1/2}$ in the plain vacuum, $n = 1$, or a denser substance.

The straight-forward inference parallels the familiar experience of relativity: an umbrella had better be tilted when the rain comes with the wind or when running in the rain.

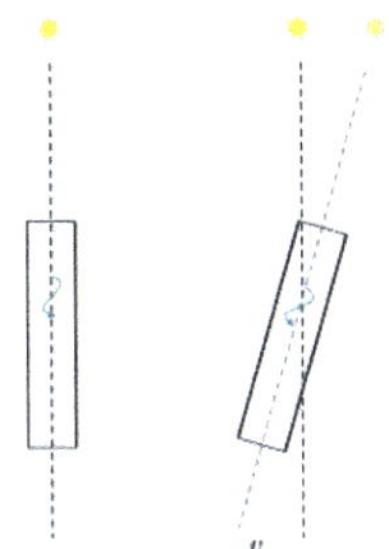

Telescope has to be tilted away from the true zenith position of a distant star to compensate for the Earth's orbital velocity around the Sun.

FIZEAU EXPERIMENT

Pondering upon the existence of ether in 1851, Hippolyte Fizeau followed Augustin-Jean Fresnel in reasoning that the speed of light squared $(c/n)^2$ is inversely proportional to the medium's density ρ_n. Thus, the speed of light increases from c/n in a medium moving at the speed v along with light to

$$\frac{c}{n} + v\frac{\rho_n - \rho}{\rho_n} = \frac{c}{n} + v\frac{(c/n)^2 - c^2}{c^2} = \frac{c}{n} + v\left(1 - \frac{1}{n^2}\right) \tag{E4}$$

neglecting the Doppler shift $-(\lambda/n)(dn/d\lambda)$ on the wavelength λ, proposed by Hendrik Lorentz and proved by Pieter Zeeman.

The reasoning, essentially that of Daniel Bernoulli, is in line with the flow of a ponderable substance, such as water, immersed in the paired-photon plenum. Also, as the vacuum density diminishes due to the expansion, the speed of light increases.

MICHELSON–MORLEY EXPERIMENT

The passage of light along an interferometer arm moving along with the Earth is akin to the ray through a moving telescope. From its own perspective, as an inseparable part of the paired-photon vacuum, a photon makes a round trip at the speed of $c = (c_x^2 + c_y^2)^{1/2}$ through the two arms, perpendicular and parallel to the motion, say, along the x-direction. Since the displacements are equal for the two arms, two coincident rays invariably arrive at the semi-transparent mirror simultaneously. Thus, the interference pattern remains intact, apart from the second-order effects on vacuum density.

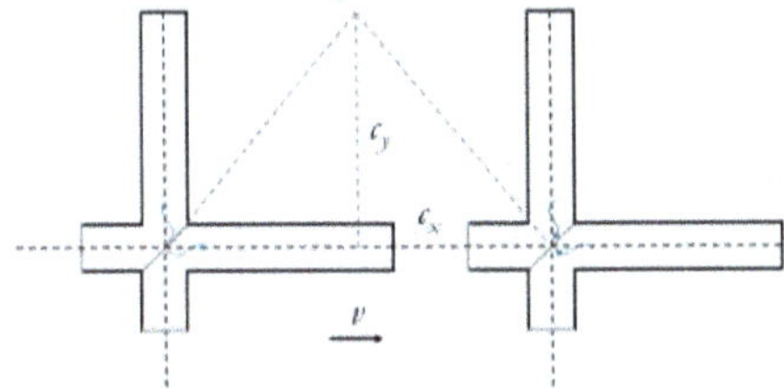

Coincident photons propagating along the arms of a moving interferometer arrive at the semi-transparent mirror simultaneously since both have undergone the same displacement.

Sagnac experiment

In 1920 Max von Laue concluded that two opposite rays in a rotating ring develop a time difference $\Delta t = t_+ - t_- = 2vl/c^2 = 4\omega A/c^2$.[33] As every part of the ring moves with velocity $\mathbf{v} = \boldsymbol{\omega} \times \mathbf{r}$ away from one ray and toward the other, Georges Sagnac confirmed the period $t_\pm = cl/n \pm \oint \mathbf{v} \cdot d\mathbf{l}/c^2$ for the ray to loop the path l that closes an area $A = \tfrac{1}{2}\oint r\,dl$. Thus, a phase difference $\Delta\phi = 2\pi c\Delta t/\lambda$ develops regardless of the refraction index n, e.g., in the rotating vacuum.

A light-carrying medium failed to explain the array of observations that relativistic coordinate transformations reproduced to first-order. However, the paired-photon vacuum makes sense of the results, for no photon is truly free but an integral part of either a plenum or a particle. This tenet complies with relativity theory in the continuum limit of Eq. A5. An infinitesimal change in kinetic energy density $d(2k)$ $\rightarrow ds^2$ balances those in potential energy density $du \rightarrow c^2d\tau^2$ and dissipation $dq \rightarrow c^2dt^2$, i.e., $ds^2 = -c^2d\tau^2 + c^2dt^2$ and yields the famous factor $\rightarrow d\tau/dt = \gamma = (1 - v^2/c^2)^{1/2}$.

Pioneer anomaly

Radio signals to Pioneer 10 and 11 spacecraft trekking out of the solar system shift in frequency inexplicably by $6 \cdot 10^{-9}$ Hz/s. The unaccountable shift in communication, operating at uplink 2.113 and downlink 2.295 GHz, relates by $d_tf/f = d_tv/c$ to an acceleration $d_tv = a$ of about $8 \cdot 10^{-10}$ m/s^2 toward the Sun. A conventional cause is not apparent; hence the acceleration is deemed anomalous.[34]

In addition to accelerating in the solar system gravitation, the spacecraft accelerates in the galactic and universal gravitation as much as its motion deviates from the Sun's orbital motion. Thus, the craft accelerates along the steepest gradient in density to attain balance in the least time (Eq. A5). The index $n = (1 - \Sigma_i GM_i/c^2r_i)^{-1/2}$ decreases from the Earth's surface, astronomical unit away from the Sun, to the faraway galactic density. So, does $\Delta n = 5.34 \cdot 10^{-9}$ relate to d_tf and by c/f to anomalous $a = 7.6 \cdot 10^{-10}$ m/s^2? Is the spacecraft accelerating $a = c/t$ toward the Sun also due to spatial effluxes of the expansion at the Hubble rate $H = 1/t$?

As for the stellar aberration, annual and daily variation in the frequency is accounted for by adding the corresponding components of the speed of light.

Appendix F: Motions of galaxies

Star systems are among the oldest structures in the universe. So, they must be in balance with their surroundings. However, the mass of a galaxy is not enough to explain the high velocities of stars. Unlike the orbital velocities of planets that decrease from the center of the solar system toward the outskirts, the orbital velocities of stars and gas clouds increase from the galaxy center and attain an approximately constant level far away at the rims. Thus, more mass is required to explain this velocity curve than is observed in the stars and gas clouds. So, what is this dark matter?

GALAXY ROTATION

The void's energy density, $\rho = c^2/4\pi G t^2 \approx 0.55\cdot10^{-9}\,\text{J/m}^3$, equals the average energy density of the matter (Eq. B2). In other words, the void embodies the gravity of the universe,[35] increasing from the sparse present toward the dense past. The gradual change is seen at the galactic scale as a tiny acceleration, $a_R = GM/R^2 = c/t = cH$, where the universe's expanding radius, $R = ct$, is expressed as the product of the speed of light, c, and the age of the universe, t, or equivalently the Hubble parameter, $H = 1/t$. This cosmic acceleration forces distant galaxies away from us and local galaxies toward us and affects galaxy rotation. Since forces tally where fluxes tally, the mass, M_o, of the local galaxy group, relates to the mass, M, of the universe as the zero-velocity surface radius, $R_o = GM_o/c^2$, beyond which the Hubble flow begins,[36] relates to the radius of the universe, $R = GM/c^2$. Influx and efflux of space tally where their driving forces $F = GMM/R^2 = Mc^2/R = c^4/G$ balance.

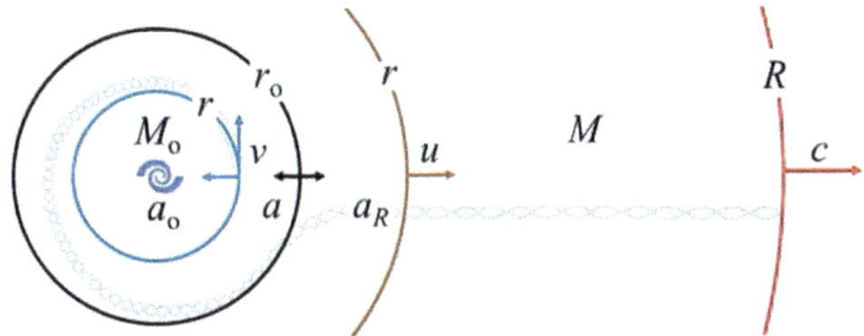

Within a radius r_o, graviton efflux emerging from local mass M_o exceeds influx from distant sources; thus, a body spirals in until v^2/r equates the universal a_t and local a_o acceleration. Beyond r_o, the influx exceeds the efflux; thus, the body recedes at a speed $u < c$. At the rim R, the total flux emerging from all matter M within R powers the expansion at the speed of light c.

Since the influx and efflux develop hand in hand, graviton by graviton, the surrounding space lengthens along r as much as the circumference $2\pi r$ shortens, i.e., $a_R = 2\pi a_t$. Eventually, the body attains an orbital velocity v corresponding to v^2/r balancing the universal acceleration a and local acceleration ao. As per Newtonian dynamics, the inward pulling efflux and the outward pushing influx tally.

A star like the Sun experiences not only acceleration, $a_o = GM_o(r)/r^2$, due to the mass, mass, $M_o(r)$, of the galaxy that resides inside its orbit but also acceleration due to the mass, M, of the expanding universe, a_t,

$$a = a_o + a_t = a_o\left(1 + \frac{a_t}{a_o}\right) = \frac{GM_o(r)}{r^2}\left(1 + \frac{1}{2\pi}\frac{M}{M_o(r)}\frac{r^2}{R^2}\right) \tag{F1}$$

where the tiny universal acceleration is $a_t = a_R/2\pi = cH/2\pi \approx 10^{-10}\,\text{m/s}^2$ per revolution. Nevertheless, it is a significant part of the total acceleration experienced by the Sun, $a = v^2/r \approx 2.32\cdot10^{-10}\,\text{m/s}^2$, at a distance, $r = 2.47\cdot10^{20}$ m, from the center of the Milky Way. The orbital velocity of the Sun, v, is thus high 240 km/s. The residual acceleration, $a - a_t = 1.23\cdot10^{-9}\,\text{m/s}^2$, due to mass inside the Sun's orbit, $M_o(r) = a_o r^2/G \approx 5.6\cdot10^{10} M_\odot$, in units of the solar mass, $M_\odot = 1.99\cdot10^{30}$ kg, accounts for most of the Milky Way's mass, estimated to be $4.6 - 6.4 \cdot 10^{10} M_\odot$ without dark matter.[37] Such a hypothetical substance is thus not needed. The acceleration of the expanding universe explains the galaxy rotation.

When the universal acceleration dominates, $a_t/a_o >> 1$, the equation of acceleration F1 reduces to the Tully-Fisher relation, $v^4 \approx a_t G M_o$.[38] Thus, we can estimate the galaxy mass from its rotational velocity asymptote. For example, the orbital speed of Andromeda, $v \approx 250$ km/s, implies a mass comparable to the Milky Way $4.6 \cdot 10^{10} M_\odot$ without dark matter.[35]

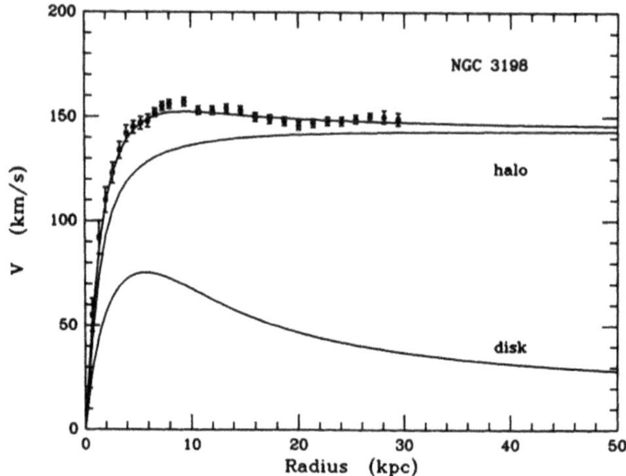

Galaxy NGC 3198 is located about 47 million light years away. The orbital velocity of its stars increases with increasing radius, reaching an approximately constant value $v \approx 150$ km/s at about 10 kpc ($3.1 \cdot 10^{20}$ m). The figure shows a model that sums the orbital velocity from the spiral galaxy's visible (disk) and dark matter (halo).[36] However, according to equation F1, the data can be explained without dark matter, using the universal acceleration, a_t. Based on the local acceleration, $a - a_t = 0.3 \cdot 10^{-10}$ m/s^2, the mass of NGC 3198, $M_o(r) = a_o r^2/G \approx 2.3 \cdot 10^{10} M_\odot$, is about half of the Milky Way.

Dwarf galaxies seem exceptionally rich in dark matter. In truth, even they house no dark matter. This is because the local acceleration due to the small mass of a dwarf galaxy is just exceptionally small compared to the universal acceleration.

All galaxies experience the universal environment much alike. Therefore, regardless of galaxy size, the orbital period of the peripheral stars, $t = 2\pi r/v$, is approximately constant, about one billion years.[39] Thus, by equation B2, the average density within the galaxy's outer circumference is about 200 times higher than the vacuum density. The escape velocities of stars in the Milky Way and Andromeda do not indicate dark matter either,[40] in agreement with equation A6 that defines the steady-state ratio of the orbital, $v^2 = GM/r$, to the escape velocity of stars, $v_{esc}^2 = 2GM/r$.

Analogously to the Tully-Fisher relation, the velocities of stars at the galactic core are in a power-law relation to the mass of its supermassive black hole.[41]

Galaxy velocity dispersion

Just as stars in galaxies orbit around, galaxies in clusters move about to balance the local and universal gravitation. The Faber–Jackson relation, ranging from tiny dwarf satellite galaxies through giant spiral galaxies to groups and rich clusters of galaxies, even encompassing the whole expanding universe, discloses the universal acceleration, $a_R = c^2/R = u^2/r = 2\pi a_t = 2\pi v^2/r$, in terms of speed of light, c, radial, u, and orbital, v, velocity. As $u^2 = 2\pi v^2$, the power-law slope for the mass vs. radial velocity is offset from orbital velocity by the factor $\sqrt{2\pi} \approx 2.5$.

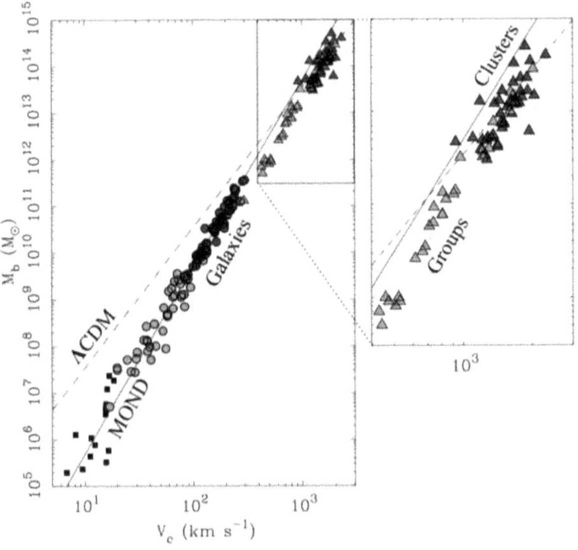

Baryonic mass, M_b, vs. velocity, v_c, follows a power law, $M \propto v^4$, from dwarf spheroidals (squares) through gas-rich (light gray circles) and star-dominated (dark gray circles) spiral galaxies to groups (light gray triangles) and clusters (dark gray triangles) of galaxies, in line with MOND (solid line) contrasting ΛCDM (dashed line). The cluster zoom shows the slope offset $\sqrt{2\pi}$.[42]

Appendix G: The fundamental forces

According to our own experience, any structure will break down when the force applied to it is strong enough. For example, granite ruptures when the pressure exceeds the characteristic strength of granite (> 150 MPa). Similarly, the four fundamental forces of nature relate to characteristic structures. The strong force rips two quarks apart. The weak force cuts the electron torus open, forming the W⁻ boson. The electromagnetic force pulls apart the paired photons. The gravitational force alters the vacuum density embodied in the photon pairs.

The electromagnetic force relates to the strong force by fine-structure constant

$$\alpha = \frac{e^2}{4\pi\varepsilon_0 c\hbar} \approx \frac{1}{137.036}. \tag{G1}$$

The electromagnetic action, i.e., the effect of the electron on the quantum pairs of the vacuum, depends on the electrical and magnetic properties of the vacuum, i.e., the permittivity, ε_o, and the permeability, μ_o, which determine the speed of light squared, $c^2 = 1/\varepsilon_o\mu_o$. The quantum of action, $\hbar = h/2\pi$, in a circular form is the neutrino (Appendix B).

Electromagnetism's and gravity's inverse proportionality to the distance squared implies that the photon carries both, as assumed for long.[43] The only difficulty has been identifying the photon pair as the carrier of gravity and realizing that the electron torus separates the paired quanta from each other, creating an electromagnetic field.

Based on the ratio of the electrostatic and gravitational fields

$$\frac{\alpha}{\alpha_G} = \frac{e^2}{4\pi\varepsilon_o r} \bigg/ \frac{Gm_e^2}{r} \approx 4.17\cdot10^{42} \tag{G2}$$

the ratio of the strong force to gravity is $1/\alpha_G \approx 5.71\cdot10^{44}$. This gives us an estimate of the neutron breakdown density, $\rho_c = \rho_M/\alpha_G \approx 3.3\cdot10^{18}$ kg/m³, where the vacuum density, $\rho_M = 1/4\pi Gt^2 \approx 0.6\cdot10^{-26}$ kg/m³, (Eq. B2) of the universe with age, $t = 1/H$, the inverse of the Hubble parameter, is inferred from the speeds of receding galaxies. Assuming that the neutrons are as packed in a neutron star as in the atomic nucleus, the pressure, equal to the density, 10^{17} kg/m³, is not enough to break the neutron's tetrahedral structure (Chapter 4). The critical pressure is exceeded first in the abyss of a black hole.

Based on equation B2, the coupling constant

$$\alpha_G = \frac{Gm_e^2}{\hbar c} = \frac{R}{r}\frac{m_e}{M}\frac{m_e cr}{\hbar} \approx 1.75\cdot10^{-45} \tag{G3}$$

expresses the ratio of the universe's radius, R, to the electron radius, r, in its first factor, multiplied by the second factor, the ratio of the electron mass, m_e, to the universe's mass, M, just as Weyl, Eddington, and Dirac assumed. The last factor can be understood, in units of h, as the internal motion of the electron, the spin. As gravity relates to the vast energy-sparse universe, it is very weak compared with the other fundamental forces related to the tiny energy-dense particles.

We can estimate the total number of quanta in the universe from the total action

$$nh = Mc^2t = \frac{c^4Rt}{G} = \frac{c^5t^2}{G} \Leftrightarrow n = \frac{c^5t^2}{Gh} \approx 10^{121}. \tag{G4}$$

This amount of quanta was at the beginning, and the same amount will also be at the end. The quanta only unfold from the curved particles into the flat void as the universe evolves.

Appendix H: From matter into the void

At the current age of the universe, 14 billion years, the void emerges from all matter at the rate

$$\frac{dV}{dt} = 4\pi R^2 \frac{dR}{dt} = 4\pi c^3 t^2 \tag{H1}$$

of $6.61 \cdot 10^{61}$ m³/s. Based on the energetic equivalence between matter and the vacuum, $E = Mc^2 = GM^2/R$, the power of the universe

$$P = \frac{dE}{dt} = \frac{c^5}{G} \tag{H2}$$

is about $3.64 \cdot 10^{52}$ W, and the force of expansion is

$$F = \frac{P}{c} = \frac{c^4}{G} = \frac{GM}{R^2}\sum_i M_i = a_R \sum_i M_i = \sum_i \frac{v_i^4}{G}, \tag{H3}$$

where $M = \Sigma M_i$ is the sum of all mass. According to equations H1 and H2, the void's energy density is about 0.55 nJ/m³. Only 0.1 ‰ is light, i.e., about 400 photons in cm³ of average energy 10^{-22} J, since the cosmic microwave background temperature is 2.725 K, equivalent to $3.76 \cdot 10^{-23}$ J.[44] This unpaired-photon fraction decreases along with decreasing matter.

As the void expands, its average temperature decreases. Deviations, such as the hydrogen gas 21-cm line imprinted in the cosmic spectrum at an early stage of the universe,[45] smooth out. The flight of the farthest galaxies, almost at the speed of light, sums up all transformations of matter into the void.

The vacuum spectrum displays photons of energy ε_i relative to the average energy $k_B T$ distributed in numbers n_i among rays of paired photons. The number of ways the photons populate, in-phase or antiphase, the numerous rays, crisscrossing in degenerate directions g_i, is the product of n_i combinations of the sets with $n_i + 2g_i - 1$ elements

$$W = \prod_i \frac{(n_i + 2g_i - 1)!}{n_i!(2g_i - 1)!}\, e^{-n_i \varepsilon_i / k_B T} \approx \prod_i \frac{(n_i + 2g_i)!}{n_i! 2g_i!}\, e^{-n_i \varepsilon_i / k_B T}. \tag{H4}$$

Taking logarithm and using Stirling's approximation for the factorials yields

$$\ln W = \Sigma_i (n_i + 2g_i)\ln(n_i + 2g_i) - n_i \ln n_i - 2g_i \ln 2g_i - n_i \varepsilon_i / k_B T. \tag{H5}$$

At a stationary state,

$$\frac{d\ln W}{dt} = \sum_i \frac{d\ln W}{dn_i}\frac{dn_i}{dt} = \sum_i \frac{dn_i}{dt}\left(\ln \frac{n_i + 2g_i}{n_i} - \varepsilon_i / k_B T\right) = 0, \qquad (H6)$$

the most probable, energetically optimal number of photons over their total number

$$n_i = \frac{2g_i}{e^{\varepsilon_i/k_B T} - 1} \qquad (H7)$$

complies with the Bose-Einstein distribution with two states for each photon, in-phase or antiphase, per the ray locus.[46]

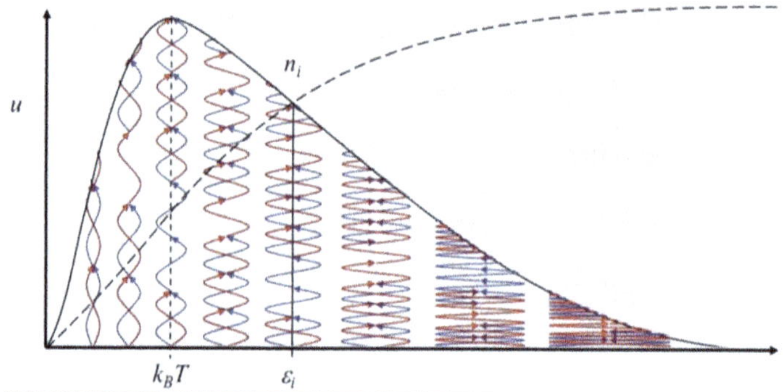

The vacuum spectral density comprises numerous rays of photons (blue-red waves). The paired photons cannot be seen as light but are sensed as inertia and gravitation by their coupling to matter. On the other hand, the odd quanta distributed in-phase or antiphase (blue or red) among the paired rays are seen as light and manifest as electromagnetism. The modular structure displays itself also in the Casimir and dynamic Casimir effects.

Then it is enough to assume that the ultimate elementary region in the phase-space has the content h^3, as Bose wrote to Einstein,[47] and the paired-photon structure of the vacuum defines the Bose-Einstein statistics and explains Planck's radiation law. In line with Planck's law, the pairs open up with increasing temperature, and conversely, the photons pair up with decreasing temperature. Likewise, the photon unpairing counteracts an imposed electric field, producing vacuum polarization rather than theorized short-lived virtual electron-positron pairs. And the paired-photon vacuum responds to an accelerating charge by unpairing and radiating photons, i.e., adjusting occupancies of the energy levels (Eq. H7).

Characteristic sources of light add characteristics to the background spectrum of the vacuum. For example, in 1868, helium was identified in the Sun's spectrum.

While a train of photons in a ray may pair up with another train in another ray, the two rays of paired photons cannot overlap. Thus, space forms from matter rather than light.

The paired-photon strings can be expressed as a fiber bundle in algebraic topology. Its connection is the vector potential, the source of the electromagnetic field.

Its action is the curvature, manifesting as gravitation. Akin to the quantum chromodynamics vacuum, the paired rays constitute a lattice that embraces everything. From this viewpoint, the Lorenz gauge is not just a gimmick to deal with mathematical redundancy in the field variables but also a continuity condition that equates flows of density with changes in density.

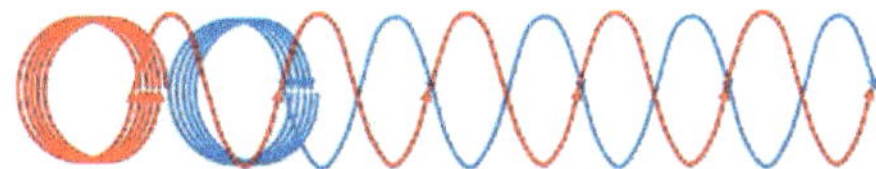

Close-up of quark segments of opposite chirality (blue and red coils) unwinding, i.e., annihilating into a ray of co-propagating photons with opposite phases (blue and red waves). As the emitted pairs of photons are without net electromagnetic fields, they are detectable only through their energy density relative to the surrounding vacuum density.

Geometrizing gravitation on the largest scale with the metric of the cosmological principle, $ds^2 = a(t)^2 dx^2 - c^2 dt^2$, sums up by Pythagoras' theorem the infinitesimal distance, ds, from the spatial, dx, and temporal, dt, coordinates that correspond to the photon wavelength and period. While the scale factor $a(t)$ parameterizes the expansion quite well, the thermodynamic balance between all matter and space implies that the universe expands physically rather than [para]metrically. As the quanta of matter transform into the paired photons of the vacuum, the energy density decreases. Photon redshifts document this density change from the past to the present.

From ancient times until now, the substance of space has been an immense mystery. Newton considered the void as a remarkably mobile and elastic essence. Indeed, the photon wavelengths have extended tremendously over the eons as ponderable matter—fundamentally, photons too—has transformed into the vacuum.

The National Library of Finland (Wikipedia)

NOTES

PREFACE

1. Boltzmann, L., *The Second Law of Thermodynamics* (D. Reidel, 1974), p. 13–14.
2. Anderson, P. W., *More and Different: Notes from a Thoughtful Curmudgeon.* (World Scientific, 2011), p. 236.
3. Shaw, B. G., Saint Joan: A Chronicle Play in 6 Scenes and an Epilogue. (Constable & Co., Ltd., 1924).
4. Husserl, E., The Crisis of European Sciences and Transcendental Phenomenology: An Introduction to Phenomenological Philosophy. (Northwestern University Press, 1970).

1. WHY?

1. Ball, P., *Shapes: Nature's Patterns: a Tapestry in Three Parts.* (Oxford University Press, 2011).
2. Maupertuis, P.-L. M. d., Les Loix du mouvement et du repos déduites d'un principe métaphysique. Hist. Acad. Roy. Sci. Belleslett., 267 (1746).
3. Terrall, M., The Man Who Flattened the Earth: Maupertuis and the Sciences in the Enlightenment. (University of Chicago Press, 2002).
4. Carnot, S., *Reflections on the Motive Power of Fire.* (Dover Publications, 1960).
5. Merritt, D., Cosmology and convention. Stud. Hist. Philos. Sci. B 57, 41 (2017).
6. Gregory, T. R., *The Evolution of the Genome.* (Elsevier Science, 2011).
7. Albert, R., Barabási, A.-L., Statistical mechanics of complex networks. Rev. Mod. Phys. 74, 47 (2002); Clauset, A., Shalizi, C. R., Newman, M. E. J., Power-law distributions in empirical data. SIAM Rev. 51, 661 (2009); Newman, M. E. J., Power laws, Pareto distributions and Zipf's law. Contemp. Phys. 46, 323 (2005); Sornette, D., *Critical Phenomena in Natural Sciences: Chaos, Fractals, Self-organization and Disorder: Concepts and Tools.* (Springer, 2006); Buchanan, M., *Ubiquity: Why Catastrophes Happen.* (Three Rivers Press, 2002).
8. Bak, P., *How Nature Works.* (Oxford University Press, 1997).
9. Lander, E. S., Linton, L. M., Birren, B. et al., Initial sequencing and analysis of the human genome. Nature 409, 860 (2001).
10. Biller, O., El-Sana, J., Kedem, K., The influence of language orthographic characteristics on digital word recognition. Lit. Linguist Comput. 30, 495 (2015).
11. Postman, M., Clusters as Tracers of Large-Scale Structure. in Evolution of Large-Scale Structure. (Print Partners Ipskamp, 1999), p. 270
12. Zhao, K., Musolesi, M., Hui, P. et al., Explaining the power-law distribution of human mobility through transportation modality decomposition. Sci. Rep. 5, 9136 (2015).
13. Herrada, E. A., Tessone, C. J., Klemm, K. et al., Universal scaling in the branching of the tree of life. PLoS ONE 3, e2757 (2008).
14. Ruegsegger, C., Stucki, D. M., Steiner, S. et al., Impaired mTORC1-dependent expression of homer-3 influences SCA1 pathophysiology. Neuron 89, 129 (2016) neurosciencenews.com/sca1-purkinje-cells-3365
15. Paragorgia Arborea (May 21, 2002) en.wikipedia.org/wiki/Deep-water_coral
16. Hargittai, I., Pickover, C. A., *Spiral Symmetry.* (World Scientific, 1992); McNally, J., Earth's Most Stunning Natural Fractal Patterns. Science (Oct. 9, 2010).
17. Rossi, P., *The Birth of Modern Science.* (Wiley, 2001), p. 325–328.
18. Pinterest, Logarithmic spiral in Nature

pinterest.com/pin/39631683594555 6187

19 Gleick, J., *Chaos: Making a New Science.* (Open Road Media, 2011).

20 Goulielmakis, E., Uiberacker, M., Kienberger, R. et al., Direct measurement of light waves. Science 305, 1267 (2004).

21 Abbott, B. P., Abbott, R., Abbott, T. D. et al., Observation of gravitational waves from a binary black hole merger. Phys. Rev. Lett. 116, 061102 (2016).

22 Huska, J., 2016 Stock market update: key chart insights and indicators (Mar. 7, 2016) www.secitmarket.com/2016-stock-market-update-key-chart-insights-and-indicators-march-15409

23 Kinnunen, J., Gabriel Tarde as a founding father of innovation diffusion research. Acta Sociol. 39, 431 (1996).

24 Ball, P., *Critical Mass.* (Random House, 2014), p. 240-247.

25 Verhulst, P.-F., Notice sur la loi que la population suit dans son accroissement. Corres. Math. et Phys. 10, 113 (1838).

26 Gould, S. J., Eldredge, N., Punctuated equilibria: the tempo and mode of evolution reconsidered. Paleobiol. 3, 115 (1977).

27 Limpert, E., Stahel, W., Abbt, M., Lognormal distributions across the sciences: keys and clues. BioScience 51, 341 (2001).

28 Poincaré, H., *Calcul des probabilités.* (Gauthier-Villars, 1912), p. 171.

29 Mäkelä, T., Annila, A., Natural patterns of energy dispersal. Phys. Life Rev. 7, 477 (2010).

30 Kapteyn, J. C., Skew frequency curves in biology and statistics. Recueil des travaux botaniques néerlandais 13, 105 (1916).

31 Pareto, V., *Cours d'Économie Politique.* (Librairie Droz, 1964), p. 299–345.

32 Zipf, G. K., Human Behavior and the Principle of Least Effort: An Introduction to Human Ecology. (Addison-Wesley Press, 1949).

33 Clauset, A., Structure strangeness. www.cs.unm.edu/~aaron/blog/archives /2011/01/labyrinths_cave; Lopes, A. M., Machado, J. A. T., Power law behavior and self-similarity in modern industrial accidents. Int. J. Bifurc. Chaos 25, 1550004 (2015).

34 Benford, F., The law of anomalous num-bers. Proc. Am. Phil. Soc. 78, 551 (1938).

35 Schmidt-Nielsen, K., *Scaling: Why is Animal Size So Important?* (Cambridge University Press, 1984), p. 43.

36 Snell, O., Die Abhängigkeit des Hirngewichts von dem Körpergewicht und den geistigen Fähigkeiten. Arch. Psychiatr. 23, 436 (1892).

37 Thompson, D. A. W., *On Growth and Form.* (Cambridge University Press, 1992).

38 Huxley, J. S., *Problems of Relative Growth.* (Dover, 1972).

39 Lotka, A. J., The frequency distribution of scientific productivity. J. Wash. Acad. Sci. 16, 317 (1926).

40 Solla Price, D. J. d., Networks of scientific papers. Science 149, 510 (1965).

41 Richardson, L. F., *Statistics of Deadly Quarrels* (Boxwood Press, 1960).

42 Bak, P., Tang, C., Wiesenfeld, K., Self-organized criticality: an explanation of 1/f noise. Phys. Rev. Lett. 59, 381 (1987).

43 Wolfram, S., *A New Kind of Science.* (Wolfram Media, 2002).

44 Gardner, M., Mathematical Games – The fantastic combinations of John Conway's new solitaire game "Life". Sci. Am. 223, 120 (1970).

45 Watts, D. J., *Small Worlds: The Dynamics of Networks between Order and Randomness.* (Princeton University Press, 1999); Strogatz, S. H., Exploring complex networks. Nature 410, 268 (2001); Bak, P., Duncan J. Watts: Small Worlds: The dynamics of networks between order and randomness. Physics Today 53, 54 (2000).

46 Rocha, L. A. O., Lorente, S., Bejan, A., *Constructal Law and the Unifying Principle of Design.* (Springer New York, 2012); Bejan, A., *The Physics of Life: The Evolution of Everything.* (St. Martin's Press, 2016).

47 Waldrop, M. M., Complexity: The Emerging Science at the Edge of Order and Chaos. (Simon and Schuster, 1993).

48 Gibrat, R., Les Inégalités Economiques: Applications: Aux Inégalités des Richesses, à la Concentration des Entreprises, aux Populations des Villes, aux Statistiques des Familles, etc., d'une loi Nouvelle, la loi de L'effect Proportionnel. (Recueil Sirey, 1931).

49 Stanley, M. H. R., Amaral, L. A. N., Buldyrev, S. V. et al., Scaling behaviour in the growth of companies. Nature 379, 804 (1996).

50 West, G., Scale: The Universal Laws of Life, Growth, and Death in Organisms, Cities, and Companies. (Penguin Publishing Group, 2017), p. a:13–22, b:109, c:434.

51 Camazine, S., Deneubourg, J.-L., Franks, N. R. et al., *Self-organization in Biological Systems.* (Princeton University Press, 2003); Schwager, M., Rus, D., Slotine, J.-J., Decentralized, adaptive coverage control for networked robots. Int. J. Robot. Res. 28, 357 (2009); Gross, M., Reading the hive mind. Curr. Biol. 29, R1055 (2019).

52 Richard, G., Sebastian, R., Ben, F. et al., Structural organisation and dynamics in king penguin colonies. J. Phys. D 51, 164004 (2018).

53 Husserl, E., The Crisis of European Sciences and Transcendental Phenomenology: An Introduction to Phenomenological Philosophy. (Northwestern University Press, 1970), p. 37.

54 Popper, K. R., The Myth of the Framework: In Defence of Science and Rationality. (Psychology Press, 1996), p. Author's note.

55 Armstrong, D. M., *What is a Law of Nature?* (Cambridge University Press, 1983).

56 Strogatz, S., Usain Bolt's split times and the power of calculus. Quanta Magazine (Apr. 3, 2019).

2. WHAT IS TIME?

1 Annila, A., The matter of time. Entropy 23, 943 (2021).

2 Leighton, R. B., Feynman, R. P., *Surely You're Joking, Mr. Feynman!* (Norton, 1985), p. 343.

3 Newton, I., *Philosophiæ Naturalis Principia Mathematica.* (Daniel Adee, 1846), p. 77.

4 Peacock, J. A., *Cosmological Physics.* (Cambridge University Press, 1998), p. 66.

5 Eddington, A. S., *The Nature of the Physical World.* (Cambridge University Press, 1948), p. a: 34.

6 Dyson, F. W., Eddington, A. S., Davidson, C. R., A Determination of the deflection of light by the Sun's gravitational field, from observations made at the solar eclipse of May 29, 1919. Phil. Trans. Roy. Soc. A. 220, 291 (1920).

7 Wheeler, J. A., Hermann Weyl and the unity of knowledge. Am. Sci. 74, 366 (1986).

8 Monk, R., *Time Reborn* by Lee Smolin, The Guardian (Jun. 6, 2013).

9 Husserl, E., The Crisis of European Sciences and Transcendental Phenomenology: An Introduction to Phenomenological Philosophy. (Northwestern University Press, 1970), p. 37.

10 Dyson, F., *Disturbing the Universe.* (Harper & Row, 1981), p. 193.

11 Schrödinger, E., Nobel Lecture: The fundamental idea of wave mechanics www.nobelprize.org/prizes/physics/1933/schrodinger/lecture

12 Laughlin, R. B., A Different Universe: Reinventing Physics from the Bottom Down. (Basic Books, 2005), p. d:50.

13 Rossi, P., *The Birth of Modern Science.* (Wiley, 2001), p. 42–55.

14 Einstein, A., Born, M., *The Born-Einstein Letters.* (MacMillan, 1971), p. 91.

15 Hoffmann, B., Dukas, H., *Albert Einstein: Creator and Rebel.* (Viking Press, 1973), p. 257.

16 Pruss, A. R., *The Principle of Sufficient Reason: A Reassessment.* (Cambridge University Press, 2006), p. 3.

17 Carroll, S., Even physicists don't understand quantum mechanics: Worse, they don't seem to want to understand it., New York Times (Sept. 7, 2019).

18 Byrne, P., The many worlds of Hugh Everett. Sci. Am. (Oct. 21, 2008).

19 Maxwell, N., Does orthodox quantum theory undermine, or support, scientific realism? Philos. Q. 43, 139 (1993).

20 Feynman, R. P., The relation of science and religion (May 2, 1956) calteches.library.caltech.edu/49/2/Religion

21 Baggott, J., Farewell to Reality: How Modern Physics Has Betrayed the Search for Scientific Truth. (Pegasus Books, 2013), p. a:ix–xiv, b:e148–151, 266–272.

22 Sagan, C., The Demon-Haunted World:

Science as a Candle in the Dark. (Random House, 2011).

23 Zeh, H. D., *The Physical Basis of The Direction of Time*. (Springer Science & Business Media, 2001).

24 Rovelli, C., *The Order of Time*. (Penguin Books Limited, 2018).

25 Maxwell, J. C., A dynamical theory of the electromagnetic field. Phil. Trans. R. Soc. A 155, 459 (1865).

26 Popper, K. R., *The Logic of Scientific Discovery*. (Psychology Press, 2002), p. a:75, b:280.

27 Einstein, A., *Ideas And Opinions*. (Crown Publishers Inc., 1954), p. 271.

28 Markie, P., The Stanford Encyclopedia of Philosophy: Rationalism vs. Empiricism. (2015); Nersessian, N. J., Faraday to Einstein: Constructing Meaning in Scientific Theories. (Martinus Nijhoff Publishers, 1984).

29 Tinsley, J. N., Molodtsov, M. I., Prevedel, R. et al., People can sense single photons. Nat. Commun. 7, 12172 (2016).

30 Hasan, A., Fumerton, R., The Stanford Encyclopedia of Philosophy: Knowledge by Acquaintance vs. Description. (2014).

31 Annila, A., Natural thermodynamics. Physica A 444, 843 (2016).

32 Wittgenstein, L., *Philosophical Investigations*. (Basil Blackwell, 1958), p. 109.

33 Annila, A., Salthe, S., Threads of time. ISRN Therm. 850957, 7 (2012).

34 Raman, C. V., Bhagavantam, S., Experimental proof of the spin of the photon. Nature 129, 22 (1932).

35 Kuhn, T. S., *Black-Body Theory and the Quantum Discontinuity, 1894–1912*. (Oxford University Press, 1978), p. 170–187.

36 Heisenberg, W., Physics and Philosophy: The Revolution in Modern Science. (HarperCollins, 2007).

37 Minev, Z. K., Mundhada, S. O., Shankar, S. et al., To catch and reverse a quantum jump mid-flight. Nature (2019).

38 Bernstein, J., Quantum leaps. Inference Review 4 (2018).

39 Unsöld, A., Baschek, B., *The New Cosmos*. (Springer-Verlag, 1983).

40 Nernst, W., Über einen Versuch, von quantentheoretischen Betrachtungen zur Annahme stetiger Energieänderungen zurückzukehren. Verhandlungen der Deutschen Physikalischen 18, 83 (1916).

41 Curd, P., Graham, D. W., *The Oxford Handbook of Presocratic Philosophy*. (Oxford University Press, 2008).

42 Davies, P., Superforce: The Search for a Grand Unified Theory of Nature. (Simon and Schuster, 1985), p. 104.

43 Bacon, F., *Novum Organum*. (Cambridge University Press, 2000), p. 39.

44 Whitaker, A., *Einstein, Bohr and the Quantum Dilemma*. (Cambridge University Press, 1996), p. 177.

45 Smolin, L., *Time Reborn*. (Houghton Mifflin Harcourt, 2013), p. 110–113.

46 Einstein, A., Relativity: the Special and the General Theory. (Methuen & Co Ltd, 1920).

47 McGrew, W. F., Zhang, X., Fasano, R. J. et al., Atomic clock performance enabling geodesy below the centimetre level. Nature 564, 87 (2018).

48 Berry, M. V., *Principles of Cosmology and Gravitation*. (Cambridge University Press, 1976), p. 91.

49 Tuisku, P., Pernu, T. K., Annila, A., In the light of time. Proc. R. Soc. A. 465, 1173 (2009).

50 Pound, R. V., Rebka Jr., G. A., Gravitational red-Shift in nuclear resonance. Phys. Rev. Lett. 3, 439 (1959).

51 Unruh, W. G., Notes on black-hole evaporation. Phys. Rev. D 14, 870 (1976).

52 Hafele, J. C., Keating, R. E., Around-the-world atomic clocks: predicted relativistic time gains. Science 177, 166 (1972).

53 Rossi, B., Hall, D. B., Variation of the rate of decay of mesotrons with momentum. Phys. Rev. 59, 223 (1941).

54 Annila, A., Least-time paths of light. Mon. Not. R. Astron. Soc. 416, 2944 (2011).

55 Carroll, S., From Eternity to Here: The Quest for the Ultimate Theory of Time. (Penguin, 2010).

56 Reichl, L. E., *A Modern Course in Statistical Physics*. (John Wiley & Sons, 2016).

57 Reich, E. S., Quantum paradox seen in diamond. Nature Magazine (Aug. 20, 2013).

58 Annila, A., Salthe, S., Physical foundations of evolutionary theory. J. Non-Equil. Therm. 35, 301 (2010).

59 Annila, A., Baverstock, K., Discourse on

order vs. disorder. Comm. Integr. Biol. 9, e1187348 (2016).

60 Ball, P., *Critical Mass.* (Random House, 2014), p. 85-87.

61 Boltzmann, L., *The Second Law of Thermodynamics* (D. Reidel, 1974), p. a:13–14, b:14–32.

62 Annila, A., Kuismanen, E., Natural hierarchy emerges from energy dispersal. Biosystems 95, 227 (2009).

63 Annila, A., Annila, E., Why did life emerge? Int. J. Astrobio. 7, 293 (2008).

64 Locke, S., How does solar power work? Sci. Am. (Oct. 20, 2008).

65 Rashed, R., A pioneer in anaclastics: Ibn Sahl on burning mirrors and lenses. Isis 81, 464 (1990).

66 Mahoney, M. S., *The Mathematical Career of Pierre de Fermat 1601–1665.* (Princeton University Press, 1994).

67 Leibniz, G. W., *Theodicy.* (Lightning Source, 2010), p. 128.

68 Laplace, P.-S., *Essai Philosophique Sur Les Probabilités.* (H. Remy, 1829), p. 3.

69 Rodych, V., The Stanford Encyclopedia of Philosophy, Wittgenstein's Philosophy of Mathematics. (2011).

70 Mäkelä, T., Annila, A., Natural patterns of energy dispersal. Phys. Life Rev. 7, 477 (2010).

71 Kimura, M., *The Neutral Theory of Molecular Evolution.* (Cambridge University Press, 1983).

72 Sharma, V., Annila, A., Natural process – natural selection. Biophys. Chem. 127, 123 (2007).

73 Einstein, A., Theorie der Opaleszenz von homogenen Flüssigkeiten und Flüssigkeitsgemischen in der Nähe des kritischen Zustandes. Ann. Phys. 338, 1275 (1910).

74 Gibbs, J. W., *The Collected Works of J. Willard Gibbs.* (Yale University Press, 1948).

75 Maupertuis, P.-L. M. d., Les Loix du mouvement et du repos déduites d'un principe métaphysique. Hist. Acad. Roy. Sci. Belleslett., 267 (1746).

76 Terrall, M., The Man Who Flattened the Earth: Maupertuis and the Sciences in the Enlightenment. (University of Chicago Press, 2002).

77 Kaila, V. R. I., Annila, A., Natural selection for least action. Proc. R. Soc. A. 464, 3055 (2008).

78 Annila, A., The 2nd law of thermodynamics delineates dispersal of energy. Int. Rev. Phys. 4, 29 (2010).

79 Megginson, L. C., Lessons from Europe for American business. Soc. Sci. Q. 44, 3 (1963).

80 Einstein, A., *Autobiographical Notes.* (Open Court Publishing Company, 1979), p. 31.

81 Tyack, D. B., Ways of Seeing: An essay on the history of compulsory schooling. Harvard Educ. Rev. 46, 367 (1976).

82 Hugo, V., *Choses vues 1830–1871.* (Le cercle du livre de France, 1951), p. 155.

83 Stubhaug, A., *Gösta Mittag-Leffler: A Man of Conviction.* (Springer Science & Business Media, 2010), p. 312.

84 Kuhn, T. S., The Essential Tension: Selected Studies in Scientific Tradition and Change. (Chicago University Press, 1977), p. 21–30.

85 Feyerabend, P., *Science in a Free Society.* (New Left Books, 1978), p. 52.

86 Einstein, A., *The Collected Papers of Albert Einstein, Geometry and Experience.* (Princeton University Press, 2002), p. 123–124.

87 Ekeberg, B., Metaphysical Experiments: Physics and the Invention of the Universe. (University of Minnesota Press, 2019), p. 75.

88 Zinsser, J. P., Emilie du Chatelet: Daring Genius of the Enlightenment. (Penguin, 2007).

89 Clausius, R., On a mechanical theorem applicable to heat. Phil. Mag. 40, 122 (1870).

90 Annila, A., Rotation of galaxies within gravity of the universe. Entropy 18, 191 (2016).

91 Annila, A., The meaning of mass. Int. J. Theor. Math. Phys. 2, 67 (2012).

92 Penzias, A. A., Wilson, R. W., A Measurement of excess antenna temperature at 4080 MC/S. Astrophys. J. 142, 419 (1965).

93 Smolin, L., Unger, R. M., *The Singular Universe and the Reality of Time: A Proposal in Natural Philosophy.* (Cambridge University Press, 2014), p. 391. Cortês, M., Smolin, L., The universe as a process of unique events. Phys. Rev. D 90,

084007 (2014).

94 Berra, Y., When You Come to a Fork in the Road, Take It!: Inspiration and Wisdom from One of Baseball's Grandest Heroes. (Thorndike Press, 2001), p. 159.

95 Connes, A., Rovelli, C., Von Neumann algebra automorphisms and time-thermodynamics relation in Generally covariant quantum theories. Classical and Quantum Gravity 11, 2899 (1994).

96 Grönholm, T., Annila, A., Natural distribution. Math. Biosci. 210, 659 (2007).

97 Neubert, T. N., A Critique of Pure Physics: Concerning the Metaphors of New Physics. (Xlibris, 2009), p. 56.

98 Musser, G., A Defense of the reality of time. Quanta Magazine (May 16, 2017).

99 Cowen, R., "Time Crystals" Could be a legitimate form of perpetual motion. Sci. Am. (Feb. 27, 2012)

100 Forrest, P., The Stanford Encyclopedia of Philosophy: The Identity of Indiscernibles. (2016).

101 Butler, T., Plague History: Yersin's discovery of the causative bacterium in 1894 enabled, in the subsequent century, scientific progress in understanding the disease and the development of treatments and vaccines. Clin. Microbiol. Infect. 20, 202 (2014).

102 Mauskopf, S. H., The atomic structural theories of Ampère and Gaudin: molecular speculation and Avogadro's hypothesis. Isis 60, 61 (1969).

103 Wheeler, J. A., Albert Einstein in The World Treasury of Physics, Astronomy, and Mathematics. (Little, Brown, & Co, 1991), p. 567–569.

104 Wheeler, J. A., *Biographic Memoirs* (National Academy Press, 1980), p. 102.

105 Hintikka, J., *Inquiry as Inquiry: A Logic of Scientific Discovery*. (Springer Science & Business Media, 2013).

106 Jaakkola, S., Sharma, V., Annila, A., Cause of chirality consensus. Curr. Chem. Biol. 2, 53 (2008) arXiv:0906.0254.

107 Jaakkola, S., El-Showk, S., Annila, A., The driving force behind genomic diversity. Biophys. Chem. 134, 136, 232 (2008) arXiv:0807.0892.

108 Würtz, P., Annila, A., Roots of diversity relations. J. Biophys., 654672 (2008).

109 Smolin, L., The Trouble with Physics: The Rise of String Theory, the Fall of a Science, and What Comes Next. (Mariner Books, 2007), p. 256.

110 Neumann, J. v., Method in the Physical Sciences in The Unity of Knowledge. (Doubleday & Co., 1955), p. 157.

111 Bush, V., As we may think. The Atlantic Monthly (July 1945).

112 Mermin, D., What's wrong with this pillow? Physics Today 42, 9 (1989).

113 Salam, A., Unification of Fundamental Forces: The First 1988 Dirac Memorial Lecture. (Cambridge University Press, 2005), p. 99.

3. WHAT IS EVERYTHING MADE OF?

1 Albert, R., Barabási, A.-L., Statistical mechanics of complex networks. Rev. Mod. Phys. 74, 47 (2002).

2 Melsen, A. G. v., From Atomos to Atom: The History and Concept of the Atom. (Dover Phoenix Editions, 1952).

3 Freeman, K., Ancilla to the Pre-Socratic Philosophers: A Complete Translation of the Fragments in Diels Fragmente Der Vorsokratiker. (Forgotten Books, 1948).

4 Rossi, P., *The Birth of Modern Science*. (Wiley, 2001), p. a:63–66, b:140.

5 LeBel, M.-M., Physics stops where natural metaphysics starts (2009) fqxi.org/community/forum/topic/506

6 Smolin, L., *Time Reborn*. (Houghton Mifflin Harcourt, 2013), p. a:110–113, b:141.

7 Forrest, P., The Stanford Encyclopedia of Philosophy: The Identity of Indiscernibles. (2016).

8 Sambursky, S., *The Physical World of the Greeks*. (Princeton University Press, 2014), p. 22.

9 Kuhn, T. S., The Essential Tension: Selected Studies in Scientific Tradition and Change. (Chicago University Press, 1977), p. 21–30.

10 Newton, I., Opticks: Or, a Treatise of the Reflexions, Refractions, Inflexions and Colours of Light. (William Innys at the West-End of St. Paul's, 1730).

11 Kuhlmann, M., Physicists debate whether the world is made of particles or fields – or something else entirely. Sci. Am. (Aug.

1, 2013).

12 Hobson, A., There are no particles, there are only fields. Am. J. Phys. 81, 211 (2013).

13 Sciamanda, R. J., There are no particles, and there are no fields. Am. J. Phys. 81, 645 (2013).

14 Dalton, J., *A New System of Chemical Philosophy*. (William Dawson & Sons, 1808).

15 Lewis, G. N., The Conservation of photons. Nature 118, 874 (1926).

16 Heeck, J., How stable is the photon? Phys. Rev. Lett. 111, 021801 (2013).

17 Stone, A. D., *Einstein and the Quantum: The Quest of the Valiant Swabian*. (Princeton University Press, 2013); Cormier-Delanoue, C., From relativity to the photon. Found. Phys. 19, 1171 (1989).

18 Neuenschwander, D. E., *Emmy Noether's Wonderful Theorem*. (JHU Press, 2017), p. 344.

19 Annila, A., Rotation of galaxies within gravity of the universe. Entropy 18, 191 (2016).

20 Lakatos, I., The role of crucial experiments in science. Stud. Hist. Philos. Sci. A 4, 309 (1974).

21 Einstein, A., On the method of theoretical physics. Philos. Sci. 1, 163 (1934).

22 Pruss, A. R., *The Principle of Sufficient Reason: A Reassessment*. (Cambridge University Press, 2006), p. 3.

23 Unsöld, A., Baschek, B., *The New Cosmos*. (Springer-Verlag, 1983).

24 Feynman, R. P., *Feynman Lectures on Gravitation*. (Addison-Wesley, 1995), p. 10.

25 Conselice, C. J., Wilkinson, A., Duncan, K. et al., The evolution of galaxy number density at z < 8 and its implications. Astrophys. J. 830, 83 (2016).

26 Mach, E., *The Science of Mechanics*. (Cambridge University Press, 2013).

27 Sciama, D. W., On the origin of inertia. Mon. Not. R. Astron. Soc. 113, 34 (1953).

28 Barbour, J. B., Pfister, H., *Mach's Principle: From Newton's Bucket to Quantum Gravity*. (Springer Science & Business Media, 1995).

29 Feynman, R. P., 'What Do You Care What Other People Think?': Further Adventures of a Curious Character. (Penguin, 2007).

30 Michelson, A. A., Morley, E. W., On the relative motion of the Earth and the luminiferous ether. Am. J. Sci. 34, 333 (1887).

31 Crawford Jr., F. S., *Waves Berkeley Physics course*. (McGraw-Hill 1968), p. Chapter 5.

32 Misner, C. W., Thorne, K. S., Wheeler, J. A., *Gravitation*. (Freeman, 1973).

33 Nersessian, N. J., *Faraday to Einstein: Constructing Meaning in Scientific Theories*. (Martinus Nijhoff Publishers, 1984), p. a:65–66, b:115, c:146.

34 Maxwell, J. C., A dynamical theory of the electromagnetic field. Phil. Trans. R. Soc. A 155, 459 (1865).

35 Poynting, J. H., On the transfer of energy in the electromagnetic field. Phil. Trans. R. Soc. 175, 343 (1884).

36 Heaviside, O., *Electromagnetic Theory*. (American Mathematical Society, 2003), p. 10.

37 Trouton, F. T., Noble, H. R., The mechanical forces acting on a charged electric condenser moving through space. Phil. Trans. Royal Soc. A 202, 165 (1903).

38 Grahn, P., Annila, A., Kolehmainen, E., On the carrier of inertia. AIP Advances 8, 035028 (2018).

39 Boyle, R., *The Works of the Honourable Robert Boyle*. (J. and F. Rivington, 1772), p. cxii.

40 Greene, B., The Fabric of the Cosmos: Space, Time, and the Texture of Reality. (Knopf Doubleday Publishing Group, 2007).

41 Feynman, R. P., Leighton, R. B., Sands, M. L., The Feynman Lectures on Physics, Desktop Edition Volume III: The New Millennium Edition. (Basic Books, 2013), p. 1–1.

42 Arndt, M., Nairz, O., Vos-Andreae, J. et al., Wave-particle duality of C60 molecules. Nature 401, 680 (1999).

43 Eibenberger, S., Gerlich, S., Arndt, M. et al., Matter–wave interference of particles selected from a molecular library with masses exceeding 10 000 amu. Phys. Chem. Chem. Phys. 15, 14696 (2013).

44 Trutz, B., Double-slit experiment (Jun. 11, 2009)

commons.wikimedia.org/wiki/File:Doub
leslitexperiment.svg

45 Smolin, L., Unger, R. M., The Singular
Universe and the Reality of Time: A
Proposal in Natural Philosophy.
(Cambridge University Press, 2014), p.
362.

46 Jacques, V., Wu, E., Grosshans, F. et al.,
experimental realization of Wheeler's
delayed-choice gedanken experiment.
Science 315, 966 (2007).

47 Dirac, P. A. M., *The Principles of Quantum
Mechanics.* (Clarendon Press, 1981), p. a:9,
b:266.

48 Magyar, G., Mandel, L., Interference
fringes produced by superposition of two
independent maser light beams. Nature
198, 255 (1963); Pfleegor, R. L., Mandel,
L., Interference of independent photon
beams. Phys. Rev. 159, 1084 (1967).

49 Hong, C. K., Ou, Z. Y., Mandel, L.,
Measurement of subpicosecond time
intervals between two photons by
interference. Phys. Rev. Lett. 59, 2044
(1987).

50 Press, T. A., Ping-pong makes physics
come alive, The Deseret News (Mar. 19,
1989).

51 Phys.Org., Researchers chart the 'secret'
movement of quantum particles (Dec. 22,
2017) phys.org/news/2017-12-secret-
movement-quantum-particles

52 Griffiths, D., *Introduction to Quantum
Mechanics.* (Pearson Prentice Hall, 2005),
p. 468.

53 Kahneman, D., *Thinking, Fast and Slow.*
(Farrar, Straus and Giroux, 2011), p. 70.

54 Kocsis, S., Braverman, B., Ravets, S. et
al., Observing the average trajectories of
single photons in a two-slit interferometer. Science 332, 1170 (2011); Mahler, D.
H., Rozema, L., Fisher, K. et al., Experimental nonlocal and surreal Bohmian
trajectories. Sci. Adv. 2, e1501466 (2016).

55 Aharonov, Y., Bohm, D., Significance of
electromagnetic potentials in quantum
theory. Phys. Rev. D 115, 485 (1959).

56 Overstreet, C., Asenbaum, P., Curti, J. et
al., Observation of a gravitational
Aharonov-Bohm effect. Science 375, 226
(2022).

57 Freire, O., The Quantum Dissidents: Re-
building the Foundations of Quantum

Mechan-ics (1950–1990). (Springer,
2014), p. 41.

58 Becker, A., What Is Real?: The
Unfinished Quest for the Meaning of
Quantum Physics. (Basic Books, 2018).

59 Magaña-Loaiza, O. S., De Leon, I.,
Mirhosseini, M. et al., Exotic looped
trajectories of photons in three-slit
interference. Nature Communications 7,
13987 (2016).

60 Tuisku, P., Pernu, T. K., Annila, A., In
the light of time. Proc. R. Soc. A. 465,
1173 (2009).

61 Laughlin, R. B., *A Different Universe:
Reinventing Physics from the Bottom Down.*
(Basic Books, 2005), p. a:48, b:120–121.

62 Jackson, J. D., Okun, L. B., Historical
roots of gauge invariance. Rev. Mod.
Phys. 73, 663 (2001).

63 Tonomura, A., (BBC and Royal
Institution, 1994).

64 Wilson, C. M., Johansson, G.,
Pourkabirian, A. et al., Observation of
the dynamical Casimir effect in a
superconducting circuit. Nature 479, 376
(2011).

65 Leon, C. C., Rosławska, A., Grewal, A. et
al., Photon superbunching from a generic
tunnel junction. Sci. Adv. 5, eaav4986
(2019).

66 Heisenberg, W., Physics and Philosophy:
The Revolution in Modern Science.
(Penguin, 2000), p. 18.

67 Riek, C., Sulzer, P., Seeger, M. et al.,
Subcycle quantum electrodynamics.
Nature 541, 376 (2017).

68 Partanen, M., Häyrynen, T., Oksanen, J.
et al., Photon mass drag and the
momentum of light in a medium. Phys.
Rev. A 95, 063850 (2017).

69 Fulling, S. A., Davies, P. C. W., Radiation
from a moving mirror in two
dimensional space-time: conformal
anomaly. Proc. R. Soc. A. 348, 393
(1976).

70 McEvoy, P., *Niels Bohr: Reflections on
Subject and Object.* (Microanalytix, 2001), p.
135.

71 Smolin, L., Einstein's Unfinished
Revolution: The Search for What Lies
Beyond the Quantum. (Penguin
Publishing Group, 2019), p. 94.

72 Feyerabend, P., *Against Method.* (Verso,

1993), p. 52.

73 Bohm, D., A Suggested Interpretation of the quantum theory in terms of "hidden" variables. Phys. Rev. D 85, 166 (1952).

74 Neumann, J. v., *Mathematical Foundations of Quantum Mechanics*. (Princeton University Press, 1955).

75 Bell, J., *Speakable and Unspeakable in Quantum Mechanics*. (Cambridge University Press, 1987), p. 65.

76 Shen, J.-T., Fan, S., Strongly correlated two-photon transport in a one-dimensional waveguide coupled to a two-level system. Phys. Rev. Lett. 98, 153003 (2007).

77 Riek, C., Seletskiy, D. V., Moskalenko, A. S. et al., Direct sampling of electric-field vacuum fluctuations. Science 350, 420 (2015).

78 Couder, Y., Fort, E., Single-Particle Diffraction and Interference at a Macroscopic Scale. Phys. Rev. Lett. 97, 154101 (2006); Wolchover, N., Have we been interpreting quantum mechanics wrong this whole time? Quanta Magazine (Jun. 30, 2014).

79 Wolchover, N., Famous experiment dooms alternative to quantum weirdness. Quanta Magazine (Oct. 11, 2018).

80 Bush, J. W. M., The new wave of pilot-wave theory. Physics Today 68 (2015).

81 Weinberg, S., *The Quantum Theory of Fields*. (Cambridge University Press, 1995), p. 31–38.

82 Popper, K. R., The Myth of the Framework: In Defence of Science and Rationality. (Psychology Press, 1996), p. Author's note.

83 Sagnac, G., L'éther lumineux démontré par l'effet du vent relatif d'éther dans un interféromètre en rotation uniforme. Comptes Rendus 157, 708 (1913); Restuccia, S., Toroš, M., Gibson, G. M. et al., Photon bunching in a rotating reference frame. Phys. Rev. Lett. 123, 110401 (2019).

84 Massey, H. S. W., Lord Blackett. Phys. Today 27, 69 (1974).

85 Pati, J. C., Salam, A., Lepton number as the fourth "color". Phys. Rev. D 10, 275 (1974).

86 Ball, P., Splitting the quark. Nature (2007).

87 Annila, A., Least-time paths of light. Mon. Not. R. Astron. Soc. 416, 2944 (2011); Annila, A., Space, time and machines. Int. J. Theor. Math. Phys. 2, 16 (2012) arXiv:0910.2629; Annila, A., Kallio-Tamminen, T., Tangled in entanglements. Physics Essays 20, 495 (2010) arXiv:1006.0463; Koskela, M., Annila, A., Least-action perihelion precession. Mon. Not. R. Astron. Soc. 417, 1742 (2011); Annila, A., Probing Mach's principle. Mon. Not. R. Astron. Soc. 423, 1973 (2012).

88 Annila, A., The meaning of mass. Int. J. Theor. Math. Phys. 2, 67 (2012); Annila, A., All in action. Entropy 12, 2333 (2010).

89 Horgan, J., In physics, telling cranks from experts ain't easy. Sci. Am. (Dec. 11, 2011).

90 Wilczek, F., Materiality of a Vacuum www.youtube.com/watch?v=BBXDrNn 6pUg&feature=youtu.be

91 Whitaker, A., *Einstein, Bohr and the Quantum Dilemma*. (Cambridge University Press, 1996), p. 177.

92 Arago, F., Biographies of Distinguished Scientific Men. (Ticknor and Fields, 1859).

93 Einstein, A., Infeld, L., The Evolution of Physics: The Growth of Ideas From Early Concepts to Relativity and Quanta. (Simon and Schuster, 1938).

94 Dicke, R. H., Gravitation without a principle of equivalence. Rev. Mod. Phys. 29, 363 (1957).

95 Smolin, L., The Trouble with Physics: The Rise of String Theory, the Fall of a Science, and What Comes Next. (Mariner Books, 2007), p. a:vii–xxiii, b:3–17, c:349.

96 Greene, B., The Elegant Universe: Superstrings, Hidden Dimensions, and the Quest for the Ultimate Theory. (W. W. Norton, 2003), p. 448

97 Griffiths, D., *Introduction to Elementary Particles*. (Wiley-VCH, 2008).

98 Kedia, H., Bialynicki-Birula, I., Peralta-Salas, D. et al., Tying knots in light fields. Phys. Rev. Lett. 111, 150404 (2013).

99 Wheeler, J. A., Hermann Weyl and the unity of knowledge. Am. Sci. 74, 366 (1986).

100 Wikipedia, Physics beyond the Standard

Model (Aug. 15, 2019)
en.wikipedia.org/wiki/Physics_beyond_t
he_Standard_Model

[101] Bauer, T. H., Spital, R. D., Yennie, D. R. et al., The hadronic properties of the photon in high-energy interactions. Rev. Mod. Phys. 50, 261 (1978).

[102] Pais, A., Subtle is the Lord: The Science and the Life of Albert Einstein. (Oxford University Press, 1982), p. 178–179.

[103] Francis, C., *Light after Dark II: The Large and the Small.* (Troubador Publishing, 2016), p. 180.

[104] Strassler, M., OPERA: what went wrong (Apr. 2, 2012)
profmattstrassler.com/articles-and-posts/particle-physics-basics/neutrinos/neutrinos-faster-than-light/opera-what-went-wrong

[105] Raman, C. V., Bhagavantam, S., Experimental proof of the spin of the photon. Nature 129, 22 (1932).

[106] Kronig, R. D. L., The neutrino theory of radiation and the emission of β-rays. Nature 137, 149 (1936).

[107] Grupen, C., *Astroparticle Physics* (Springer 2005), p. 95.

[108] Higgs, P., Broken symmetries and the masses of gauge bosons. Phys. Rev. Lett. 13, 508 (1964).

[109] Englert, F., Brout, R., Broken symmetry and the mass of gauge vector mesons. Phys. Rev. Lett. 13, 321 (1964).

[110] Guralnik, G., Hagen, C. R., Kibble, T. W. B., Global conservation laws and massless particles. Phys. Rev. Lett. 13, 585 (1964).

[111] Rovelli, C., *Seven Brief Lessons on Physics.* (Penguin Publishing Group, 2016).

[112] Geim, A. K., Kim, P., Carbon wonderland, graphene, a newly isolated form of carbon, provides a rich lode of novel fundamental physics and practical applications. Sci. Am. 298, 90 (2008).

[113] Roche, J., What is mass? Eur. J. Phys. 26, 1 (2005).

[114] Thackray, A. W., The origin of Dalton's chemical atomic theory: Daltonian doubts resolved. Isis 57, 35 (1966).

[115] Annila, A., Kolehmainen, E., Atomism revisited. Physics Essays 29, 532 (2016).

[116] Lee, T. D., Yang, C. N., Question of parity conservation in weak interactions. Phys. Rev. 104, 254 (1956).

[117] Wu, C. S., Ambler, E., Hayward, R. W. et al., Experimental test of parity conservation in beta decay. Phys. Rev. 105, 1413 (1957).

[118] Lederman, L., Teresi, D., *The God Particle: If the Universe Is the Answer, What Is the Question?* (Houghton Mifflin Harcourt, 2006), p. Interlude C.

[119] Enz, C. P., *Of Mind and Spirit, Selected Essays of Charles Enz.* (World Scientific, 2009), p. 95.

[120] Mäkelä, T., Annila, A., Natural patterns of energy dispersal. Phys. Life Rev. 7, 477 (2010).

[121] Schrödinger, E., Über die kräftefreie Bewegung in der relativistischen Quantenmechanik. Sitz. Preuss. Akad. Wiss. Phys.-Math. 24, 418 (1930).

[122] Eddington, A. S., *New Pathways in Science.* (Cambridge University Press, 1935), p. 235.

[123] Feynman, R. P., *QED: The Strange Theory of Light and Matter.* (Universities Press, 1985), p. 152.

[124] Hestenes, D., The zitterbewegung interpretation of quantum mechanics. Found. Physics., 20, 1213 (1990).

[125] Griffiths, D. J., *Introduction to Electrodynamics.* (Cambridge University Press, 2017), p. 358–359.

[126] Bender, D., Derrick, M., Fernandez, E. et al., Tests of QED at 29 GeV center-of-mass energy. Phys. Rev. D 30, 515 (1984).

[127] Parson, A. L., A Magneton theory of the structure of the atom. Smithsonian Miscellaneous Collections 65, 1 (1915); Duffield, J., physicsdetective.com/a-worble-embracing-itself

[128] Compton, A. H., Scientific Papers of Arthur Holly Compton: X-Ray and Other Studies. (Chicago University Press, 1973), p. 777.

[129] Wood, C., The strange numbers that birthed modern algebra. Quanta Magazine (Sept. 6, 2018).

[130] Deaver, B. S., Fairbank, W. M., Experimental evidence for quantized flux in superconducting cylinders. Phys. Rev. Lett. 7, 43 (1961).

[131] Casimir, H. B. G., On the attraction between two perfectly conducting plates.

Proc. K. Ned. Akad. Wet. 51, 793 (1948); Somers, D. A. T., Garrett, J. L., Palm, K. J. et al., Measurement of the Casimir torque. Nature 564, 386 (2018).

132 Paalanen, M. A., Tsui, D. C., Gossard, A. C., Quantized Hall Effect at Low Temperatures. Phys. Rev. B 25, 5566 (1982).

133 Legros, A., Benhabib, S., Tabis, W. et al., Universal T-linear resistivity and Planckian dissipation in overdoped cuprates. Nat. Phys. 15, 142 (2018).

134 Kwon, D., Ten things you might not know about antimatter. Symmetry Magazine (Apr. 28, 2015).

135 O'Connor, T., Wong, H. Y., The Stanford Encyclopedia of Philosophy: Emergent Properties. (2015).

136 Pernu, T. K., Annila, A., Natural emergence. Complexity 17, 44 (2012).

137 Lamb, W. E., Nobel Lecture: Fine structure of the hydrogen atom www.nobelprize.org/prizes/physics/195 5/lamb/lecture

138 Lincoln, D., The inner life of quarks. Sci. Am. 307, 36 (2012).

139 Charitos, P., Interview with George Zweig (Dec 13, 2013) ep-news.web.cern.ch/content/interview-george-zweig

140 Georgi, H., Glashow, S. L., Unity of all elementary-particle forces. Phys. Rev. Lett. 32, 438 (1972), p. 1.

141 Glashow, S., Unification – Then and now indico.cern.ch/event/70765/contributio ns/2071844/attachments/1032322/1470 455/Glashow.pdf

142 Hansson, J., The "Proton spin crisis" – a quantum query. Prog. Phys. 3, 51 (2010).

143 Peskin, M. E., Schroeder, D. V., *An Introduction To Quantum Field Theory*. (Avalon Publishing, 1995).

144 Pohl, R., Nez, F., Fernandes, L. M. et al., Laser spectroscopy of muonic deuterium. Science 353, 669 (2016); Bezginov, N., Valdez, T., Horbatsch, M. et al., A measurement of the atomic hydrogen Lamb shift and the proton charge radius. Science 365, 1007 (2019); Matveev, A., Yost, D. C., Maisenbacher, L. et al., Two-photon frequency comb spectroscopy of atomic hydrogen. Science 370, 1061 (2020).

145 Street, J., Stevenson, E., New Evidence for the existence of a particle of mass intermediate between the proton and electron. Phys. Rev. D 52, 1003 (1937).

146 Bartusiak, M., Who ordered the muon?, (Sept. 27, 1987).

147 Abel, C., Afach, S., Ayres, N. J. et al., Measurement of the permanent electric dipole moment of the neutron. Phys. Rev. Lett. 124, 081803 (2020).

148 Mannel, T., Theory and phenomenology of CP violation. Nucl, Phys. B 167, 170 (2007).

149 Smolin, L., Three Roads to Quantum Gravity: A New Understanding of Space, Time and the Universe. (Basic Books, 2001), p. a:111, b:113.

150 Neubert, T. N., A Critique of Pure Physics: Concerning the Metaphors of New Physics. (Xlibris, 2009), p. 219.

151 Misra, B., Sudarshan, E. C. G., The Zeno's paradox in quantum theory. J. Math. Phys. 18, 756 (1977).

152 Jammer, M., The Philosophy of Quantum Mechanics: The Interpretations of Quantum Mechanics in Historical Perspective. (Wiley, 1974), p. 151.

153 Blackett, P. M. S., Nobel lecture: Cloud chamber researches in nuclear physics and cosmic radiation www.nobelprize.org/prizes/physics/194 8/blackett/lecture

154 Cole, K. C., Sympathetic Vibrations: Reflections on Physics as a Way of Life. (Bantam Books, 1985), p. 228.

155 Ball, P., Quantum theory rebuilt from simple physical principles. Quanta Magazine (Aug. 30, 2017).

156 Griffiths, R. B., *Consistent Quantum Theory*. (Cambridge University Press, 2002), p. 408.

157 Aaij, R., Abellán Beteta, C., Adeva, B. et al., Observation of a narrow pentaquark state, Pc4312+, and of the two-peak structure of the Pc4450+. Phys. Rev. Lett. 122, 222001 (2019).

158 Finnish Society of Sciences and Letters, Section for Mathematics and Physics: Randomness and order in the exact sciences.

159 Veltman, M. J. G., *Facts and Mysteries in Elementary Particle Physics*. (World Scientific, 2003), p. 348.

160 Alduino, C., Alessandria, F. et al., First results from CUORE: A search for lepton number violation via 0vbb decay of 130Te. Phys. Rev. Lett. 120, 132501 (2018).

161 Handbook of LHC Higgs cross sections: 2. differential distributions. (2012).

162 Blucher, E., *Properties of the Tau Lepton*. (Atlantica Séguier Frontières, 1993), p. 241–243.

163 Ciezarek, G., Franco Sevilla, M., Hamilton, B. et al., A challenge to lepton universality in B-meson decays. Nature 546, 227 (2017).

164 Tomita, A., Chiao, R. Y., Observation of Berry's topological phase by use of an optical fiber. Phys. Rev. Lett. 57, 937 (1986).

165 Wolfenstein, L., Neutrino oscillations and stellar collapse. Phys. Rev. D 20, 2634 (1979).

166 Berry, M. V., *Principles of Cosmology and Gravitation*. (Cambridge University Press, 1976), p. 91.

167 Euler, L., De la force de percussion et sa veritable mesure. Mem. Acad. R. Sci. B-Lett. Berlin 24–5, 21 (1745).

168 Woit, P., Not Even Wrong: The Failure of String Theory and the Search for Unity in Physical Law. (Basic Books, 2006).

169 Lakatos, I., *Criticism and the Growth of Knowledge*. (Cambridge University Press, 1970), p. 92.

170 Lindley, D., What's wrong with quantum mechanics? Phys. Rev. Focus 16, 10 (2005).

171 Einstein, A., Podolsky, B., Rosen, N., Can quantum-mechanical description of physical reality be considered complete? Phys. Rev. Lett. 47, 777 (1935).

172 Pais, A., Einstein and the quantum theory. Rev. Mod. Phys. 51, 863 (1979).

173 Mercier de Lépinay, L., Ockeloen-Korppi, C. F., Woolley, M. J. et al., Quantum mechanics–free subsystem with mechanical oscillators. *Science* 372, 625 (2021).

174 Hesse, M. B., Action at a distance in classical physics. Isis 46, 337 (1955).

175 Megidish, E. et al., Entanglement swapping between photons that have never coexisted. ev. Phys. Rev. Lett. A 110, 210403 (2013).

176 Vailati, E., *Leibniz and Clarke: A Study of Their Correspondence*. (Oxford University Press, 1997).

177 Terrall, M., The Man Who Flattened the Earth: Maupertuis and the Sciences in the Enlightenment. (University of Chicago Press, 2002), p. 48.

178 Tesla, N., Radio power will revolutionize the world. Modern Mechanics and Inventions, 40 (Jul. 1, 1934).

179 Husserl, E., The Crisis of European Sciences and Transcendental Phenomenology: An Introduction to Phenomenological Philosophy. (Northwestern University Press, 1970), p. 46.

180 Illy, J., Einstein teaches Lorentz, Lorentz teaches Einstein their collaboration in general relativity, 1913–1920. Arch. Hist. Exact Sci. 39, 247 (1989).

181 Dirac, P. A. M., Is there an aether? Nature 168, 906 (1951).

182 Giovagnoli, M., Angels in the Workplace: Stories and Inspirations for Creating a New World of Work. (Wiley, 1998).

183 Powell, C. S., *God in the Equation: How Einstein Transformed Religion*. (Simon and Schuster, 2003), p. 29.

4. WHY DOES THE UNIVERSE EXPAND?

1 Baryshev, Y., Teerikorpi, P., *Fundamental Questions of Practical Cosmology, Exploring the Realm of Galaxies*. (Springer Netherlands, 2012).

2 Slipher, V. M., The radial velocity of the Andromeda nebula. Lowell Observatory Bulletin 1, 56 (1913).

3 Slipher, V. M., Spectrographic observations of nebulae. Popular Astronomy 23, 21 (1915).

4 Annila, A., Least-time paths of light. Mon. Not. R. Astron. Soc. 416, 2944 (2011).

5 Hubble, E., A Relation between distance and radial velocity among extra-galactic nebulae. Proc. Natl. Acad. Sci. USA 15, 168 (1929).

6 Einstein, A., Kosmologische Betrachtungen zur allgemeinen Relativitaetstheorie. Sitzungsberichte der Königlich Preussischen Akademie der

Wissenschaften Berlin 1, 142 (1917).

7 Gamow, G., *My World Line*. (Viking Press, 1970), p. 44.

8 Friedman, A., Über die Möglichkeit einer Welt mit konstanter negativer Krümmung des Raumes. Z. Phys. 21, 326 (1924).

9 Friedman, A., Über die Krümmung des Raumes. Z. Phys. 10, 377 (1922).

10 Lemaître, G., Un univers homogène de masse constante et de rayon croissant rendant compte de la vitesse radiale des nébuleuses extra-galactiques. Ann. Soc. Sci. Bruxelles 47, 49 (1927).

11 Kuhn, T. S., *The Structure of Scientific Revolutions*. (University of Chicago Press, 1970), p. 78–88.

12 Weinberg, S., *Cosmology*. (Oxford University Press, 2008), p. 34–44.

13 Baryshev, Y., Teerikorpi, P., *Discovery of Cosmic Fractals*. (World Scientific Publishing Company, 2002); Hertzprung, E., Über die Sterne der Unterabteilung c und ac nach der Spektralklassifikation von A. C. Maury. Astron. Nachr. 179, 373 (1908); Salpeter, E., The luminosity function and stellar evolution. Astrophys. J. 121, 161 (1955); Faber, S. M., Jackson, R. E., Velocity dispersions and mass-to-light ratios for elliptical galaxies. Astrophys. J. 204, 668 (1976); Liu, X. D., Jones, B. J. T., Evolution of LYα clouds: observational biases due to line crowding. Mon. Not. R. Astron. Soc. 230, 481 (1988); Parnell, H. C., Carswell, R. F., Effects of blending on estimates of the redshift evolution of QSO LY-Alpha absorbers. Mon. Not. R. Astron. Soc. 230, 491 (1988).

14 Kearns, E., Kajita, T., Totsuka, Y., Detecting massive neutrinos. Sci. Am., 281, 64 (1999); Bahcall, J. N., Nobel lecture: Solving the mystery of the missing neutrinos (Apr. 28, 2004) www.sns.ias.edu/~jnb

15 Postman, M., Clusters as Tracers of Large-Scale Structure. In Evolution of Large-Scale Structure. (Print Partners Ipskamp, 1999), p. 270

16 Albert, R., Scale-free networks in cell biology. J. Cell Science 118, 4947 (2005).

17 Mortlock, D. J., Warren, S. J., Venemans, B. P. et al., A luminous quasar at a Redshift of Z = 7.085. Nature 474, 616 (2011); Schmidt, M., Schneider, D. P., Gunn, J. E., Spectroscopic CCD surveys for quasars at large redshift. IV. Evolution of the luminosity function from quasars detected by their Lyman-alpha emission. Astron. J. 110, 68 (1995).

18 Unsöld, A., Baschek, B., *The New Cosmos*. (Springer-Verlag, 1983).

19 Anderson, M. E., Gaspari, M., White, S. D. M. et al., Unifying X-ray scaling relations from galaxies to clusters. Mon. Not. R. Astron. Soc. 449, 3806 (2015).

20 West, G. B., Brown, J. H., Basal metabolic rate and Cellular Energetics: The origin of allometric scaling laws in biology from genomes to ecosystems: towards a quantitative unifying theory of biological structure and organization. J. Exp. Biol. 208, 1575 (2005).

21 Ostriker, J. P., Mitton, S., Heart of Darkness: Unraveling the Mysteries of the Invisible Universe. (Princeton University Press, 2013); Liddle, A., *An Introduction to Modern Cosmology*. (Wiley, 2007); Ryden, B., *Introduction to Cosmology*. (Addison Wesley, 2002).

22 Barrow, J. D., Tiple, F. J., *The Anthropic Cosmological Principle*. (Clarendon Press, 1986); Collins, C. B., Hawking, S., Why is the universe isotropic? Astrophys. J. 180, 317 (1973).

23 Annila, A., Evolution of the universe by the principle of least action. Physics Essays 30, 248 (2017).

24 Annila, A., Space, time and machines. Int. J. Theor. Math. Phys. 2, 16 (2012) arXiv:0910.2629.

25 Pruss, A. R., *The Principle of Sufficient Reason: A Reassessment*. (Cambridge University Press, 2006).

26 Maxwell, J. C., A dynamical theory of the electromagnetic field. Phil. Trans. R. Soc. A 155, 459 (1865).

27 Janek, V., What shape is the universe? (Dec. 23, 2015) phys.org/news/2015-05-universe

28 Turner, M. S., Wilczek, F., Is our vacuum metastable? Nature 298, 633 (1982).

29 Thomson, W. B. K., On the dynamical theory of heat: with numerical results deduced from Mr. Joule's equivalent of a thermal unit and M. Regnault's obser-vations on steam. Proc. R. Soc. Edinb. 3,

48 (1851); Dyson, F. J., Time without end: physics and biology in an open universe. Rev. Mod. Phys. 51, 447 (1979).

30 Einstein, A., Über den Äther. Verh. Schweiz. Naturforsch. Ges. 105, 85 (1924).

31 Anderson, P. W., Brainwashed by Feynman? Phys. Today 53, 11 (2000).

32 Stachel, J., *Early History of Quantum Gravity*. (Kluwer, 1999); Rovelli, C., Notes for a Brief History of Quantum Gravity. arXiv:gr-qc/0006061 (2001).

33 Annila, A., Rotation of galaxies within gravity of the universe. Entropy 18, 191 (2016).

34 Annila, A., All in action. Entropy 12, 2333 (2010).

35 Nersessian, N. J., *Faraday to Einstein: Constructing Meaning in Scientific Theories*. (Martinus Nijhoff Publishers, 1984), p. a:37–39, b: 69–70.

36 Feynman, R. P., Leighton, R. B., Sands, M., *The Feynman Lectures on Physics*. (Addison–Wesley, 1964).

37 Annila, A., Cosmic rays report from the structure of space. Advan. Astron. 135025, 11 (2015).

38 Annila, A., Flyby anomaly via least action. Prog. Phys. 13, 92 (2017).

39 Annila, A., The meaning of mass. Int. J. Theor. Math. Phys. 2, 67 (2012).

40 Annila, A., Probing Mach's principle. Mon. Not. R. Astron. Soc. 423, 1973 (2012).

41 Feynman, R. P., *Feynman Lectures on Gravitation*. (Addison-Wesley, 1995), p. a:10, b:23–28.

42 Euler, L., *Briefe an eine deutsche Prinzessin*. (1776), p. 173–176.

43 Riemann, B., *Neue mathematische Prinzipien der Naturphilosophie*. (B. G. Teubner 1892), p. 528–538.

44 Yarkovsky, I. O., Hypothese cinetique de la Gravitation universelle et connexion avec la formation des elements chimiques. (Yarkovsky, I. O., 1888).

45 Griffiths, D., *Introduction to Elementary Particles*. (Wiley-VCH, 2008).

46 Heaviside, O., A Gravitational and electromagnetic analogy. The Electrician 31, 281 (1893).

47 Kepler, J., *New Astronomy*. (Cambridge University Press, 1992), p. 55.

48 Berkeley, G., The Works of George Berkeley. Vol. 1 of 4: Philosophical Works, 1705–1721. (Project Gutenberg, 1721).

49 McMullin, E., The Origins of the field concept in physics. Phys. Persp. 4, 13 (2002).

50 Faraday, M., Experimental researches in Electricity. 1164 (1831).

51 Faraday, M., On the possible relation of gravity to electricity. Phil. Trans. R. Soc. A 141, 1 (1851).

52 Laughlin, R. B., Nobel lecture: Fractional quantization. Rev. Mod. Phys. 71, 863 (1999).

53 Misner, C. W., Thorne, K. S., Wheeler, J. A., *Gravitation*. (Freeman, 1973).

54 Mäkelä, T., Annila, A., Natural patterns of energy dispersal. Phys. Life Rev. 7, 477 (2010).

55 Grahn, P., Annila, A., Kolehmainen, E., On the carrier of inertia. AIP Advances 8, 035028 (2018); Annila, A., The substance of gravity. Physics Essays 28, 208 (2015).

56 Sandage, A., The redshift-distance Relation. IX – Perturbation of the very nearby velocity field by the mass of the local group. Astrophys. J. 307, 1 (1986).

57 Wulf, T., Observations on the radiation of high penetration power on the Eiffel tower. Physik. Z. 11, 811 (1910).

58 Hess, V. F., Über Beobachtungen der durchdringenden Strahlung bei sieben Freiballonfahrten. Physik. Z. 13, 1084 (1912).

59 Drury, L. O. C., Origin of cosmic rays. Astropart. Phys. 39, 52 (2012); Blasi, P., The origin of galactic cosmic rays. Astron. Astrophys. Rev. 21, 70 (2013).

60 Nagano, M., Search for the end of the energy spectrum of primary cosmic rays. New J. Phys. 11, 065012 (2009).

61 Zatsepin, G. T., Kuzmin, V. A., Upper limit of the spectrum of cosmic rays. J. Exp. Theor. Phys. 4, 78 (1966).

62 Newton, I., Original Letter from Isaac Newton to Richard Bentley. (1692).

63 Berra, Y., When You Come to a Fork in the Road, Take It!: Inspiration and Wisdom from One of Baseball's Greatest Heroes. (Thorndike Press, 2001), p. a:53, b:159.

64 Agullo, I., Del Rio, A., Navarro-Salas, J., Electromagnetic duality anomaly in curved spacetimes. Phys. Rev. Lett. 118, 111301 (2017); Francis, M. R., The hunt for the truest north. Symmetry Magazine (Sept. 13, 2016).

65 Weinberg, S., Nobel Lectures: Conceptual foundations of the unified theory of weak and electromagnetic interactions. (World Scientific Publishing Co, 1992).

66 Laughlin, R. B., *A Different Universe: Reinventing Physics from the Bottom Down.* (Basic Books, 2005), p. a:xv, b:120–121, c:125.

67 Kragh, H., *Dirac: A Scientific Biography.* (Cambridge University Press, 1990), p. 184.

68 Barrow, J. D., *The Constants Of Nature.* (Random House, 2010), p. 92.

69 Giudice, G. F., The dawn of the post-naturalness era. arXiv:1710.07663 (2017).

70 Carter, B., Large number coincidences and the anthropic principle in cosmology. In *Confrontation of cosmological theories with observational data; Proceedings of the Symposium.* ed. Ellis, G. (1973), Vol. 43, p. 3225.

71 Baggott, J., Farewell to Reality: How Modern Physics Has Betrayed the Search for Scientific Truth. (Pegasus Books, 2013), p. 148–151, 266–272.

72 Smolin, L., *Time Reborn.* (Houghton Mifflin Harcourt, 2013), p. 104.

73 Melamed, Y. Y., Lin, M., The Stanford Encyclopedia of Philosophy: Principle of Sufficient Reason. (2017).

74 Rugh, S., Zinkernagel, H., The quantum vacuum and the cosmological constant problem. Stud. Hist. Philos. Sci. B 33 663 (2001).

75 Smolin, L., Unger, R. M., *The Singular Universe and the Reality of Time: A Proposal in Natural Philosophy.* (Cambridge University Press, 2014), p. a:42, b:259.

76 Guth, A. H., *The Inflationary Universe.* (Perseus Books, 1997).

77 Ijjas, A., Steinhardt, P. J., Loeb, A., Cosmic inflation theory faces challenges. Sci. Am., 316, 32 (2017).

78 Tryon, E. P., Is the universe a vacuum fluctuation? Nature 246, 396 (1973).

79 Assis, A. K. T., History of the 2.7 K temperature prior to Penzias and Wilson. Apeiron 2, 79 (1995).

80 Penzias, A., Nobel lecture: The origin of elements 1978 www.nobelprize.org/prizes/physics/1978/penzias/lecture

81 Aron, J., Planck shows almost perfect cosmos – plus axis of evil. New Sci. (Mar. 21, 2013).

82 Land, K., Magueijo, J., Examination of evidence for a preferred axis in the cosmic radiation anisotropy. Phys. Rev. Lett. 95, 071301 (2005).

83 Cruz, M., Martinez-Gonzalez, E., Vielva, P. et al., Detection of a non-Gaussian spot in WMAP. Mon. Not. R. Astron. Soc. 356, 29 (2004).

84 New survey hints at exotic origin for the Cold Spot (Apr. 25, 2017) phys.org/news/2017-04-survey-hints-exotic-cold.html#jCp

85 Wolchover, N., A Jewel at the heart of quantum physics. Quanta Magazine (Sept. 17, 2019).

86 Mackenzie, R., Shanks, T., Bremer, M. N. et al., Evidence against a supervoid causing the CMB Cold Spot. Mon. Not. R. Astron. Soc. 470, 2328 (2017).

87 Charitos, P., Interview with George Zweig (Dec. 13, 2013) ep-news.web.cern.ch/content/interview-george-zweig

88 Charley, S., What's left to learn about antimatter? Symmetry Magazine (Apr. 11, 2017).

89 Ahmadi, M., Alves, B. X. R., Baker, C. J. et al., Characterization of the 1S–2S transition in antihydrogen. Nature (2018).

90 Ulmer, S., Smorra, C., Mooser, A. et al., High-precision comparison of the antiproton-to-proton charge-to-mass ratio. Nature 524, 196 (2015); Hori, M., Aghai-Khozani, H., Sótér, A. et al., Buffer-gas cooling of antiprotonic helium to 1.5 to 1.7 K, and antiproton-to–electron mass ratio. Science 354, 610 (2016); Nagahama, H., Smorra, C., Sellner, S. et al., Sixfold improved single particle measurement of the magnetic moment of the antiproton. Nat. Commun. 8, 14084 (2017).

91 Raynova, I., Paving the way for a new antimatter experiment (Mar. 17, 2017) home.cern/cern-people/updates/2017/03/paving-way-

new-antimatter-experiment

92 Feynman, R. P., Nobel lecture: The development of the space-time view of quantum electrodynamics (2014) www.nobelprize.org/prizes/physics/196 5/feynman/lecture

93 Christenson, J. H., Cronin, J. W., Fitch, V. L. et al., Evidence for the 2π decay of the K02 meson system. Phys. Rev. Lett. 13, 138 (1964).

94 Mannel, T., Theory and phenomenology of CP violation. Nucl. Phys. B 167, 170 (2007).

95 Peccei, R. D., Quinn, H. R., CP conservation in the presence of pseudoparticles. Phys. Rev. Lett. 38, 1440 (1977).

96 Jaakkola, S., Sharma, V., Annila, A., Cause of chirality consensus. Curr. Chem. Biol. 2, 53 (2008) arXiv:0906.0254.

97 Seaton, K. A., Hackett, L. M., Stations, trains and small-world networks. Physica A 339, 635 (2004); Sen, P., Dasgupta, S., Chatterjee, A. et al., Small-world properties of the Indian railway network. Phys. Rev. E 67, 036106 (2003).

98 Sabulsky, D. O., Dutta, I., Hinds, E. A. et al., Experiment to detect dark energy forces using atom interferometry. Phys. Rev. Lett. 123, 061102 (2019).

99 Howell, E., Euclid: ESA's Search for dark matter & dark energy (Mar. 23, 2017) www.space.com/36195-euclid-esa-facts

100 Seife, C., "Tired-light" hypothesis gets re-tired. Science (Jun. 28, 2001).

101 Berry, M. V., *Principles of Cosmology and Gravitation.* (Cambridge University Press, 1976), p. 91.

102 The SCP "Union2.1" SN Ia compilation supernova.lbl.gov/Union; Riess, A. G. et al., Type Ia Supernova Distances at Redshift >1.5 from the Hubble Space Telescope Multi-cycle Treasury Programs: The Early Expansion Rate. Astrophys. J. 853, 126 (2018).

103 Whitbourn, J. R., Shanks, T., The galaxy luminosity function and the local hole. Mon. Not. R. Astron. Soc. 459, 496 (2016).

104 Bowman, J. D., Rogers, A. E. E., Monsalve, R. A. et al., An absorption profile centred at 78 megahertz in the sky-averaged spectrum. Nature 555, 67 (2018).

105 Sokol, J., New tactics clash on speed of expanding universe. Science 365, 306 (2019).

106 Merritt, D., Cosmology and convention. Stud. Hist. Philos. Sci. B 57, 41 (2017).

107 Kogut, A., Fixsen, D., Fixsen, S. et al., Arcade: Absolute radiometer for cosmology, astrophysics, and diffuse emission. New Astron. Rev. 50, 925 (2006).

108 Morandi, A., Sun, M., Probing dark energy via galaxy cluster outskirts. Mon. Not. R. Astron. Soc. 457, 3266 (2016).

109 Rácz, G., Dobos, L., Beck, R. et al., Concordance cosmology without dark energy. Mon. Not. R. Astron. Soc. Letters 469, L1 (2017).

110 Einstein, A., Relativity: the Special and the General Theory. (Methuen & Co Ltd, 1920).

111 Illy, J., Einstein teaches Lorentz, Lorentz teaches Einstein their collaboration in general relativity, 1913–1920. Arch. Hist. Exact Sci. 39, 247 (1989).

112 Laugwitz, D., Bernhard Riemann 1826–1866 Turning Points in the Conception of Mathematics. (Birkhäuser Basel, 1999), p. 263.

113 Annila, A., The 2nd law of thermodynamics delineates dispersal of energy. Int. Rev. Phys. 4, 29 (2010).

114 Ekeberg, B., Metaphysical Experiments: Physics and the Invention of the Universe. (University of Minnesota Press, 2019).

115 Sandage, A. R., Current problems in the extragalactic distance scale. Astrophys. J. 127, 513 (1958).

116 Freedman, W. L., Madore, B. F., Gibson, B. K. et al., Final results from the Hubble Space Telescope key project to measure the Hubble constant. Astrophys. J. 553, 47 (2001).

117 Mattig, W., Über den Zusammenhang zwischen Rotverschiebung und scheinbarer Helligkeit. Astron. Nachr. 284, 109 (1958).

118 Terrell, J., The luminosity distance equation in Friedmann cosmology. Am. J. Phys. 45, 869 (1977).

119 Rossi, P., *The Birth of Modern Science.* (Wiley, 2001), p. 297–300.

[120] Lubin, L. M., Sandage, A., The Tolman surface brightness test for the reality of the expansion. IV. A measurement of the Tolman signal and the luminosity evolution of early-type galaxies. Astron. J. 122, 1084 (2001).

[121] Zwicky, F., On the red shift of spectral lines through interstellar space. Proc. Natl. Acad. Sci. USA 15, 773 (1929).

[122] Suntola, T., *The Dynamic Universe*. (CreateSpace Independent Publishing Platform, 2011).

[123] Pound, R. V., Rebka Jr., G. A., Gravitational red-shift in nuclear resonance. Phys. Rev. Lett. 3, 439 (1959).

[124] Wojtak, R., Hansen, S. H., Hjorth, J., Gravitational redshift of galaxies in clusters as predicted by general relativity. Nature 477, 567 (2011).

[125] Michelson, A. A., Morley, E. W., On the relative motion of the Earth and the luminiferous ether. Am. J. Sci. 34, 333 (1887).

[126] Rubin, V., *Bright Galaxies, Dark Matters*. (American Institute of Physics, 1996).

[127] Zwicky, F., Die Rotverschiebung von extragalaktischen Nebeln. Helv. Phys. Acta 6, 110 (1933).

[128] White, L. A., Wimps in the dark matter wind. Symmetry Magazine (Apr. 4, 2017).

[129] Aprile, E., Aalbers, J., Agostini, F. et al., First dark matter search results from the XENON1T experiment. Phys. Rev. Lett. 119, 181301 (2017).

[130] Ouellet, J. L., Salemi, C. P., Foster, J. W. et al., Design and implementation of the ABRACADABRA-10 cm axion dark matter search. Physical Review D 99, 052012 (2019).

[131] Dattaro, L., What could dark matter be? Symmetry Magazine (Dec. 1, 2015).

[132] Wolchover, N., The man who's trying to kill dark matter. Wired (Jan. 14, 2017).

[133] Drukier, A. K., Freese, K., Spergel, D. N., Detecting cold dark-matter candidates. Phys. Rev. D 33, 3495 (1986).

[134] Castelvecchi, D., Beguiling dark-matter signal persists 20 years on. Nature (28 March 2018).

[135] Lelli, F., McGaugh, S. S., Schombert, J. M. et al., One law to rule them all: the radial acceleration relation of galaxies. Astrophys. J. 836, 152 (2017).

[136] Nicastro, F., Kaastra, J., Krongold, Y. et al., Observations of the missing baryons in the warm–hot intergalactic medium. Nature 558, 406 (2018).

[137] Lakatos, I., *The Methodology of Scientific Research Programmes: Philosophical Papers*. (Cambridge University Press, 1978), p. a:1, b:32, c:35.

[138] Traunmüller, H., in *Zeitschrift für Naturforschung A* (2018), Vol. 73, p. 1005.

[139] Kosowsky, A., Viewpoint: Connecting the bright and dark sides of galaxies. Phys. Persp. 9, 130 (2016).

[140] Ibata, R., Nipoti, C., Sollima, A. et al., Do globular clusters possess dark matter haloes? A case study in Ngc 2419. Mon. Not. R. Astron. Soc. 428, 3648 (2013).

[141] Naab, T., Ostriker, J. P., Theoretical challenges in galaxy formation. Ann. Rev. Astron. Astrophys. 55, 59 (2017).

[142] Wittenburg, N., Kroupa, P., Famaey, B., The formation of exponential disk galaxies in MOND. arXiv:2002.01941 (2020).

[143] Harvey, D., Courbin, F., Kneib, J. P. et al., A detection of wobbling brightest cluster galaxies within massive galaxy clusters. Mon. Not. R. Astron. Soc. 472, 1972 (2017).

[144] Contenta, F., Balbinot, E., Petts, J. A. et al., Probing dark matter with star clusters: A dark matter core in the ultra-faint dwarf Eridanus II. Mon. Not. R. Astron. Soc. 476, 3124 (2018).

[145] Stengers, I., *Cosmopolitics*. (University of Minnesota Press, 2010), p. 22.

[146] Bacon, F., *Novum Organum*. (Cambridge University Press, 2000), p. 38.

[147] Annila, A., Wikström, M., Dark matter and dark energy denote the gravitation of the expanding universe. Front. Phys. 10, 995977 (2022).

[148] Banik, I., Zhao, H., Dynamical history of the local group in ΛCDM – II. Including external perturbers in 3D. Mon. Not. R. Astron. Soc. 467, 2180 (2017).

[149] Pawlowski, M., Running away from Einstein phys.org/news/2017-03-einstein

[150] Dokkum, P. v., Danieli, S., Cohen, Y. et al., A galaxy lacking dark matter. Nature 555, 629 (2018).

[151] Famaey, B., McGaugh, S., Milgrom, M., MOND and the dynamics of NGC 1052-DF2. Mon. Not. R. Astron. Soc. 480, 473

(2018).

152 Kafle, P. R., Sharma, S., Lewis, G. F. et al., The need for speed: Escape velocity and dynamical mass measurements of the Andromeda galaxy. Mon. Not. R. Astron. Soc. 475, 4043 (2018).

153 Criss, R., Hofmeister, A., Galactic density and evolution based on the virial theorem, energy minimization, and conservation of Angular Momentum. Galaxies 6, 115 (2018).

154 Meurer, G. R., Obreschkow, D., Wong, O. I. et al., Cosmic Clocks: A tight radius–velocity relationship for H I-selected galaxies. Mon. Not. R. Astron. Soc. 476, 1624 (2018).

155 Clemence, G. M., The relativity effect in planetary motions. Rev. Mod. Phys. 19, 361 (1947).

156 Airy, G. B., Account of some circumstances historically connected with the discovery of the planet exterior to Uranus. Mon. Not. R. Astron. Soc. 7, 121 (1846).

157 Koskela, M., Annila, A., Least-action perihelion precession. Mon. Not. R. Astron. Soc. 417, 1742 (2011).

158 Dicke, R. H., Gravitation without a Principle of Equivalence. Rev. Mod. Phys. 29, 363 (1957).

159 Everitt, C. W. F., DeBra, D. B., Parkinson, B. W. et al., Gravity Probe B: Final results of a space experiment to test general relativity. Phys. Rev. Lett. 106, 221101 (2011).

160 Frontiers, Collaborative peer review www.frontiersin.org/about/review-system

161 Feyerabend, P., *Against Method*. (Verso, 1993), p. 52.

162 Dongen, J. v., *Einstein's Unification*. (Cambridge University Press, 2010), p. 44.

163 Clowe, D., Bradač, M., Gonzalez, A. H. et al., A direct empirical proof of the existence of dark matter. Astrophys. J. Lett. 648, L109 (2006).

164 Bozza, V., Jetzer, P., Mancini, L. et al., Microlensing by compact objects associated with gas clouds. Astron. Astrophys. 382, 6 (2002).

165 Massey, R., Harvey, D., Liesenborgs, J. et al., Dark matter dynamics in Abell 3827: New data consistent with standard cold dark matter. Mon. Not. R. Astron. Soc. 477, 669 (2018).

166 Milgrom, M., A Modification of the Newtonian dynamics as a possible alternative to the hidden mass hypothesis. Astrophys. J. 270, 365 (1983).

167 Scott, D., White, M., Cohn, J. D. et al., Cosmological difficulties with Modified Newtonian Dynamics (Or: La fin du MOND?). arXiv:astro-ph/0104435 (2001).

168 Sanders, R. H., Cosmology with Modified Newtonian Dynamics (MOND) Mon. Not. R. Astron. Soc. 296, 1009 (1998); Bekenstein, J. D., Relativistic gravitation theory for the MOND paradigm. Phys. Rev. D 70, 083509 (2004).

169 Finnish Society of Sciences and Letters, Section for Mathematics and Physics: Randomness and order in the exact sciences

170 Lakatos, I., The role of crucial experiments in science. Stud. Hist. Philos. Sci. A 4, 309 (1974).

171 Popper, K. R., *The Myth of the Framework: In Defence of Science and Rationality*. (Psychology Press, 1996), p. Author's note.

172 Michell, J., On the means of discovering the distance, magnitude, &c. of the fixed stars. Phil. Trans. R. Soc. 74, 35 (1784).

173 News, S., Einstein's Gravity: One Big Idea Forever Changed How We Understand the Universe. (Diversion Books, 2016).

174 Blundell, K., *Black Holes: A Very Short Introduction*. (Oxford University Press, 2015), p. 1.

175 Anderson, R. W., *The Cosmic Compendium: Black Holes*. (LULU Press, 2015), p. 42.

176 Longair, M. S., *The Cosmic Century: A History of Astrophysics and Cosmology*. (Cambridge University Press, 2006), p. 65.

177 Valtonen, M. J., Zola, S., Ciprini, S. et al., Primary black hole spin in OJ 287 as determined by the general relativity centenary flare. Astrophys. J. Lett. 819, L37 (2016).

178 Thorne, K., Black Holes & Time Warps: Einstein's Outrageous Legacy (Norton & Company, 1995).

179 Lehmonen, L., Annila, A., Fron. Phys. 954439, 10 (2022).

180 Oppenheimer, J. R., Volkoff, G. M., On massive neutron cores. Phys. Rev. 55, 374 (1939).

181 Profumo, S., Fundamental physics from the sky: Cosmic rays, gamma rays and the hunt for dark matter. J. Phys. Conf. Ser. 485, 012007 (2014).

182 Sanders, J. S., Fabian, A. C., Taylor, G. B. et al., A very deep chandra view of metals, sloshing and feedback in the Centaurus cluster of galaxies. Mon. Not. R. Astron. Soc. 457, 82 (2016); Hlavacek-Larrondo, J., Allen, S. W., Taylor, G. B. et al., Probing the extreme realm of active galactic nucleus feedback in the massive galaxy cluster, RX J1532.9+3021. Astrophys. J. 777, 163 (2013); Strickland, D. K., Heckman, T. M., Iron line and diffuse hard X-ray emission from the starburst galaxy M82. Astrophys. J. 658, 258 (2007).

183 Giacintucci, S., Markevitch, M., Johnston-Hollitt, M. et al., Discovery of a giant radio fossil in the Ophiuchus galaxy cluster. Astrophys. J. 891, 1 (2020).

184 Su, M., Fermi Bubble: Giant Gamma-Ray Bubbles in the Milky Way. Cosmic Rays in Star-Forming Environments. (Springer, 2013).

185 Ayala Solares, H. A., Hui, C. M., Hüntemeyer, P., Fermi Bubbles with HAWC. arXiv:1508.06592 (2015).

186 Chapline, G., Dark Energy Stars. arXiv:astro-ph/0503200 (2004).

187 Sauter, F., Über das Verhalten eines Elektrons im homogenen elektrischen Feld nach der relativistischen Theorie Diracs. Z. Phys. 69, 742 (1931); Heisenberg, W., Euler, H., Folgerungen aus der Diracschen Theorie des Positrons. Z. Phys. 98, 714 (1936).

188 Mignani, R. P., Testa, V., González Caniulef, D. et al., Evidence for vacuum birefringence from the first optical-polarimetry measurement of the isolated neutron Star RXJ1856.5−3754. Mon. Not. R. Astron. Soc. 465, 492 (2017).

189 Celotti, A., Miller, J. C., Sciama, D. W., Astrophysical evidence for the existence of black holes. Class. Quantum Grav. 16, A3 (1999).

190 Alexander, T., Bar-Or, B., A universal minimal mass scale for present-day central black holes. Nat. Astron. 1, 0147 (2017).

191 Pacucci, F., Ferrara, A., Grazian, A. et al., First identification of direct collapse black hole candidates in the early universe in CANDELS/GOODS-S. Mon. Not. R. Astron. Soc. 459, 1432 (2016).

192 Castelvecchi, D., The black-hole collision that reshaped physics. Nature 531, 428 (2016).

193 Chen, H. Y., Essick, R., Vitale, S. et al., Observational selection effects with ground-based gravitational wave detectors. Astrophys. J. 835, 31 (2017).

194 Cho, A., Merging neutron stars generate gravitational waves and a celestial light show. Science (Oct. 16, 2007).

195 Pavan, L., Bordas, P., Pühlhofer, G. et al., The long helical jet of the Lighthouse nebula, IGR J11014-6103. A&A 562, A122 (2014); Halpern, J. P., Tomsick, J. A., Gotthelf, E. V. et al., Discovery of X-ray pulsations from the integral source IGR J11014−6103. Astrophys. J. 795, L27 (2014).

196 Becker, W., Old pulsars still have new tricks to teach us. ESA (Jul. 26, 2006) www.esa.int/Science_Exploration/Space _Science/Old_pulsars_still_have_new_tr icks_to_teach_us

197 Loisel, G. P., Bailey, J. E., Liedahl, D. A. et al., Benchmark experiment for photoionized plasma emission from accretion-powered X-ray sources. Phys. Rev. Lett. 119, 075001 (2017); Dallilar, Y., Eikenberry, S. S., Garner, A. et al., A precise measurement of the magnetic field in the corona of the black hole binary V404 Cygni. Science 358, 1299 (2017); Gandhi, P., Bachetti, M., Dhillon, V. S. et al., An elevation of 0.1 light-seconds for the optical jet base in an accreting galactic black hole system. Nat. Astron. 1, 859 (2017).

198 Smolin, L., The Trouble with Physics: The Rise of String Theory, the Fall of a Science, and What Comes Next. (Mariner Books, 2007), p. 3–17.

199 Heisenberg, W., Physics and Philosophy: The Revolution in Modern Science.

(Penguin Books, 2000), p. 12.

200 Poincare, H., *Science and Hypothesis*. (Read Books Limited, 2016), p. xxii–xxiii.

201 Husserl, E., The Crisis of European Sciences and Transcendental Phenomenology: An Introduction to Phenomenological Philosophy. (Northwestern University Press, 1970), p. 46.

202 Russell, B., Theory of Knowledge: The 1913 Manuscript. (Routledge, 1984).

203 Lemaître, G., The beginning of the world from the point of view of quantum theory. Nature 127, 706 (1931).

5. WHAT DOES MATHEMATICS MEAN?

1 Xu, P., Sasmito, A. P., Yu, B. et al., Transport phenomena and properties in treelike networks. Appl. Mech. Rev. 68, 040802 (2016).

2 Etheridge, D. M., Steele, L. P., Langenfelds, R. L. et al., Natural and anthropogenic changes in atmospheric CO_2 over the last 1000 years from air in Antarctic ice and firn. J. Geophys. Res. 101, 4115 (1996).

3 Abrams, D. M., Strogatz, S. H., Linguistics: Modelling the dynamics of language death. Nature 424, 900 (2003).

4 Pinterest, Logarithmic spiral in Nature fi.pinterest.com/pin/3963168359455561 87

5 Galilei, G., *The Assayer*. (1618).

6 Wigner, E. P., The unreasonable effectiveness of mathematics in the natural sciences. Comm. Pure Appl. Math. 13, 1 (1960).

7 Hamming, R. W., The unreasonable effectiveness of mathematics. Am. Math. Monthly 87, 81 (1980).

8 Tegmark, M., Our Mathematical Universe, My Quest for the Ultimate Nature of Reality. (Penguin Books, 2014), p. a:243, b:341, c:361.

9 Smolin, L., The Trouble with Physics: The Rise of String Theory, the Fall of a Science, and What Comes Next. (Mariner Books, 2007), p. 46.

10 Husserl, E., The Crisis of European Sciences and Transcendental Phenomenology: An Introduction to Phenomenological Philosophy. (Northwestern University Press, 1970), p. 37.

11 Neuenschwander, D. E., *Emmy Noether's Wonderful Theorem*. (JHU Press, 2017), p. 344.

12 Keto, J., Annila, A., The capricious character of nature. Life 2, 165 (2012).

13 Landau, L. D., Lifshitz, E. M., *Fluid Mechanics* (Pergamon, 1959), p. 45.

14 Annila, A., Space, time and machines. Int. J. Theor. Math. Phys. 2, 16 (2012) arXiv:0910.2629.

15 ClayMathematicsInstitute, Navier–Stokes equation (Apr. 6, 2017) www.claymath.org/millennium-problems/navier-stokes-equation

16 Schroeder, M. R., Number Theory in Science and Communication: With Applications in Cryptography, Physics, Digital Information, Computing, and Self-Similarity. (Springer Berlin Heidelberg, 2005), p. 9.

17 Wiles, A., Modular elliptic curves and Fermat's last theorem. Ann. Math. 141, 443 (1995).

18 Melamed, Y. Y., Lin, M., The Stanford Encyclopedia of Philosophy: Principle of Sufficient Reason. (2017).

19 Annila, A., The meaning of mass. Int. J. Theor. Math. Phys. 2, 67 (2012); Annila, A., Kolehmainen, E., Atomism revisited. Physics Essays 29, 532 (2016).

20 Yang, C. N., Mills, R., Conservation of isotopic spin and isotopic gauge invariance. Phys. Rev. Lett. 96, 191 (1954).

21 Annila, A., All in action. Entropy 12, 2333 (2010).

22 ClayMathematicsInstitute, Yang–Mills and mass gap (Apr. 6, 2017) www.claymath.org/millennium-problems/yang-mills-and-mass-gap

23 Griffiths, D., *Introduction to Elementary Particles*. (Wiley-VCH, 2008).

24 Wikipedia, Logarithmic Spiral en.wikipedia.org/wiki/Logarithmic_spiral

25 Muriel, A., Reversibility and irreversibility from an initial value formulation. Phys. Lett. A 377, 1161 (2013).

26 Heaviside, O., *Electromagnetic Theory*. (Am. Math. Soc., 2003), p. 10.

27 Annila, A., Salthe, S., Physical foundations of evolutionary theory. J. Non-Equil. Therm. 35, 301 (2010).

28 Weyl, H., Gravitation and the electron. Proc. Natl. Acad. Sci. USA 15, 323

(1929).

29 Griffiths, D., *Introduction to Quantum Mechanics*. (Pearson Prentice Hall, 2005), p. 468.

30 Cavanagh, J., Fairbrother, W. J., Palmer III, A. G. et al., *Protein NMR Spectroscopy, Principles and Practice*. (Elsevier, 2007).

31 Tuisku, P., Pernu, T. K., Annila, A., In the light of time. Proc. R. Soc. A. 465, 1173 (2009).

32 Baggott, J., Farewell to Reality: How Modern Physics Has Betrayed the Search for Scientific Truth. (Pegasus Books, 2013), p. ix–xiv.

33 Hartle, J. B., Gravity: An Introduction to Einstein's General Relativity. (Addison-Wesley, 2003), p. 15.

34 Riemann, B., On the Hypotheses Which Lie at the Bases of Geometry. (Birkhäuser Basel, 2016).

35 Carroll, S., Spacetime and Geometry: An Introduction to General Relativity. (Pearson, 2003).

36 Hossenfelder, S., Lost in Math: How Beauty Leads Physics Astray. (Basic Books, 2018).

37 Weinberg, S., *Cosmology*. (Oxford University Press, 2008), p. 34–44.

38 Schumayer, D., Hutchinson, D. A. W., Physics of the Riemann hypothesis. Rev. Mod. Phys. 83, 307 (2011).

39 Woollam, R. C., Value of the Zeta function along the critical line from the Wolfram demonstrations project demonstrations.wolfram.com/ValueOfTheZetaFunctionAlongTheCriticalLine

40 du Sautoy, M., Nature's hidden prime number code. BBC New Magazine (Jun. 27, 2011).

41 Woltman, G., Great Internet Mersenne Prime Search (Jan. 19, 2016) www.mersenne.org

42 Annila, A., Rotation of galaxies within gravity of the universe. Entropy 18, 191 (2016).

43 Lehto, A., On the Planck scale and properties of matter. Int. J. Astrophys. Space Sci. 2, 57 (2014).

44 Wallis, J., Stedall, J. A., *The Arithmetic of Infinitesimals: John Wallis 1656*. (Springer, 2004).

45 Rossi, P., *The Birth of Modern Science*. (Wiley, 2001), p. 297–300.

46 Weinberg, S., *The Quantum Theory of Fields*. (Cambridge University Press, 1995), p. 31–38.

47 Kragh, H., *Dirac: A Scientific Biography*. (Cambridge University Press, 1990), p. 184.

48 Rugh, S., Zinkernagel, H., The quantum vacuum and the cosmological constant problem. Stud. Hist. Philos. Sci. B 33 663 (2001).

49 Ekeberg, B., Metaphysical Experiments: Physics and the Invention of the Universe. (University of Minnesota Press, 2019).

50 Woit, P., Not Even Wrong: The Failure of String Theory and the Search for Unity in Physical Law. (Basic Books, 2006).

51 Wolchover, N., As supersymmetry fails tests, physicists seek new ideas. Quanta Magazine (Nov. 20, 2012).

52 Steinhardt, P. J., The inflation debate: Is the theory at the heart of modern cosmology deeply flawed? Sci. Am., (Apr. 2, 2011).

53 Hunter, J., *The Works of John Hunter, FRS*. (Longman, 1835).

54 Deutsch, D., The Beginning of Infinity: Explanations that Transform the World. (Allen Lane, 2011), p. 496

55 Leifer, M. S., Pusey, M. F., Is a time symmetric interpretation of quantum theory possible without retrocausality? Proc. R. Soc. A. 473, 20160607 (2017).

56 Applegate, D. L., Bixby, R. M., Chvátal, V. et al., *The Traveling Salesman Problem*. (Princeton University Press, 2006).

57 ClayMathematicsInstitute, P vs NP problem (Apr. 6, 2017) www.claymath.org/millennium-problems/p-vs-np-problem

58 Annila, A., Physical portrayal of computational complexity. ISRN Comp. Math. 321372, 15 (2012).

59 Rittel, H. W. J., Webber, M. M., Dilemmas in a general theory of planning. Policy Sci. 4, 155 (1973).

60 Bayes, T., An Essay towards solving a Problem in the Doctrine of Chances. Phil. Trans. Roy. Soc. A. 53, 370 (1763).

61 Hartnett, K., To build truly intelligent machines, teach them cause and effect. Quanta Magazine (May 15, 2018).

[62] Rosenfeld, A., Zemel, R., Tsotsos, J. K., The elephant in the room. arXiv:1808.03305 (2018).

[63] Kuhn, T. S., The Essential Tension: Selected Studies in Scientific Tradition and Change. (Chicago University Press, 1977), p. 21–30.

[64] Strogatz, S. H., Nonlinear Dynamics and Chaos: With Applications to Physics, Biology, Chemistry, and Engineering. (Avalon Publishing, 2014).

[65] Thomas Unden, Daniel Louzon, Michael Zwolak et al., Revealing the emergence of classicality in nitrogen-vacancy centers. arXiv:1809.10456 (2018).

[66] Ackerman, L., Buckley, M. R., Carroll, S. M. et al., Dark matter and dark radiation. Phys. Rev. D 79, 023519 (2009).

[67] Erren, T. C., Cullen, P., Erren, M. et al., Ten simple rules for doing your best research, according to Hamming. Plos Comput. Biol. 3, e213 (2007).

[68] Hamming, R. W., Mathematics on a distant planet. Am. Math. Monthly 105, 640 (1998).

[69] Lee, J. A. N., *International Biographical Dictionary of Computer Pioneers*. (Fitzroy Dearborn, 1995), p. a:313, b:363.

6. HOW DID LIFE ORIGINATE?

[1] Annila, A., Kolehmainen, E., On the divide between animate and inanimate. J. Syst. Chem. 6, 2 (2015).

[2] West, G., Scale: The Universal Laws of Life, Growth, and Death in Organisms, Cities, and Companies. (Penguin Publishing Group, 2017).

[3] Mäkelä, T., Annila, A., Natural patterns of energy dispersal. Phys. Life Rev. 7, 477 (2010); Annila, A. Statistical physics of evolving systems. Entropy 23, 1590 (2021).

[4] Xu, J., Naturally and anthropogenically accelerated sedimentation in the lower Yellow River, China, over the past 13,000 years. Geogr. Ann. Ser. A 80, 67 (1998).

[5] Tennekes, H., The spectrum of animal flight: Insects to Pterosaurs. Prog. Aerosp. Sci. 36, 393 (1996).

[6] Hall, W. P., Physical basis for the emergence of autopoiesis, cognition and knowledge. Kororoit Institute Working Papers, 1 (2011).

[7] Rosenzweig, M. L., *Species Diversity in Space and Time*. (Cambridge University Press, 1995).

[8] Axelsen, J. B., Manrubia, S., River density and landscape roughness are universal determinants of linguistic diversity. Proc. R. Soc. B 281, 20133029 (2014).

[9] Darwin, C., On the Origin of Species by Means of Natural Selection. (Murray, 1859), p. 48.

[10] Poincare, H., *Science and Hypothesis*. (Read Books Limited, 2016), p. xxiv.

[11] Courtland, R., IEEE Spectrum. spectrum.ieee.org/nanoclast/semiconduc tors/processors/intel-now-packs-100-million-transistors-in-each-square-millimeter

[12] Milo, R., What is the total number of protein molecules per cell volume? A call to rethink some published values. Bioessays 35, 1050 (2013).

[13] Kleiber, M., Body size and metabolism. Hilgardia 6, 315 (1932).

[14] Keto, J., Annila, A., The capricious character of nature. Life 2, 165 (2012).

[15] Hesse, R., Untersuchungen über die Organe der Lichtempfindung bei niederen Thieren. II. Die Augen der Plathelminthen. Z. Wiss. Zool. 62, 527 (1897).

[16] Gehring, W. J., New perspectives on eye development and the evolution of eyes and photoreceptors. J. Hered. 96, 171 (2005).

[17] Jobling, J. A., *The Helm Dictionary of Scientific Bird Names*. (Bloomsbury Publishing, 2010), p. 33.

[18] Ball, P., *Critical Mass*. (Random House, 2014), p. 85-87.

[19] Gould, S. J., Punctuated equilibrium - A different way of seeing. New Sci. 94, 137 (1982).

[20] Maupertuis, P. L. M. d., *The Earthly Venus*. (Johnson Reprint Corp., 1966).

[21] Dobzhansky, T., Nothing in biology makes sense except in the light of evolution. Am. Biol. Teach. 35, 125 (1973).

[22] Terrall, M., The Man Who Flattened the Earth: Maupertuis and the Sciences in the Enlightenment. (University of Chicago Press, 2002), p. a:86, b:329, c:369.

[23] Popper, K. R., *The Logic of Scientific Discovery*. (Psychology Press, 2002), p. 75.

24 Noble, D., Physiology is rocking the foundations of evolutionary biology. Exp. Physiol. 98, 1235 (2013).

25 Wallace, A. R., On the Tendency of Species to form Varieties. Zool. J. Linn. Soc. 3, 45 (1858); Thompson, D. A. W., *On Growth and Form.* (Cambridge University Press, 1992).

26 Shapiro, J. A., *Evolution: A View from the 21st Century.* (Pearson Education, 2011).

27 Olkkonen, V. M., Gottlieb, P., Strassman, J. et al., In vitro assembly of infectious nucleocapsids of bacteriophage Phi 6: formation of a recombinant double-stranded RNA virus. Proc. Natl. Acad. Sci. USA 87, 9173 (1990).

28 Hull, D. L., in *Philosophy of Science.* Contemporary Philosophy: A New Survey. 2. ed. Fløistad, G. (Nijhoff, 1982), p. 280.

29 Xing, Y., Ree, R. H., Uplift-driven diversification in the Hengduan mountains, a temperate biodiversity hotspot. Proc. Natl. Acad. Sci. USA 114, E3444 (2017).

30 Brouwers, L., Thoughtomics. Sci. Am. (Feb. 16, 2012).

31 Oparin, A. I., Morgulis, S., *The Origin of Life.* (Dover Publications, 2003).

32 Michaelian, K., Thermodynamic dissipation theory for the origin of life. Earth Syst. Dynam. 2, 37 (2011).

33 GRC, Origin of Life, Gordon Research Conference (Jan. 20-25, 2008)

34 Dill, K. A., MacCallum, J. L., The protein-folding problem, 50 years on. Science 338, 1042 (2012).

35 Sharma, V., Kaila, V. R. I., Annila, A., Protein folding as an evolutionary process. Physica A 388, 851 (2009).

36 Boltzmann, L., *Populäre Schriften.* (Barth, 1905).

37 Meierhenrich, U., Kagan, H. B., Amino Acids and the Asymmetry of Life: Caught in the Act of Formation. (Springer, 2008).

38 Wikipedia, Chirality with hands (Oct. 22, 2011) en.wikipedia.org/wiki/Chirality

39 Peterson, H., A Treasury of the World's Great Speeches: Each Speech Prefaced with Its Dramatic and Biographical Setting and Placed in Its Full Historical Perspective. (Simon and Schuster, 1965), p. 469.

40 Jaakkola, S., Sharma, V., Annila, A., Cause of chirality consensus. Curr. Chem. Biol. 2, 53 (2008) arXiv:0906.0254.

41 Würtz, P., Annila, A., Roots of diversity relations. J. Biophys., 654672 (2008).

42 Annila, A., Annila, E., Why did life emerge? Int. J. Astrobio. 7, 293 (2008).

43 Sivonen, K., Jones, G., Cyanobacterial Toxins in Toxic Cyanobacteria in Water. (WHO, 1999), p. 41.

44 Georgiev, G. Y., Henry, K., Bates, T. et al., Mechanism of organization increase in complex systems. Complexity 21, 18 (2015).

45 Getling, A. V., Rayleigh-Benard Convection: Structures and Dynamics. (World Scientific, 1998).

46 Kovač, L., Life's Arrow: The epistemic singularity. EMBO reports 17, 1083 (2016).

47 Kováč, L., in *Closing Human Evolution: Life in the Ultimate Age.* ed. Kováč, L. (Springer International Publishing, 2015), p. 1.

48 Moore, D. S., The Developing Genome: An Introduction to Behavioral Epigenetics. (Oxford University Press, 2015).

49 Waddington, C. H., *The Strategy of the Genes.* (Taylor & Francis, 2014), p. 30–36.

50 Fisher, S. E., Ridley, M., Culture, genes, and the human revolution. Science 340, 929 (2013).

51 Brenner, S., Turing centenary: Life's code script. Nature 482, 461 (2012).

52 Watson, J. D., Crick, F. H. C., A structure for deoxyribose nucleic acid. Nature 171, 737 (1953).

53 Annila, A., Kolehmainen, E., Atomism revisited. Physics Essays 29, 532 (2016).

54 Crick, F., *What Mad Pursuit.* (Basic Books, 2008).

55 Ohno, S., So Much 'Junk DNA' in Our Genome. In Evolution of Genetic Systems. (Gordon & Breach, 1972), p. 366–370.

56 Ridley, M., Genome: The Autobiography of a Species in 23 Chapters. (HarperCollins Publishers, 2017).

57 Gregory, T. R., *The Evolution of the Genome.* (Elsevier Science, 2011).

58   Cullen, J. H., *Barbara McClintock*. (Facts On File, 2003), p. 96.

59   Jaakkola, S., El-Showk, S., Annila, A., The driving force behind genomic diversity. Biophys. Chem. 134, (136) 232 (2008) arXiv:0807.0892.

60   Annila, A., Baverstock, K., Genes without prominence: a reappraisal of the foundations of biology. J. R. Soc. Interface 11, 20131017 (2014).

61   Boyle, E. A., Li, Y. I., Pritchard, J. K., An expanded view of complex traits: from polygenic to omnigenic. Cell 169, 1177 (2017).

62   Huang, S., The tension between big data and theory in the "omics" era of biomedical research. Perspect Biol. Med. 61, 472 (2018).

63   Bettencourt, L. M. A., Lobo, J., Strumsky, D. et al., Urban scaling and its deviations: Revealing the structure of wealth, innovation and crime across cities. PLOS ONE 5, e13541 (2010).

64   Xu, P., Sasmito, A. P., Yu, B. et al., Transport phenomena and properties in treelike networks. Appl. Mech. Rev. 68, 040802 (2016).

65   Meymandi, A., The science of epigenetics. Psychiatry (Edgmont) 7, 40 (2010).

66   Darwin, C., The Descent of Man, and Selection in Relation to Sex. (John Murray, 1871).

67   Dawkins, R., The Blind Watchmaker: Why the Evidence of Evolution Reveals a Universe without Design. (W. W. Norton, 2015).

68   Williams, G. C., Mitton, J. B., Why reproduce sexually? J. Theor. Biol. 39, 545 (1973); Hartfield, M., Keightley, P. D., Current hypotheses for the evolution of sex and recombinationn. Integr. Zool. 7, 192 (2012).

69   Laplace, P.-S., *Essai Philosophique Sur Les Probabilités*. (H. Remy, 1829), p. 3.

70   Annila, A., Annila, E., The significance of sex. Biosystems 110, 156 (2012).

71   Frank, S. A., Puzzles in modern biology. I. Male sterility, failure reveals design. F1000Research 5, 2288 (2016).

72   Becker, G. S., *A Treatise on the Family, Enlarged Edition*. (Harvard University Press, 2009).

73   Feynman, R. P., Nobel lecture: The development of the space-time view of quantum electrodynamics (2014) www.nobelprize.org/prizes/physics/1965/feynman/lecture

74   Annila, A., Kuismanen, E., Natural hierarchy emerges from energy dispersal. BioSystems 95, 227 (2009).

75   Koskela, M., Annila, A., Looking for the Last Universal Common Ancestor. Genes 3, 81 (2012).

76   Cepelewicz, J., Fossil DNA reveals new twists in modern human origins. Quanta Magazine (Aug. 29, 2019).

77   Matson, J., Meteorite that fell in 1969 still revealing secrets of the early solar system. Sci. Am. (Feb. 15, 2010).

78   Warmflash, D., Weiss, B., Did life come from another world? Sci. Am. 293, 64 (2005).

79   Lovelock, J., *Gaia: A New Look at Life on Earth*. (Oxford University Press, 1987).

80   Doolittle, W. F., Is nature really motherly. Coevol. Q. Spring, 58 (1981).

81   Doolittle, W. F., Making evolutionary sense of Gaia. Trends Ecol. Evol. 34, 889 (2019).

82   Watson, A. J., Lovelock, J. E., Biological homeostasis of the global environment: The Pparable of Daisyworld. Tellus B 35B, 284 (1983).

83   Karnani, M., Annila, A., Gaia again. Biosystems 95, 82 (2009).

84   Green, J. K., Konings, A. G., Alemohammad, S. H. et al., Regionally strong feedbacks between the atmosphere and terrestrial biosphere. Nat. Geosci. 10, 410 (2017).

85   Doughty, C. E., Santos-Andrade, P. E., Shenkin, A. et al., Tropical forest leaves may darken in response to climate Change. Nat. Ecol. Evol. 2, 1918 (2018).

86   Diamond, J. M., Guns, Germs, and Steel: The Fates of Human Societies. (Norton, 1997); Diamond, J., Collapse: How Societies Choose to Fail or Succeed: Revised Edition. (Penguin, 2011).

87   Steffen, W., Rockström, J., Richardson, K. et al., Trajectories of the earth system in the Anthropocene. Proc. Natl. Acad. Sci. USA (2018).

88   Seager, S., Bains, W., The search for signs of life on exoplanets at the interface of

chemistry and planetary science. Sci. Adv. 1, e1500047 (2015).

89 Hart, M. H., Explanation for the absence of extraterrestrials on Earth. Q. J. R. Astron. Soc. 16, 128 (1975).

90 Marx, G., The myth of the Martians and the golden age of Hungarian science. Sci. Educ. 5, 225 (1996).

91 Drake, N., How my dad's equation sparked the search for extraterrestrial intelligence. Natl. Geogr. Mag. (Jun. 30, 2014).

92 Boer, M. A. d., Lammertsma, K., Scarcity of rare earth elements. Chem. Sus. Chem. 6, 2045 (2013).

93 Peirce, C. S., *Collected Papers of Charles Sanders Peirce.* (Belknap Press of Harvard University Press, 1974), p. 226.

7. WHAT IS CONSCIOUSNESS?

1 Reid, T., *Essays on the Intellectual Powers of Man.* (Cambridge University Press, 2011).

2 James, W., Does 'consciousness' exist? J. Philos. 1, 477 (1904).

3 Damasio, A. R., Self Comes to Mind: Constructing the Conscious Brain. (Pantheon Books, 2010), p. 40.

4 Goldstein, K., The Organism: A Holistic Approach to Biology Derived from Pathological Data in Man. (Zone Books, 2000), p. 361.

5 Marshall, N., Timme, N. M., Bennett, N. et al., Analysis of power laws, shape collapses, and neural complexity: New techniques and Matlab support via the NCC toolbox. Front. Physiol. 7, 250 (2016).

6 Braha, D., Global civil unrest: Contagion, self-organization, and prediction. PLoS One 7, e48596 (2012).

7 Schweitzer, F., Mach, R., The epidemics of donations: Logistic growth and power-Laws. PLoS One 3, e1458 (2008).

8 Karsai, M., Kaski, K., Barabási, A.-L. et al., Universal features of correlated bursty behaviour. Sci. Rep. 2, 397 (2012).

9 Bak, P., *How Nature Works.* (Oxford University Press, 1997).

10 Handel, P. H., Chung, A. L., *Noise in Physical Systems and 1/"f" Fluctuations.* (American Institute of Physics, 1993).

11 Bak, P., Tang, C., Wiesenfeld, K., Self-organized criticality: an explanation of 1/f noise. Phys. Rev. Lett. 59, 381 (1987).

12 Jensen, M. H., Obituary: Per Bak (1947–2002). Nature 420, 284 (2002).

13 Annila, A., On the character of consciousness. Front. Syst. Neurosci. 10, 27 (2016); Annila, A., The fundamental nature of motives. Front. Neurosci. 16, (2022).

14 Varpula, S., Annila, A., Beck, C., Thoughts about thinking: Cognition according to the second law of thermodynamics. Adv. Stud. Biol 5, 135 (2013).

15 He, B. J., Zempel, J. M., Snyder, A. Z. et al., The temporal structures and functional significance of scale-free brain activity. Neuron 66, 353 (2010).

16 Christensen, K., Danon, L., Scanlon, T. et al., Unified scaling law for earthquakes. Proc. Natl. Acad. Sci. USA 99, 2509 (2002).

17 Huberman, B. A., Adamic, L. A., Internet: growth dynamics of the World-Wide Web. Nature 401, 131 (1999).

18 Epstein, I. R., Pojman, J. A., An Introduction to Nonlinear Chemical Dynamics: Oscillations, Waves, Patterns, and Chaos. (Oxford University Press, 1998).

19 Marintchev, A., Fidelity and Quality Control in Gene Expression. (Academic, 2012).

20 Gutowitz, H., Cellular Automata: Theory and Experiment. (MIT Press, 1991).

21 Taleb, N. N., *The Black Swan: The Impact of the Highly Improbable.* (Random House Publishing Group, 2010), p. 147.

22 Mathews, P., Mitchell, L., Nguyen, G. T. et al., The nature and origin of heavy tails in retweet activity. arXiv:1703.05545 (2017).

23 Beggs, J. M., The criticality hypothesis: how local cortical networks might optimize information processing. Phil. Trans. R. Soc. A 366, 329 (2008).

24 Zohar, Y. C., Yochelis, S., Dahmen, K. A. et al., Controlling avalanche criticality in 2d nano arrays. Sci. Rep. 3, 1845 (2013).

25 Caldarelli, G., Scale-Free Networks: Complex Webs in Nature and Technology. (Oxford University Press, 2013).

26 Hartonen, T., Annila, A., Natural networks as thermodynamic systems.

Complexity 18, 53 (2012).

27 Matsubara, T., Statistics and dynamics in the large-scale structure of the universe. J. Phys. Conf. 31, 27 (2006); Coil, A. L., in *Planets, Stars and Stellar Systems: Volume 6: Extragalactic Astronomy and Cosmology*. ed. Oswalt, T.D. et al. Keel, W.C. (Springer Netherlands, 2013), p. 387.

28 Koch, C., Is consciousness universal? Sci. Am. (Jan. 1, 2014).

29 Kahneman, D., *Thinking, Fast and Slow*. (Farrar, Straus and Giroux, 2011), p. a:59, b:86, c:417.

30 Wolpert, L., You have to believe in something. The Telegraph (Jan. 25, 2001).

31 McGilchrist, I., The Master and His Emissary: The Divided Brain and the Making of the Western World. (Yale University Press, 2009).

32 Anderson, A., Mind: Multitasking by brain wave. Sci. Am. (Jan. 27, 2016).

33 Atmanspacher, H., The Stanford Encyclopedia of Philosophy: Quantum Approaches to Consciousness. (2015).

34 Forrest, P., The Stanford Encyclopedia of Philosophy: The Identity of Indiscernibles. (2016).

35 Horgan, J., David Chalmers thinks the dard problem is really hard. Sci. Am. (Apr. 16, 2017).

36 Chawla, D. S., To remember, the brain must actively forget. QuantaMagazine (Jul. 24, 2018).

37 Bargh, J. A., Chen, M., Burrows, L., Automaticity of social behavior: direct effects of trait construct and stereotype-activation on action. J. Pers. Soc. Psychol. 71, 230 (1996); Doyen, S., Klein, O., Pichon, C.-L. et al., Behavioral priming: It's all in the mind, but whose mind? PLoS One 7, e29081 (2012).

38 Pernu, T. K., Annila, A., Natural emergence. Complexity 17, 44 (2012).

39 Maupertuis, P.-L. M. d., *Essay de cosmologie*. (De l'Imp. d'Elie Luzac, 1751).

40 Nagel, T., What is it like to be a bat? Philos. Rev. 83, 435 (1974).

41 Gell-Mann, M., The Quark and the Jaguar: Adventures in the Simple and the Complex. (St. Martin's Press, 1995).

42 Nagel, T., Mind and Cosmos: Why the Materialist Neo-Darwinian Conception of Nature is Almost Certainly False. (Oxford University Press, 2012), p. 93.

43 Georgiev, G. Y., Henry, K., Bates, T. et al., Mechanism of organization increase in complex systems. Complexity 21, 18 (2015).

44 Capri, A. Z., Quips, Quotes and Quanta: An Anecdotal History of Physics. (World Scientific, 2011), p. 67.

45 Hardin, G., The tragedy of the commons. Science 162, 1243 (1968).

46 Würtz, P., Annila, A., Ecological succession as an energy dispersal process. Biosystems 100, 70 (2010).

47 Mäkelä, T., Annila, A., Natural patterns of energy dispersal. Phys. Life Rev. 7, 477 (2010).

48 Annila, A., Salthe, S., On intractable tracks. Physics Essays 25, 233 (2012).

49 Simons, D. J., Chabris, C. F., Gorillas in our midst: Sustained inattentional blindness for dynamic events. Perception 28, 1059 (1999).

50 Graziano, M. S. A., Webb, T. W., The attention schema theory: A mechanistic account of subjective awareness. Front. Psychol. 6, 500 (2015).

51 Merton, R. K., *Social Theory and Social Structure*. (Free Press, 1968), p. 157.

52 Dawkins, R., The Blind Watchmaker: Why the Evidence of Evolution Reveals a Universe without Design. (W. W. Norton, 2015).

53 Goodfellow, I. J., Pouget-Abadie, J., Mehdi Mirza, B. X. et al., Generative adversarial networks. arXiv:1406.2661 (2014).

54 Domingos, P., The Master Algorithm: How the Quest for the Ultimate Learning Machine Will Remake Our World. (Penguin Books Limited, 2015).

55 Shannon, C. E., A mathematical theory of communication. Bell Syst. Tech. J. 27, 379 (1948).

56 Douglas, M., Deciphering a meal. Daedalus 101, 61 (1972).

57 Karnani, M., Pääkkönen, K., Annila, A., The physical character of information. Proc. R. Soc. A 465, 2155 (2009).

58 Mcluhan, M., Understanding Media: The Extensions of Man. (Mentor, 1964), p. 7.

59 Clauset, A., Shalizi, C. R., Newman, M. E. J., Power-law distributions in empirical

data. SIAM Rev. 51, 661 (2009).

60 Landauer, R., Irreversibility and heat generation in the computing process. IBM J. Res. Dev. 5, 183 (1961).

61 Annila, A., Physical portrayal of computational complexity. ISRN Comp. Math. 321372, 15 (2012). arxiv/0906.1084

62 Neumann, J. v., Zur Theorie der Gesellschaftsspiele. Math. Ann. 100, 295 (1928).

63 Ball, P., *Critical Mass*. (Random House, 2014), p. 439.

64 Samuelson, P., *Foundations of Economic Analysis*. (Harvard University Press, 1947).

65 Nash, J. F., *Non-Cooperative Games*. (Princeton University, 1950).

66 Anttila, J., Annila, A., Natural games. Phys. Lett. A 375, 3755 (2011).

67 Leibniz, G. W., *Theodicy*. (Lightning Source, 2010), p. 128.

68 Shannon, C. E., Weaver, W., *The Mathematical Theory of Communication*. (Illinois University Press, 1971), p. 144

69 Wikipedia, Hypogram

70 Hirata, H., Yoshiura, S., Ohtsuka, T. et al., Oscillatory expression of the bHLH factor Hes1 regulated by a negative feedback loop. Science 298, 840 (2002).

71 Hobson, J. A., Pace-Schott, E. F., The Cognitive Neuroscience of sleep: Neuronal systems, consciousness and learning. Nat. Rev. Neurosci. 3, 679 (2002).

72 Nash, S., Making sense of Mount St. Helens. BioScience 60, 571 (2010).

73 Betzel, R. F., Bassett, D. S., Specificity and robustness of long-distance connections in weighted, interareal connectomes. Proc. Natl. Acad. Sci. USA 115, E4880 (2018).

74 Janata, P., Brain networks that track musical structure. Ann. N.Y. Acad. Sci. 1060, 111 (2005).

75 Wilson, B., I Am Brian Wilson: The Genius behind the Beach Boys. (Hodder & Stoughton, 2016).

76 Melamed, Y. Y., Lin, M., The Stanford Encyclopedia of Philosophy: Principle of Sufficient Reason. (2017).

77 Keto, J., Annila, A., The capricious character of nature. Life 2, 165 (2012).

78 Libet, B., Gleason, C. A., Wright, E. W. et al., Time of conscious intention to act in relation to onset of cerebral activity (readiness-potential). Brain 106, 623 (1983).

79 Dennett, D. C., *Consciousness Explained*. (Little, Brown, 2017).

80 Trevena, J., Miller, J., Brain preparation before a voluntary action: Evidence against unconscious movement initiation. Conscious. Cogn. 19, 447 (2010).

81 Penrose, R., Shadows of the Mind: A Search for the Missing Science of Consciousness. (Oxford University Press, 1994).

8. WHAT IS OUR DESTINY?

1 Alvarez, L. W., Alvarez, W., Asaro, F. et al., Extraterrestrial cause for the Cretaceous–Tertiary extinction. Science 208, 1095 (1980).

2 Zalasiewicz, J., Williams, M., Haywood, A. et al., The Anthropocene: A new epoch of geological time? Phil. Trans. R. Soc. A 369, 835 (2011).

3 West, G., Scale: The Universal Laws of Life, Growth, and Death in Organisms, Cities, and Companies. (Penguin Publishing Group, 2017).

4 Bettencourt, L. M. A., Lobo, J., Helbing, D. et al., Growth, innovation, scaling, and the pace of life in cities. Proc. Natl. Acad. Sci. USA 104, 7301 (2007).

5 Kelly, R., Lundy, M. G., Mineur, F. et al., Historical data reveal power-law dispersal patterns of invasive aquatic species. Ecography 37, 581 (2014).

6 Strogatz, S. H., Exploring complex networks. Nature 410, 268 (2001).

7 Annila, A., Kuismanen, E., Natural hierarchy emerges from energy dispersal. Biosystems 95, 227 (2009).

8 Clauset, A., Young, M., Gleditsch, K. S., On the frequency of severe terrorist events. J. Confl. Resolut. 51, 58 (2007) arXiv:physics/0606007.

9 Songa, W., Wanga, J., Satohb, K. et al., Three types of power-law distribution of forest fires in Japan. Ecol. Model. 196, 527 (2006).

10 Ruck, D. J., Bentley, R. A., Lawson, D. J., Religious change preceded economic change in the 20th century. Sci. Adv. 4 (2018).

11 Harris, M., *Cannibals and Kings: Origins of Cultures*. (Knopf Doubleday Publishing Group, 2011), p. 229.

12 Tavris, C., A sketch of Marvin Harris. Psychol. Today 8, 64 (1975).

13 Annila, A., Salthe, S., Cultural naturalism. Entropy 12, 1325 (2010).

14 Nafeez, M. A., A User's Guide to the Crisis of Civilization: And How to Save It. (Pluto Press, 2010), p. 312.

15 Westermarck, E., Sociology as a university study, inauguration of the Martin White Professorship of Sociology filosofia.fi/tallennearkisto/tekstit/5669

16 Rossi, P., *The Birth of Modern Science*. (Wiley, 2001), p. 206.

17 Ball, P., *Critical Mass*. (Random House, 2014), p. a:85-87, b:373-377, c:389-396.

18 Schimper, A. F. W., Über die Entwicklung der Chlorophyllkörner und Farbkörper. Bot. Zeitung. 41, 105 (1883); Sagan, L., On the Origin of Mitosing Cells. J. Theor. Biol. 14, 255 (1967).

19 Whitman, W., Coleman, D., Wiebe, W., Prokaryotes: The unseen majority. Proc. Natl. Acad. Sci. USA 95, 6578 (1998).

20 Samuelson, P. A., *Economics*. (Tata McGraw Hill, 2010).

21 Newman, M. E. J., Power laws, Pareto distributions and Zipf's law. Contemp. Phys. 46, 323 (2005).

22 Lovelock, J., *Gaia: a new look at life on earth*. (Oxford University Press, 1987).

23 Gould, S. J., Wonderful Life: The Burgess Shale and the Nature of History. (W. W. Norton, 1990).

24 Annila, A., Salthe, S., Economies evolve by energy dispersal. Entropy 11, 606 (2009).

25 Estola, M., Newtonian Microeconomics: A Dynamic Extension to Neoclassical Micro Theory. (Springer, 2017).

26 Walker, D. A., *William Jaffe's Essays on Walras*. (Cambridge University Press, 1983).

27 Morgenstern, O., Thirteen critical points in contemporary economic theory: An interpretation. J. Econ. Lit. 10, 1163 (1972).

28 May, R. M., Simple mathematical models with very complicated dynamics. Nature 261, 459 (1976).

29 Mäkelä, T., Annila, A., Natural patterns of energy dispersal. Phys. Life Rev. 7, 477 (2010).

30 Kellert, S. H., In the Wake of Chaos: Unpredictable Order in Dynamical Systems. (Chicago University Press, 1994).

31 Plerou, V., Gopikrishnan, P., Nunes Amaral, L. A. et al., Scaling of the distribution of price fluctuations of individual companies. Phys. Rev. E 60, 6519 (1999).

32 Peters, O., Hertlein, C., Christensen, K., A Complexity view of rainfall. Phys. Rev. Lett. 88, 018701 (2002).

33 Samuelson, P. A., A Note on the pure theory of consumer's behaviour. Economica 5, 61 (1938).

34 Koivu-Jolma, M., Annila, A., Epidemic as a natural process. Math. Biosci. 299, 97 (2018).

35 Obstfeld, M., Rogoff, K., The six major puzzles in international macroeconomics: Is there a common cause? in *NBER Macroeconomics Annual 2000*. (MIT Press, 2001).

36 Rodda, G. H., Fritts, T. H., Chiszar, D., The disappearance of Guam's wildlife. Bioscience 47, 565 (1997).

37 Thakur, M. P., Tilman, D., Purschke, O. et al., Climate warming promotes species diversity, but with greater taxonomic redundancy, in complex environments. Sci. Adv. (2017). Tisdell, C. A., *Competition, Diversity and Economic Performance*. (Edward Elgar, 2013).

38 Anttila, J., Annila, A., Natural games. Phys. Lett. A 375, 3755 (2011).

39 Kahneman, D., *Thinking, Fast and Slow*. (Farrar, Straus and Giroux, 2011), p. 251.

40 Bacon, F., *Novum Organum*. (Cambridge University Press, 2000), p. 39.

41 Cepelewicz, J., Goals and rewards redraw the brain's map of the world. Quanta Magazine (Mar. 28, 2019).

42 Leighton, R. B., Feynman, R. P., *Surely You're Joking, Mr. Feynman!* (Norton, 1985), p. 108.

43 Kaplan, S., Framing contests: Strategy making under uncertainty. Organ. Sci. 19, 729 (2008).

44 Foster, R., Kaplan, S., Creative Destruction: Why Companies That Are Built to Last Underperform the Market–

And How to Success fully Transform Them. (Crown Publishing Group, 2011), p. 384.

45 Nagel, T., Mind and Cosmos: Why the Materialist Neo-Darwinian Conception of Nature is Almost Certainly False. (Oxford University Press, 2012), p. 185.

46 Maxwell, N., Science, reason, knowledge, and wisdom: A critique of specialism. Inquiry 23, 19 (1980).

47 Verschuur, G. L., *Impact!: The Threat of Comets and Asteroids*. (Oxford University Press, 1997).

9. WHY WILL THE WORLDVIEW CHANGE?

1 Bird, A., The Stanford Encyclopedia of Philosophy: Thomas Kuhn. (2013).

2 Winograd, T., Flores, F., *Understanding Computers and Cognition: A New Foundation for Design*. (Addison-Wesley Longman, 1987), p. 97.

3 Kuhn, T. S., *The Essential Tension: Selected Studies in Scientific Tradition and Change*. (Chicago University Press, 1977), p. a:208–221, b:225–239, c:320–329.

4 Maxwell, N., Science, reason, knowledge, and wisdom: A critique of specialism. Inquiry 23, 19 (1980).

5 Neubert, T. N., A Critique of Pure Physics: Concerning the Metaphors of New Physics. (Xlibris, 2009), p. a:22, b:217.

6 Kuhn, T. S., *The Structure of Scientific Revolutions*. (University of Chicago Press, 1970), p. a:64, b:76, c:122–145, d:137, e:211, f:219.

7 Forbes, N., Mahon, B., Faraday, Maxwell, and the Electromagnetic Field: How Two Men Revolutionized Physics. (Prometheus Books, 2014).

8 Kuhn, T. S., *Black-Body Theory and the Quantum Discontinuity, 1894–1912*. (Oxford University Press, 1978), p. 170–187.

9 Michelson, A. A., *Light Waves and Their Uses*. (Chicago University Press, 1903), p. 23.

10 Merritt, D., Cosmology and convention. Stud. Hist. Philos. Sci. B 57, 41 (2017).

11 Ludlow, A. D., Benitez-Llambay, A., Schaller, M. et al., Mass-discrepancy acceleration relation: A natural outcome of galaxy formation in cold dark matter halos. Phys. Rev. Lett. 118, 161103 (2017).

12 Peskin, M. E., Schroeder, D. V., *An Introduction To Quantum Field Theory*. (Avalon Publishing, 1995).

13 Giudice, G. F., The dawn of the post-naturalness era. arXiv:1710.07663 (2017).

14 Rovelli, C., Notes for a brief history of quantum gravity. arXiv:gr-qc/0006061 (2000).

15 Montaigne, M. d., *The Essays of Montaigne*. (Project Gutenberg, 2006), p. 518.

16 Carter, B., Large number coincidences and the anthropic principle in cosmology. In *Confrontation of Cosmological Theories with Observational Data; Proceedings of the Symposium* (Krakow, Poland 1974), p. 3225.

17 Richter, B., presented at the AIP Conf. Proc. 2007 (unpublished).

18 Copi, C. J., Huterer, D., Schwarz, D. J. et al., On the large-angle anomalies of the microwave sky. Mon. Not. R. Astron. Soc. 367, 79 (2006).

19 Nieto, M. M., Anderson, J. D., Earth flyby anomalies. Phys. Today 62, 76 (2009); Annila, A., Flyby anomaly via least action. Prog. Phys. 13, 92 (2017).

20 Nieto, M. M., Turyshev, S. G., Anderson, J. D., The Pioneer anomaly: The data, its meaning, and a future test. AIP Conf. Proc. 758, 113 (2005) arXiv:gr-qc/0411077; Annila, A., Space, time and machines. Int. J. Theor. Math. Phys. 2, 16 (2012) arXiv:0910.2629; Greaves, E. D., Bracho, C., Gift, S., Rodriguez, A. M., A solution to the Pioneer anomalous annual and diurnal residuals. Prog. Phys. 17, 168 (2021).

21 Corneliussen, S. T., Peer-reviewed NASA "EM drive" report draws media enthusiasm and doubt. Phys. Today (2016); Grahn, P., Annila, A., Kolehmainen, E., On the exhaust of electromagnetic drive. AIP Advances 6, 065205 (2016).

22 Dawkins, R., The Blind Watchmaker: Why the Evidence of Evolution Reveals a Universe without Design. (W. W. Norton, 2015).

23 Portin, P., The development of genetics in the light of Thomas Kuhn's theory of scientific revolutions. Recent Adv. DNA Gene Seq. 9, 14 (2015).

24 Fleck, L., *Genesis and Development of a Scientific Fact.* (Chicago University Press, 2012), p. 41.

25 Terrall, M., The Man Who Flattened the Earth: Maupertuis and the Sciences in the Enlightenment. (University of Chicago Press, 2002), p. 300.

26 Terekhovich, V., Metaphysics of the principle of least action. Stud. Hist. Philos. Sci. B 62, 189 (2018).

27 Maxwell, N., Does orthodox quantum theory undermine, or support, scientific realism? Philos. Q. 43, 139 (1993).

28 Faye, J., The Stanford Encyclopedia of Philosophy: Copenhagen Interpretation of Quantum Mechanics. (2014); Goldstein, S., The Stanford Encyclopedia of Philosophy: Bohmian Mechanics. (2017).

29 Rossi, P., *The Birth of Modern Science.* (Wiley, 2001), p. 42–55.

30 Taleb, N. N., *The Black Swan: The Impact of the Highly Improbable.* (Random House Publishing Group, 2010), p. 147.

31 Smolin, L., *Time Reborn.* (Houghton Mifflin Harcourt, 2013), p. xxxi.

32 Annila, A., Least-time paths of light. Mon. Not. R. Astron. Soc. 416, 2944 (2011); Koskela, M., Annila, A., Least-action perihelion precession. Mon. Not. R. Astron. Soc. 417, 1742 (2011).

33 Annila, A., Rotation of galaxies within gravity of the universe. Entropy 18, 191 (2016).

34 Laughlin, R. B., A Different Universe: Reinventing Physics from the Bottom Down. (Basic Books, 2005), p. 26.

35 Bacon, F., *Novum Organum.* (Cambridge University Press, 2000).

36 Harari, Y. N., *Homo Deus: A Brief History of Tomorrow.* (Random House, 2016).

37 Flew, A., What is indoctrination? Stud. Philos. Educ. 4, 281 (1966).

38 Woods, R., Barrow, R., *An Introduction to Philosophy of Education.* (Taylor & Francis, 2006), p. 70–83.

39 Husserl, E., The Crisis of European Sciences and Transcendental Phenomenology: An Introduction to Phenomenological Philosophy. (Northwestern University Press, 1970).

40 Smolin, L., The Trouble with Physics: The Rise of String Theory, the Fall of a Science, and What Comes Next. (Mariner Books, 2007), p. 256.

41 Hossenfelder, S., Science needs reason to be trusted. Nat. Phys. 13, 316 (2017).

42 Dicke, R. H., Gravitation without a principle of equivalence. Rev. Mod. Phys. 29, 363 (1957).

43 Berry, M. V., *Principles of Cosmology and Gravitation.* (Cambridge University Press, 1976), p. 91; Jacoby, M., Thin film helps detect gravitational waves. ACS Cent. Sci. 3, 92 (2017).

44 Calaprice, A., Kennefick, D., Schulmann, R., *An Einstein Encyclopedia.* (Princeton University Press, 2015), p. 97.

45 Weinberg, S., *Dreams of a Final Theory.* (Panthenon Books, 1992), p. 85.

46 Maxwell, N., In Praise of Natural Philosophy: A Revolution for Thought and Life. (McGill-Queen's University Press, 2017), p. 3.

47 Wittgenstein, L., *Philosophical Investigations.* (Wiley, 2010), p. cxxvii.

48 Mould, R. F., A Century of X-Rays and Radioactivity in Medicine: With Emphasis on Photographic Records of the Early Years. (Taylor & Francis, 1993), p. 1–9.

49 Dalton, J., *A New System of Chemical Philosophy.* (William Dawson & Sons, 1808).

50 Lewis, G. N., The conservation of photons. Nature 118, 874 (1926).

51 Born, M., Nobel lecture: The statistical interpretations of quantum mechanics (Apr. 28, 2004) www.nobelprize.org/ prizes/physics/1954/born/lecture

52 Stengers, I., *Cosmopolitics.* (University of Minnesota Press, 2010), p. 24.

53 Popper, K. R., *Objective Knowledge: An Evolutionary Approach.* (Clarendon Press, 1979), p. 226.

54 Gamow, G., *My World Line.* (Viking Press, 1970), p. 44.

55 Macrae, N., John Von Neumann: The Scientific Genius Who Pioneered the Modern Computer, Game Theory, Nuclear Deterrence, and Much More. (American Mathematical Society, 2000).

56 Brush, S. G., John James Waterston and the kinetic theory of gases. Am. Sci. 49, 202 (1961).

57 Kamat, P. V., Schatz, G. C., Know the

difference: Scientific publications versus scientific reports. J. Phys. Chem. Lett. 6, 858 (2015).

58 Biscaro, C., Giupponi, C., Co-authorship and bibliographic coupling network effects on citations. PLoS ONE 9, e99502 (2014).

59 Kennefick, D., Einstein versus the Physical Review. Phys. Today 58, 43 (2005).

60 Merali, Z., ArXiv Rejections lead to spat over screening process. Nature (Jan. 29, 2016).

61 Noorden, R. V., Global scientific output doubles every nine years. Nature News Blog (May 7, 2014).

62 DailyNews, *Yogi Berra: An American Original.* (Sports Publishing, LLC, 2001), p. 58.

63 Brembs, B., Button, K., Munafò, M., Deep impact: Unintended consequences of journal rank. Front. Hum. Neurosci. 7, 291 (2013).

64 Reich, E. S., Plastic Fantastic: How the Biggest Fraud in Physics Shook the Scientific World. (St. Martin's Press, 2009).

65 Koepsell, D., Scientific Integrity and Research Ethics: An Approach from the Ethos of Science. (Springer, 2016).

66 Taleb, N. N., *Skin in the Game: Hidden Asymmetries in Daily Life.* (Random House Publishing Group, 2018), p. 29.

67 Afshar, S. S., Flores, E., McDonald, K. F. et al., Paradox in wave-particle duality. Found. Phys. 37, 295 (2007).

68 Chown, M., Editorial: Keep science fair, and keep it clean. New Sci. 2591 (2007).

69 Marmet, P., Paul Marmet newtonphysics.on.ca/info/author

70 Ekeberg, B., Metaphysical Experiments: Physics and the Invention of the Universe. (University of Minnesota Press, 2019), p. 72.

71 Sokal, A., Bricmont, J., *Fashionable Nonsense.* (Picador 1999), p. 300.

72 Nersessian, N. J., *Faraday to Einstein: Constructing Meaning in Scientific Theories.* (Martinus Nijhoff Publishers, 1984), p. 154.

73 Prusiner, S. B., Madness and Memory: The Discovery of Prions–A New Biological Principle of Disease. (Yale University Press, 2014).

74 Schopenhauer, A., Die Welt als Wille und Vorstellung: vier Bücher nebst einem Anhange, der die Kritik der Kantischen Philosophie enthält. (Brockhaus, 1819), p. xvi.

75 Darwin, C., On the Origin of Species by Means of Natural Selection. (Murray, 1859), p. 48.

76 Bell, D., Secret science. Sci. Public Policy 26, 450 (1999).

77 Horgan, J., Why there will never be another Einstein. Sci. Am. (Aug. 23, 2015).

78 Martin, B., Science and politics: An A-to-Z guide to issues and controversies. (Sage 2014), p. 145–149.

79 Blackwell, R. J., *Galileo, Bellarmine, and the Bible.* (University of Notre Dame Press, 1991).

80 Nagel, T., Mind and Cosmos: Why the Materialist Neo-Darwinian Conception of Nature is Almost Certainly False. (Oxford University Press, 2012), p. a:93, b:185, c:128.

81 Ball, P., Questioning truth, reality and the role of science. Quanta Magazine (May 24, 2018).

82 Lyotard, J. F., *The Postmodern Condition: A Report on Knowledge.* (University of Minnesota Press, 1984).

CONCLUDING WORDS

1 Heaviside, O., *Electromagnetic Theory.* (American Mathematical Society, 2003), p. 10.

2 Baggott, J., Farewell to Reality: How Modern Physics Has Betrayed the Search for Scientific Truth. (Pegasus Books, 2013), p. ix–xiv.

3 Bacon, F., *The Advancement of Learning.* (Clarendon Press, 1876), p. 41.

APPENDICES

1 Gibbs, J. W., *The Collected Works of J. Willard Gibbs.* (Yale University Press, 1948); Phillies, G. D. J., Readings and Misreadings of J. Willard Gibbs Elementary Principles in Statistical Mechanics. arXiv:1706.01500 (2017).

2 Annila, A., Annila, E., Why did life emerge? Int. J. Astrobio. 7, 293 (2008); Jaakkola, S., Sharma, V., Annila, A., Cause of chirality consensus. Curr. Chem. Biol. 2, 53 (2008)

arXiv:0906.0254; Jaakkola, S., El-Showk, S., Annila, A., The driving force behind genomic diversity. Biophys. Chem. 134, (136) 232 (2008) arXiv:0807.0892; Karnani, M., Annila, A., Gaia again. Biosystems 95, 82 (2009); Annila, A., Salthe, S., Economies evolve by energy dispersal. Entropy 11, 606 (2009).

3 Mäkelä, T., Annila, A., Natural patterns of energy dispersal. Phys. Life Rev. 7, 477 (2010).

4 Annila, A., The meaning of mass. Int. J. Theor. Math. Phys. 2, 67 (2012); Annila, A., Kolehmainen, E., Atomism revisited. Physics Essays 29, 532 (2016).

5 Tuisku, P., Pernu, T. K., Annila, A., In the light of time. Proc. R. Soc. A. 465, 1173 (2009); Griffiths, D. J., *Introduction to Electrodynamics*. (Cambridge University Press, 2017), p. 358–359.

6 May, R. M., Simple mathematical models with very complicated dynamics. Nature 261, 459 (1976).

7 Grahn, P., Annila, A., Kolehmainen, E., On the carrier of inertia. AIP Advances 8, 035028 (2018).

8 Annila, A., All in action. Entropy 12, 2333 (2010).

9 Unsöld, A., Baschek, B., *The New Cosmos*. (Springer-Verlag, 1983); Feynman, R. P., *Feynman Lectures on Gravitation*. (Addison-Wesley, 1995), p. 10.

10 Maciejko, J., Qi, X-L., Drew, H. D., Zhang, S-C., Topological Quantization in Units of the Fine Structure Constant. Phys. Rev. Lett. 105, 166803 (2010); Shuvaev, A., Pan, L., Tai, L. et al., Universal rotation gauge via quantum anomalous Hall effect. Appl. Phys. Lett. 121, 193101 (2022).

11 Compton, A. H., Scientific Papers of Arthur Holly Compton: X-Ray and Other Studies. (Chicago University Press, 1973), p. 777.

12 Griffiths, D., *Introduction to Elementary Particles*. (Wiley-VCH, 2008).

13 Irons, M. L., The curvature and geodesics of the torus www2.rdrop.com/~half/math/torus/index

14 Peskin, M. E., Schroeder, D. V., *An Introduction To Quantum Field Theory*. (Avalon Publishing, 1995).

15 Beringer, J., Arguin, J.-F., Barnett, R. M. et al., Review of particle physics. Phys. Rev. D 86, 010001 (2012).

16 Nambu, Y., An empirical mass spectrum of elementary particles. Prog. Theor. Phys. 7, 595 (1952); Mac Gregor, M. H., *Power Of Alpha, The: Electron Elementary Particle Generation With Alpha-quantized Lifetimes And Masses.* (World Scientific Publishing Company, 2007), p. 371-373.

17 Feynman, R. P., *QED: The Strange Theory of Light and Matter.* (Universities Press, 1985), p. 152.

18 Schmookler, B., Duer, M., Schmidt, A. et al., Modified structure of protons and neutrons in correlated pairs. Nature 566, 354 (2019).

19 Feynman, R. P., Leighton, R. B., Sands, M., *The Feynman Lectures on Physics.* (Addison–Wesley, 1964).

20 Afshar, S. S., Flores, E., McDonald, K. F. et al., Paradox in wave-particle duality. Found. Phys. 37, 295 (2007).

21 Hong, C. K., Ou, Z. Y., Mandel, L., Measurement of subpicosecond time intervals between two photons by inter-ference. Phys. Rev. Lett. 59, 2044 (1987).

22 Ou, Z. Y., Mandel, L., Non-Local and nonclassical effects in two-photon down-conversion. Quantum Optics: J. Eur. Opt. Soc. Part B 2, 71 (1990).

23 Vaidman, L., Past of a quantum particle. Phys. Rev. A 81, 052104 (2013).

24 Barut, A. O., Meystre, P. A classical model of EPR experiment with quantum mechanical correlations and Bell inequali-ties. Phys. Lett. A 105, 458 (1984).

25 Berry, M. V., *Principles of Cosmology and Gravitation.* (Cambridge University Press, 1976), p. 91.

26 Seife, C., "Tired-light" hypothesis gets re-tired. Science (Jun. 28, 2001).

27 Hogg, D. W., Baldry, I. K., Blanton, M. R. et al., The K correction. arXiv:astro-ph/0210394 (2002).

28 The SCP "Union2.1" SN Ia compilation supernova.lbl.gov/Union; Riess, A. G. et al., Type Ia Supernova Distances at Redshift >1.5 from the Hubble Space Telescope Multi-cycle Treasury Programs: The Early Expansion Rate. Astrophys. J. 853, 126 (2018).

29 Annila, A., Least-time paths of light. Mon. Not. R. Astron. Soc. 416, 2944

(2011).

30 Annila, A., Probing Mach's principle. Mon. Not. R. Astron. Soc. 423, 1973 (2012); Everitt, C. W. F., DeBra, D. B., Parkinson, B. W. et al., Gravity Probe B: Final results of a space experiment to test general relativity. Phys. Rev. Lett. 106, 221101 (2011).

31 Koskela, M., Annila, A., Least-action perihelion precession. Mon. Not. R. Astron. Soc. 417, 1742 (2011).

32 Hughes, J., Stellar Aberration, (2021) https://einsteinwaswrongrelativelyspeaking.wordpress.com/2021/03/18/stellar-aberration-a-stellar-aberration/

33 Laue, M. v., Zum Versuch von F. Harreß. Ann. Phys. 367, 448 (1920).

34 Nieto M. M. and Anderson J. D. Using early data to illuminate the Pioneer anomaly. Class. Quantum Gravity, 22, 5343 (2005); Greaves, E. D., Bracho, C., Gift, S., Rodriguez, A. M., A solution to the Pioneer anomalous annual and diurnal residuals. Prog. Phys. 17, 168 (2021).

35 Annila, A., Rotation of galaxies within gravity of the universe. Entropy 18, 191 (2016).

36 Karachentsev, I. D., Kashibadze, O. G., Makarov, D. I. et al., The Hubble flow around the Local Group. Mon. Not. R. Astron. Soc. 393, 1265 (2009).

37 Albada, T. S. v., Bahcall, J. N., Begeman, K. et al., Distribution of dark matter in the spiral galaxy NGC 3198. Astrophys. J. Part 1 295, 305 (1985).

38 Tully, R. B., Fisher, J. R., A new method of determining distances to galaxies. Astron. Astrophys. 54, 661 (1977).

39 Meurer, G. R., Obreschkow, D., Wong, O. I. et al., Cosmic Clocks: A tight radius–velocity relationship for H I-selected galaxies. Mon. Not. R. Astron. Soc. 476, 1624 (2018).

40 Kafle, P. R., Sharma, S., Lewis, G. F. et al., The need for speed: Escape velocity and dynamical mass measurements of the Andromeda Galaxy. Mon. Not. R. Astron. Soc. 475, 4043 (2018).

41 Ferrarese, L., Merritt, D., A fundamental relation between supermassive black holes and their host galaxies. Astrophys. J. Lett. 539, L9 (2000).

42 McGaugh, S. S., A tale of two paradigms: the mutual incommensurability of ΛCDM and MOND. Can. J. Phys. 93, 250 (2014).

43 Faraday, M., On the possible relation of gravity to electricity. Phil. Trans. R. Soc. A 141, 1 (1851).

44 Bennett, C. L. et al., Nine-year Wilkinson microwave anisotropy probe (WMAP) observations: final maps and results. Astrophys. J. Suppl. 208, 54 pp. (2013).

45 Bowman, J. D., Rogers, A. E. E., Monsalve, R. A. et al., An absorption profile centred at 78 megahertz in the sky-averaged spectrum. Nature 555, 67 (2018); Singh, S., Nambissan T. J., On the detection of a cosmic dawn signal in the radio background. Nat. Astron. (2022)

46 Griffiths, D., *Introduction to Quantum Mechanics*. (Pearson Prentice Hall, 2005), p. 468.

47 Venkataraman, G., *Bose And His Statistics*. (Universities Press, 1992), p. 14.